Human-Environment Interactions

Volume 8

Series Editor

Emilio F. Moran, Michigan State University, Bloomington, IN, USA

The *Human-Environment Interactions* series invites contributions addressing the role of human interactions in the earth system. It welcomes titles on sustainability, climate change and societal impacts, global environmental change, tropical deforestation, reciprocal interactions of population-environment-consumption, large-scale monitoring of changes in vegetation, reconstructions of human interactions at local and regional scales, ecosystem processes, ecosystem services, land use and land cover change, sustainability science, environmental policy, among others. The series publishes authored and edited volumes, as well as textbooks. It is intended for environmentalists, anthropologists, historical, cultural and political ecologists, political geographers, and land change scientists.

Human-environment interaction provides a framework that brings together scholarship sharing both disciplinary depth and interdisciplinary scope to examine past, present, and future social and environmental change in different parts of the world. The topic is very relevant since human activities (e.g. the burn of fossil fuels, fishing, agricultural activities, among others) are so pervasive that they are capable of altering the earth system in ways that could change the viability of the very processes upon which human and non-human species depend.

More information about this series at http://www.springer.com/series/8599

Thomas Weith · Tim Barkmann ·
Nadin Gaasch · Sebastian Rogga ·
Christian Strauß · Jana Zscheischler
Editors

Sustainable Land Management in a European Context

A Co-Design Approach

 Springer

Editors
Thomas Weith
Institute of Environmental Science
and Geography
University of Potsdam, Campus Golm
Potsdam, Germany

Working Group "Co-Design
of Change and Innovation"
Leibniz Centre for Agricultural
Landscape Research (ZALF)
Müncheberg, Germany

Nadin Gaasch
Science Management and Transfer
Potsdam Institute for Climate Impact
Research (PIK)
Potsdam, Germany

Christian Strauß
Research Area 'Land Use and Governance'
Leibniz Centre for Agricultural Landscape
Research (ZALF)
Müncheberg, Germany

Tim Barkmann
Research Area 'Land Use and Governance'
Leibniz Centre for Agricultural Landscape
Research (ZALF)
Müncheberg, Germany

Sebastian Rogga
Leibniz Centre for Agricultural Landscape
Research (ZALF)
Müncheberg, Germany

Jana Zscheischler
Working Group "Co-Design of Change and
Innovation"
Leibniz Centre for Agricultural Landscape
Research (ZALF)
Müncheberg, Germany

ISSN 2214-2339   ISSN 2452-1744 (electronic)
Human-Environment Interactions
ISBN 978-3-030-50840-1   ISBN 978-3-030-50841-8 (eBook)
https://doi.org/10.1007/978-3-030-50841-8

This Springer imprint is published by the registered company Springer Nature Switzerland AG
The registered company address is: Gewerbestrasse 11, 6330 Cham, Switzerland

Foreword

Sustainable land management is a key issue among the various applications of sustainable development. Humans are mobile terrestrial beings who need land as *terra firma* to support their activities and obtain vital resources. And yet land makes up only a third of the earth's surface and is finite, much of it already having been populated by humans. Other basic preconditions for ensuring the survival of humans were the availability of freshwater and a plant cover that provided food and shelter, which was used for gathering and hunting. This activity was steadily improved owing to skill and intellect, particularly after learning how to use fire as an effective energy source.

The real importance of land, however, was discovered when humans started to practice agriculture, especially crop farming. This was the very origin of land use and the first decisive step in transforming nature into a typical human environment, called culture. Humans took ownership of land and selected certain plant and animal species for crop cultivation and livestock husbandry around their settlements. These were built as solid farmsteads, marking a second type of land use and a completely artificial land cover.

During plant cultivation, farmers became aware of the uppermost layer of the land, called soil, the quality of which, above all fertility, was indicated by the plant cover. Humans soon learned to determine which sites were best suited for agriculture—deep sandy to loamy soils that were easy to till. Management began by replacing the natural plant cover with crop plants, usually in pure stands, which involved working the land using tools such as hoes and ploughs. Grain and root crops soon became the mainstay of human food supply, promoting population growth—which again required more farmland: a vicious circle evolved.

Cultural development is marked by continuous technical progress involving long-term consequences that were often impossible to foresee. Progress in agriculture resulted in bigger food quantities than farmers needed. This surplus enabled food to be supplied to a new, non-farming human population that led to urban life and civilisation, causing an irreversible division of rural and urban land use, lifestyle and mentality. Urban citizens, free from the daily toil of struggling with nature to produce food, developed new ideas and values such as landscape beauty, a love of nature, and animal welfare. However, these ideas and values are irrevocably tied to reliable daily nourishment, which depends on rural supply. As the urban population became aware of this, they started to govern farmers' activities, adjusting them to urban needs and ideas. This culminated in industrialisation and the rapid growth of cities, which induced the intensification and technical modernisation of farming, again conceived by city dwellers, whose standard of living rose to heights never seen before.

This achievement was also based on rapid advances in science and technology, which of course also included agriculture and food production. These advances created new attitudes and values towards fundamental aspects of life. Human rights and well-being took centre stage, encouraging further population growth and the greater exploitation of land and its resources. The concept of sustainable development, which has gained general consent as shown in several international resolutions, aims to overcome this dilemma.

And yet it takes precise specification to translate this broad, integrative concept into concrete measures, which also holds for land management: which components, resources or functions of land are to be managed in which way at a given location? Since land is finite, its very different qualities renders it necessary to choose sites or locations that are best suited for the various competing land uses, which are only compatible in part. These land uses include farming, forestry, settlement, urban-industrial development, mining, regenerative energy production, nature conservation, leisure and recreation. Sustainable management should seek to adapt land use to a site's qualities, rather than modifying these qualities to suit the land use. In addition, use intensities must be controlled to mitigate ecological deterioration, which can also be reduced by consciously designing the spatial arrangement of land uses.

The management of rural land is usually prioritised because it supplies the urban population with commodities such as food from grain and root crops. Such crops depend in turn on productive arable soils, and yet these are gradually being damaged by crop farming treatments that need to be applied. Since soils are the most precious and vulnerable land resource, a resource that cannot be restored, such negative impacts must be mitigated to the greatest extent possible, particularly since soils now have a new function—the sequestration of CO_2 to combat climate change.

All these management measures must be organised by adaptive and participatory governance institutions. It is an enormous challenge that may be helped by the information and proposals given in this book.

Wolfgang Haber
Technical University of Munich
Freising, Germany
e-mail: haber@wzw.tum.de

Contents

Part IV   Outlook

Chapter 1
A Knowledge-Based European Perspective on Sustainable Land Management: Conceptual Approach and Overview of Chapters

Thomas Weith, Tim Barkmann, Nadin Gaasch, Sebastian Rogga, Christian Strauß, and Jana Zscheischler

Abstract This introductory chapter, written by the editors, provides an overview of their conceptual approach, the book's line of argumentation, and an insight into the different chapters of the book "Sustainable Land Management in a European Context—a co-design approach". The synopsis highlights the various approaches and possible applications of a co-design approach.

Keywords Sustainable land management · Land governance · Co-design · Co-production of knowledge · Transdisciplinarity

T. Weith (✉)
Institute of Environmental Science and Geography, University of Potsdam, Campus-Golm, Karl-Liebknecht-Str. 24-25, 14476 Potsdam, Germany
e-mail: Thomas.Weith@zalf.de

T. Weith · J. Zscheischler
Working Group "Co-Design of Change and Innovation", Leibniz Centre for Agricultural Landscape Research, 15374 Müncheberg, Germany
e-mail: Jana.Zscheischler@zalf.de

S. Rogga
Leibniz Centre for Agricultural Landscape Research, Eberswalder Str. 84, 15374 Müncheberg, Germany
e-mail: Sebastian.Rogga@zalf.de

N. Gaasch
Science Management and Transfer, Potsdam Institute for Climate Impact Research, Telegraphenberg A31, 14471 Potsdam, Germany
e-mail: nadin.gaasch@pik-potsdam.de

T. Barkmann · C. Strauß
Research Area "Land Use and Governance", Leibniz Centre for Agricultural Landscape Research, Eberswalder Str. 84, 15374 Müncheberg, Germany
e-mail: timbarkmann@gmx.de

C. Strauß
e-mail: Christian_Strauss@gmx.de

1.1 New Conceptual Approaches for New Challenges

Urban and rural landscapes are constantly undergoing processes of change. Humankind has been influencing the natural environment on earth for centuries, resulting in different types of land use. In the nineteenth and twentieth centuries, radical changes in land use (Lambin et al. 2001; Hersperger and Bürgi 2009; Niewöhner et al. 2016; Jepsen et al. 2015; Plieninger et al. 2016) were brought about by the "agricultural revolution" and the "industrial revolution", which presented new opportunities for mobility and urban development.

Today, international challenges such as climate change, loss of biodiversity, soil degradation and food security or megatrends such as globalisation processes fuel discussions about the current land use system and land use policies, as well as future options. Significant changes have occurred, e.g. natural land has been transformed. In addition, qualitative aspects such as the loss of ecosystem services in combination with land degradation is an inconvenient truth.

International bodies such as the United Nations' Intergovernmental Panel on Climate Change (IPCC 2019) and the Intergovernmental Panel on Biodiversity and Ecosystem Services (IPBES 2018) specified changes in land use as one of the key aspects in debates about sustainability, culminating in the adoption of the Sustainable Development Goals (SDG) by the United Nations.

Several SDGs include relevant objectives in the policy field of land use. Not only SDG 15 (Life on Land) covers a wide range of recommendations—SDG 3 (Good Health and Well-Being), SDG 6 (Clean Water and Sanitation), SDG 9 (Industry, Innovation, and Infrastructure), SDG 11 (Sustainable Cities and Communities) and SDG 13 (Climate Action) also address further important aspects by highlighting a need for change across sectors. Implementation of these objectives is seen at the regional and local level, requiring changes of governance processes (Weith et al. 2019).

Looking specifically at cities and urbanised areas, HABITAT III focuses on resilient and sustainable cities and human settlements. Emphasis is placed on integrated urban and territorial development and new forms of urban governance to prevent urban sprawl and a further depletion of natural resources. In addition, there is a need for a balanced approach to the development of urban and rural areas, which are interdependent. This will indirectly promote the reduction of greenhouse gas emissions and air pollution (Liu et al. 2013). In a nutshell, there is a scientific and societal need to discuss options and solutions for land use change.

Land use is a complex human-nature interaction, generating diverse cultural landscapes influenced to a greater or lesser extent by humankind. In particular, land used for urban areas and infrastructure puts pressure on ecosystem services due to continued soil sealing and the fragmentation of landscapes following continuous land conversion. Consequences occur not only at the local or regional scale—they are globally interconnected. Consequently, human action accelerates global change, which in turn has an impact on humankind. Although various partial models have been developed for different types of land use (e.g. Siedentop et al. 2009; Siedentop and Fina 2010; Weber and Höferl 2009; Plieninger et al. 2016; EEA 2017), there is no accepted model that provides a comprehensive explanation of land use change. One important factor for land use change is interaction between population and economic dynamics (cf. Storper and Scott 2009). On a global scale, the population is increasing, particularly in economically prosperous regions and in metropolitan areas (UN 2014), putting pressure on the land by using space for settlement activities. In Europe—the spatial focus of our book—75% of the population live in urban areas, which is expected to increase to over 80% by 2020 (EEA 2017: 29). Due to urban–rural linkages, rural areas are likewise affected by this development. Land use and environmental impacts therefore depend not only on the size of the population, but also on its spatial distribution (Zasada 2011).

The persistent gap between the sustainability goals and spatial developments that are noncompliant with those goals raises questions about methods of governance to ensure compatibility with the goals. At first glance, the land use system in Europe, including Germany, seems to be organised within sophisticated land use policy frameworks, comprising land use planning and impact analysis tools. This assumption also seems to reinforce the fact that, against the backdrop of the aforementioned, the annual land use change rate is low relative to the total amount of land (EEA 2017).

However, land use is influenced by a complex interaction of more or less coordinated governance patterns, referring to a variety of sectoral administrative and disciplinary activities. Quite often, there is no integration of these complex factors. Although functional perspectives are required, territorial and sectoral powers dominate (Hooghe and Marks 2003; Blatter 2004; Sikor et al. 2013). In 2002, Young mentioned deficits with respect to fit, interplay and scale that hamper the successful implementation of sustainable land use strategies, e.g. in settlement development. He explained governance mismatches as drivers of land use conflicts by referring to incongruity between ecosystems and institutional arrangements (fit), a lack of adequate interaction between institutions (interplay), as well as a lack of interrelations between temporal and spatial scales.

As yet, there is no integrated approach to land use governance that provides applicable inter-sectoral governance to handle multi-stakeholder interrelations. In particular, current land use decisions are based on an insufficient poly-rational understanding of actors' positions (Davy 2012, 85), as well as deficits in knowledge provision (Salet 2014; Frantzeskaki and Kabisch 2015; Giebels et al. 2016). This includes knowledge gaps, also related to the complexity in land use governance, which is characterised by high levels of uncertainty (unpredictable developments and interrelations), disagreement (conflicting aims and demands) and distributed capacities (multi-actor landscape) (cf. Hummelbrunner and Jones 2013, 2).

In conclusion, contemporary land use governance systems are characterised by gaps, mismatches, or other dysfunctionalities that lead to the promotion of unsustainable land use. Above all, integrative land use regulations are lacking, particularly at the EU level. This deficit can be traced back to path dependencies (Getimis et al. 2014), based on closed disciplinary debates in science that are continued and reflected by sectoral perspectives in land use governance.

It therefore appears necessary to broaden the understanding of the system that influences land use by elucidating the interrelationships between different land use sectors, integrating diverse knowledge bases, and intermediating between stakeholders' various action patterns (Ison et al. 2013). Besides the need for a better understanding of land use systems, there is also a need for concrete future action.

We argue that this requires new governance approaches that integrate different knowledge types and perspectives, and that develop new socio-technical solutions in a continuous process to narrow the persistent gaps and contribute to more sustainable land use.

The concept of sustainable land management (SLM) is being increasingly discussed against this backdrop. "Land management" refers primarily to the procedural dimension of land use and land development for coordinating spatial, sectoral and temporal aspects in multi-level governance processes (see Engelke and Vancutsem 2010: 70). Central aspects include the integration of multi-stakeholder perspectives by linking ecological, socio-economic and political aspects as well as intertemporal dimensions (Hurni 2000; Schwilch et al. 2012). Management includes reconfigurations of the set of instruments as well as "technological, political and legal measures and activities" (Haber et al. 2010, 378–379). Land management takes into account interactions among different land use types and land-related sectors, explicitly incorporating rural and urban demands as well as economic, social and ecosystem functions of land. It is suggested as a framework to provide system-oriented solutions for dealing with land use conflicts (Repp and Weith 2015). In this regard, land management is strongly linked to debates on environmental and landscape governance (EEA 2017).

Understanding and managing land resources on a landscape level calls for systemic knowledge of the diverse actor groups that influence land use, and a coherent approach towards possible sustainable futures and adequate measures. This cannot be provided by science alone because normative aspects play a decisive role. It is therefore crucial for sustainable land management to integrate and utilise different perspectives and knowledge types from academia and practice. To that effect, we advocate a conception of SLM that refers to ongoing debates on new "modes" of knowledge production, reflecting the demand of real-world problem orientation, actor orientation and implementation (Zscheischler and Rogga 2015). In this context, co-design processes can be considered as a central instrument of SLM with different applications and purposes: starting at the co-design of just and acceptable processes, through the co-production of knowledge to the co-design of socially robust orientations and solutions (Moser 2016; Mauser et al. 2013).

Based on the challenges, state and drivers of land use trends in Europe described above, the central question is how to produce and handle new knowledge about land management to create innovative and actionable solutions towards sustainability. One special focus of this book is therefore co-design- processes and the co-production of knowledge to foster new modes of land use governance.

The book is organised following the "co-design and co-creation" concept as proposed by Mauser et al. (2013) as part of the sustainability discourse. It deals with the role and significance of different actors and their inclusion in (political) planning processes, emphasised for more than a decade in institutionalist approaches (cf. Ostrom 2011). Particular attention is paid to the knowledge stocks of these actors, including their experience-based knowledge (co-production of knowledge). This knowledge is not only used for a better understanding of problems and starting situations, but also to prepare and implement decision processes. Thus, implementation orientation plays a special role. All this takes place in the highly complex context of spatial and landscape development organised in a multi-level system (Reimer et al. 2014).

Co-design processes are increasingly assessed in science with regard to their effects in decision-making processes. It is assumed that co-design processes can raise acceptance because of their integrational and participative nature (Busse and Siebert 2018) and can support the development and implementation of social innovations. That is why they play an important role in the discussion about change and transition processes (Tschakert et al. 2016). At the same time, they are associated with developing action possibilities for stakeholders and increasing the resilience of decisions. In consequence, co-design approaches are much more than "simple" participation processes, which have been known for a long time and in some cases are applied in a formalised way (e.g. Arnstein 1969). They also go beyond the mere application of learning loops or transdisciplinary concepts.

We therefore stress the early and ongoing involvement of a broad variety of actors in solution-oriented processes that take up the role as knowledge providers alongside academic and technical experts. Ideally, the process consists of three steps: the analysis of the state and drivers of change; the co-design and co-production of knowledge; and the co-evolution and co-dissemination of that knowledge. We argue that the comprehensive approach to sustainable land management should follow this concept, and take up these three steps.

To conclude, the book is based on the assumption that successful—and hence sustainable—land management (SLM) requires routines of knowledge co-production to put sustainable land use into practice. It thus refers to ongoing debates on new "modes" of science, which stress problem orientation, action, and negotiation of research activities—as it is highlighted in the following chapters.

1.2   Overview of the Main Parts and the Single Chapters

The book starts with a foreword and this introductory overview. The main body is subdivided accordingly into three main parts:

- "Land use: state and drivers in Europe",
- "Co-production of knowledge", and
- "Co-evolution: New system solutions and governance".
- The book closes with a concluding and summarising "Outlook".

The chapters in **part one** focus on the development of different forms of land use as a result of the interplay between societal demands on land and physical-ecological conditions. Society is in a situation of ongoing changes and developments. Important framing factors are the social and economic system, population development and economic driving forces, urbanisation and spatial interrelations. These factors influence the development of regions and landscapes with their infrastructure, agriculture in its respective intensity, forestry in form and extent, the use of resources and raw materials as well as the preservation of specific landscapes and their characteristics. The chapters in part one highlight these interlinkages and provide overviews and current states and drivers of land use change, as well as describing any lack of knowledge.

The chapter on *Landscape change in Europe* by García-Martín, Quintas-Soriano, Torralba, Wolpert and Plieninger provides a first introductory overview of the evolution and change of landscapes throughout history. In this emerging research field on landscape change in Europe, the authors explain why these changes happen, and what they fully entail. They embed the evolution of landscapes in an understanding of complex social-ecological systems and the drivers behind them. In consequence, they gain a better understanding of complex, dynamic and interlinked change processes that are characterised inter alia by environmental degradation and increased land use conflicts. While considering these challenges, the authors provide guidance for co-designed sustainable landscape management. The results are based on a combination of quantitative analysis with participatory approaches on the one hand, and diverse spatial and temporal scales on the other to achieve a comprehensive understanding of past changes and future trajectories.

From an economic point of view, land is a limited and scarce natural resource that faces competing and rising demands. These demands refer to different types of land use, such as agriculture, nature and natural resource protection, industrial areas, human settlements and infrastructure. Land use conflicts have evolved and are driven by various factors. Based on a previous study, Kirschke et al. focus on the economic aspects of sustainable land management in the chapter about *New trends and drivers for agricultural land use in Germany*. This chapter provides an overview of the main economic drivers of land use change in Europe, bringing together theoretical approaches from regional economics, economic geography, agricultural economics, environmental and resource economics, and infrastructure planning. New developments in agricultural land use in Germany are analysed based on general land use trends. Indicators for agricultural land use changes that create conflicts include rising land prices, deteriorating environmental conditions and changing land use structures. The authors describe major drivers behind this development, which basically reflect market forces and new policy frameworks such as the Renewable Energy Act (REA) and particularities related to German reunification.

Besides economic pressure, demographic change is described as one of the central factors of human influence on land use change. But does demographic change really contribute directly to land use change? Do changes in population size and composition directly affect changes in land use? Hoffmann reflects in his chapter on *Demographic change and land use* the questions of (a) whether and to what degree clear correlations between demographic change and observable land use changes could be found in the existing literature, and (b) what the result of the literature review means for regional studies and regional development policies. The chapter contributes to a more evidence-based view of the topic, helping to avoid misleading simplifications and even myths on interlinkages that often influence concepts, values and decisions in politics and policies shaping land use.

Urbanisation is seen as one of the major driving forces of today's land use systems. This phenomenon involves the usually irreversible conversion of mainly agricultural land. Urban land use change can occur in quite different forms in terms of spatial layout, building density and speed of change, to name but a few aspects. The global dimension of urbanisation and related land use change is now on the agenda of policy-makers and researchers worldwide. To provide an overview of these complex processes, Nuissl and Siedentop endeavour in their chapter on *Urbanisation and land use change* to achieve conceptual clarification, highlighting drivers and impacts. They provide a systematic overview of influencing instruments and strategies for coping with urban land use change.

In contrast to the urbanisation focus of the previous chapter, Doernberg and Weith seek to expand the view towards an urban–rural interrelations perspective. Urban and rural regions are no longer seen as distinct places, but as functionally connected in the context of the sustainable use of land-related resources, quite often with blurring boundaries. In recent years, both scientific and societal discourses have given new impetus to this thematic debate. At the same time, it becomes quite obvious that there are currently no conceptual approaches available that are comprehensive in terms of content, and can capture the complexity of regional urban–rural interrelationships. In consequence, concepts are missing to provide practical support for regional planning and regional development policies. The chapter on *Urban–rural interrelations* outlines the existing approaches, and highlights opportunities for developing new concepts, especially by defining requirements of new concepts to overcome the shortcomings of models currently in place in science and practice.

The chapters in **part two** "Co-production of knowledge" explain and discuss explicit ways of knowledge co-production and co-dissemination to foster sustainable land management. In recent years, new ways of integrative knowledge generation have been developed and applied to find new solutions to complex real-world problems. This includes land use issues, starting with problem definition through identifying and describing land use conflicts to developing new land governance processes. The involvement of various actor groups, such as political and administrative decision-makers, is an immanent part of the co-design and co-production process. Especially transdisciplinary approaches, translated into the field of land use sciences, are seen as adequate methodological ways to be tested and implemented in the context of land use challenges. They also show great promise for creating social innovation processes. Various forms of simulation games and real-world labs are being discussed to support creative ways of handling new challenges in land use. Quite often, these knowledge production processes are supported by new ways of knowledge management, implementation and transfer, using the manifold opportunities offered by digitalisation processes.

The first chapter in this part explains additional values of transdisciplinary approaches for solving land use problems against the backdrop of urgent complex real-world challenges and changed societal demands on knowledge. In *Transdisciplinarity in land use sciences,* Zscheischler states that transdisciplinarity is no longer just about the production of new scientific insights, but also about the solution-oriented goal and action knowledge that support sustainable development and land management. However, the extent to which these projects have succeeded has increasingly been critically questioned in recent years. The chapter introduces the development of the concept on transdisciplinary research (TDR), describes the current criticism of TDR and presents a critical assessment based on empirical findings from research practice. The results reveal several unsolved implementation challenges that can be traced back to a misfit with academic structures and a lack of empirical knowledge.

For decades, fostering innovations was seen as a key approach for problem solving, also in sustainable development. Socio-technological change and transformations should lead to more sustainable land use practices. However, new perspectives on innovation emerge when addressing long-term societal goals, the variety of actor groups involved, multi-level governance and a lack of usual commercialisation potentials. Sustainability innovations frequently contradict social practices, regulations and existing infrastructure, and focus on social innovation processes. Since there is little understanding of how transformation and socio-technological change can be effectively governed and supported in the specific field of sustainable land use and management, the chapter contributes specifically to this knowledge gap. In *Innovations for sustainable land management,* Zscheischler and Rogga refer to nine projects that address that specific issue. The authors identify different types and degrees of innovations, approaches to manage innovation processes, and the leverage points of these solutions in the governance system of sustainable land management. The results stress the need for reflexive processes of social learning and cognitive reframing by embedding experimental innovation management approaches such as real-world laboratories into larger transdisciplinary and participatory processes.

One of the key aspects in change and innovation processes is quite often underestimated or misunderstood: how to transmit knowledge from one actor to another. In the article on *Knowledge exchange at science-policy inter-faces in spatial planning, land use and soil management*, Pütz and Brassel investigate this topic in detail in the context of land use change. Based on literature review and expert interviews, they identify six types of knowledge exchange and examine barriers and opportunities for knowledge exchange. This offers a better understanding of processes, and provides suggestions for knowledge exchange activities. However, this will be a challenging task for the future, due to actors' different expectations and experiences of how knowledge is to be exchanged.

Knowledge-driven change and innovation processes will be supported by new integrative methodological approaches. Maaß addresses this in her chapter entitled *Serious games in sustainable land management* by the using the method of experimental games. She presents an integrated approach that combines a serious game with a land use and transport model for analysing the effects of high prices for fossil fuels on land use. This approach opens up the possibility to simulate the complexity of sustainable land management and to incorporate learning aspects for decision-makers. Participants in the serious game came from the Hamburg Metropolitan Region. The decision-makers attended several meetings to develop their individual strategies on how to cope with rising energy prices.

Beyond simulation, it has always been a great challenge to test and implement sustainability-driven processes of change and innovation. Real-world laboratories have gained in importance as an adequate research format in recent years in Europe and especially in Germany. The underlying assumption is that transformation processes towards sustainability can be investigated under real-world conditions in order to gain knowledge of their dynamics, to identify characteristics of successful transformation processes, and to be able to transfer this knowledge to other cases. The chapter by Kanning, Richter-Harm, Scurrell and Yildiz entitled *Real-world laboratories initiated by practitioner stakeholders for sustainable land management* analyses existing types of real-world labs, and reflects on the possibilities generated by real-world laboratories that have been initiated by practitioner stakeholders. "Energieavantgarde Anhalt" provides evidence for such labs, which contributes to sustainable land management in the context of the energy transition in rural areas, featuring small and medium-sized towns. The practicability and variability of real-world lab approaches pave the way for future options for application.

Another dimension of transitions is highlighted in the chapter on *Knowledge management for sustainability: The spatial dimension of higher education as an opportunity for land management* by Schulz, Köhler and Weith. In this case, digitalisation offers new opportunities for knowledge creation and dissemination, which are an essential part of sustainable land management. Digitalisation will support collaborative activities related to land use, bringing together diverse interests and help detect new patterns of cooperation. These developments will also have consequences for knowledge spaces and the institutional and personnel knowledge carriers established within them. The chapter outlines these consequences in the context of higher education research. Concepts from both domains—higher education and land management—are combined in a fruitful way, facilitating new interpretations of spatial and digital artefacts as well as debates about knowledge dissemination.

In recent years, the modes of knowledge transfer from science to society, and vice versa, have been changing. Our traditional understanding of scientific knowledge transfer in the form of unidirectional modes of communication is supplemented by new ways of knowledge exchange and knowledge production. As such, the notion of transfer and implementation should be discussed with a new focus in mind. Based on a framework for transfer and implementation activities in transdisciplinary research settings, Rogga discusses specific paths of knowledge dissemination, and especially of knowledge transfer, in his article on *Transcending the loading dock paradigm—rethinking science-practice transfer and implementation in sustainable land management*. He discusses the systemic understanding of transfer and implementation regarding unidirectional approaches, and provides definitions for transfer and implementation in sustainable land management. He recommends taking a deeper look into knowledge management and considering what it means for sustainable land use management.

In **part three** *Co-evolution: new system solutions and governance,* the authors suggest and discuss various ways to integrate co-design and co-production approaches into applicable as well as integrative concepts, instruments and measures. In the cases outlined, system solutions are a result of co-design and co-production processes that demonstrate the complexity of interaction and competing targets.

Starting with embedded change projects, a description is given of small-scale system solutions that combine settlement development, water, and energy supply. Second, the new approach of urban agriculture combines different fields of action in cities, such as urban land use policy, development of open spaces, food security and social integration. Third, conceptual improvements reflect on the functional aspects of ecosystem services and the development of green infrastructure. The chapter ends with a broad discussion about upcoming challenges in land use science.

Changes towards meeting the ambitious 2030 Sustainable Development Goals (SDGs) demand new systemic approaches to resource management. Material Flow Management (MFM) is regarded as a vital tool for complex systems. Heck explains in the chapter entitled *Small-scale system solutions—material flow management in settlements* how MFM will contribute to the protection of land, the conversion of abandoned land and upcycling of degraded land. He demonstrates the usefulness of the concept, despite its relative novelty, by showing practical applications of MFM in small-scale systems characterised by decentralised material and energy flows. The chapter pays special attention to augmenting source and sink capacity, employing MFM to reduce impacts on ecosystems both upstream and downstream, i.e. on the use of resources as well as on the amount of emissions.

Specht, Schimichowski and Fox-Kämper illustrate another applicable concept. They focus on a concept that combines urban and rural potentials of co-production of goods as a (business) model in the chapter on *Multifunctional urban landscapes: The potential role of urban agriculture as an element of sustainable land management.* They explore how urban agriculture can contribute to sustainable land management and co-production. To this end, background information is given on the (re-)emerging phenomenon of urban food production and on what motivates those involved to implement collaborative practices. The functions and services provided by urban agriculture as an element of sustainable land management are explored using the three pillars of sustainability. It is shown that urban agriculture may reduce land use conflicts, and support new social activities for saving and qualifying open spaces.

In recent years, new scientific concepts for the assessment, and hence development, of ecosystems have also been established in science and land use planning. In the chapter on *Integrating ecosystem services, green infrastructure and nature-based solutions,* Haase includes three perspectives for developing new solutions to save urban nature as a habitat for humans, flora and fauna. Complementary paths to increased urban sustainability are shown, combining knowledge for action. Nonetheless, implementation is still a long way off, and there are also unsolved issues, such as the social inclusiveness of the three approaches.

Part three closes with a chapter on *Upcoming challenges in land use science—an international perspective* by Fürst. This chapter complements and extends the discussion on important concepts such as multifunctionality and social-ecological frameworks applied in land use science. The author also reflects on current political debates and challenges in terms of methodological aspects, actor involvement and project designs—in the context of sustainable land management. Future research topics related to the UN Sustainable Development Goals are raised, along with proposals for advancing land use science. As a future perspective, Fürst discusses how the co-development of knowledge and the co-design of land use system research could be conceived.

An **Outlook** is given in the final chapter of the book, entitled *Conclusions and research perspectives*. Here, the editors reflect on the different aspects of knowledge based on sustainable land management. They develop an integrative view on the topic and point out options for improving in particular the science-policy interface in the European context.

Acknowledgements This book would not have possible without the dedicated work of many committed supporters. First of all, we thank the German Funding Agency BMBF Ministry of Education and Research and the project management agency PTJ for their financial and administrative support in the context of the Scientific Accompanying Research Project "Sustainable Land Management (module B)" (FKZ 033L004). We are also very grateful to all of the contributing authors. Special thanks go to Petra Koeppe and Alice Baumgärtner as well as to Teresa Gehrs (Lingua Connect) and agrathaer for helping us finalise the book. Last but not least, we thank the Springer team for their support and patience.

References

Arnstein, S. R. (1969). The ladder of citizen participation. *Journal of the American Institute of Planners., 35*(4), 216–224.

Blatter, J. (2004). From 'spaces of place' to 'spaces of flows'? Territorial and functional governance in cross-border regions in Europe and North America. *International Journal of Urban and Regional Research, 28*(3), 530–548.

Busse, M., & Siebert, R. (2018). Acceptance studies in the field of land use—A critical and systematic review to advance the conceptualization of acceptance and acceptability. *Land Use Policy, 76,* 235–245.

Davy, B. (2012). *Land policy: Planning and the spatial consequences of property.* Farnham, Burlington: Ashgate Publishing Limited.

EEA. (2017). Landscapes in transition. An account of 25 years of land cover change in Europe. EEA Report No 10/2017: https://www.eea.europa.eu/publications/landscapes-in-transition/at_download/file.

Engelke, D., & Vancutsem, D. (2010). *Sustainable land use management in Europe.* Lyon: Providing strategies and tools for decision-makers.

Frantzeskaki, N., & Kabisch, N. (2015). Designing a knowledge co-production operating space for urban environmental governance—Lessons from Rotterdam, Netherlands and Berlin, Germany. *Environmental Science and Policy,* Online First.

Getimis, P., Reimer, M., & Blotevogel, H. (2014). Conclusion: Multiple trends of continuity and change. In: M. Reimer & H. Blotevogel (Eds.), *Spatial planning systems and practices in Europe*. London and New York: Routledge.

Giebels, D., van Buuren, A., & Edelenbos, J. (2016). Knowledge governance for ecosystem-based management: Understanding its context-dependency. *Environmental Science and Policy, 55,* 424–435.

Haber, W., Bückmann, W., & Endres, E. (2010). Anpassung des Landmanagements in Europa an den Klimawandel. *Natur und Recht, 32,* 377–383.

Hersperger, A. M., & Bürgi, M. (2009). Going beyond landscape change description: Quantifying the importance of driving forces of landscape change in a Central Europe case study. *Land Use Policy, 26,* 640–648.

Hooghe, L., & Marks, G. (2003). Unraveling the Central State, but how? Types of multi-level governance. *American Political Science Review, 97*(2), 233–243.

Hummelbrunner, R., & Jones, J. (2013). A guide for planning and strategy development in the face of complexity. ODI Background Notes. [Online] Available from: https://www.odi.org/publicati ons/7325-aid-development-planning-strategy-complexity. Accessed on May 26, 2016.

Hurni, H. (2000). Assessing sustainable land management (SLM). *Agriculture, Ecosystems and Environment, 81,* 83–92.

IPBES/Montanarella, L., Scholes, R., & Brainich, A. (Eds.). (2018). The IPBES assessment report on land degradation and restoration. Secretariat of the Intergovernmental Science-Policy Platform on Biodiversity and Ecosystem Services, Bonn, Germany, pp. 744.

IPCC. (2019). Climate change and land: An IPCC special report on climate change, desertifica-tion, land degradation, sustainable land management, food security, and greenhouse gas fluxes in terrestrial ecosystems. https://www.ipcc.ch/site/assets/uploads/2019/11/SRCCL-Full-Report-Compiled-191128.pdf.

Ison, R., Blackmore, C., & Iaquinto, B. L. (2013). Towards systemic and adaptive governance: Exploring the revealing and concealing aspects of contemporary social-learning metaphors. *Ecological Economics, 87,* 34–42.

Jepsen, M., Kuemmerle, T., Müller, D., Erb, K., Verburg, P., Haberl, H., et al. (2015). Transitions in European Land-management regimes between 1800 and 2010. *Land Use Policy, 49,* 53–64. https://doi.org/10.1016/j.landusepol.2015.07.003

Lambin, E. F., Turner, B. L., Geist, H. J., Agbola, S. J., Angelsen, A., Bruce, J. W., et al. (2001). The causes of land-use and land-cover change: Moving beyond the myths. *Global Environmental Change, 11*(4), 261–269. https://doi.org/10.1016/S0959-3780(01)00007-3

Liu, J., Hull, V., Batistella, M., DeFries, R., Dietz, T., Fu, F., et al. (2013). Framing sustainability in a telecoupled world. *Ecology and Society, 18*(2), 26. https://doi.org/10.5751/ES-05873-180226

Mauser, W., Klepper, G., Rice, M., Schmalzbauer, B. S., Hackmann, H., Leemans, R., & Moore, H. (2013). Transdisciplinary global change research: The co-creation of knowledge for sustainability. *Current Opinion in Environmental Sustainability, 5*(3–4), 420–431.

Moser, S. C. (2016). Can science on transformation transform science? Lessons from co-design. *Current Opinion in Environmental Sustainability, 20,* 106–115.

Niewöhner, J., Bruns, A., Hostert, P., Krueger, T., Nielsen, J. Ø, Haberl, H., et al. (Eds.). (2016). *Land use competition*. Switzerland: Springer.

Ostrom, E. (2011). Background on the institutional analysis and development framework. *Policy Studies Journal, 39*(1), 7–27. https://doi.org/10.1111/j.1541-0072.2010.00394.x.

Plieninger, T., Draux, H., Fagerholm, N., Bieling, C., Bürgi, M., Kizos, T., et al. (2016). The driving forces of landscape change in Europe: A systematic review of the evidence. *Land Use Policy, 57,* 204–214. https://doi.org/10.1016/j.landusepol.2016.04.040

Reimer, M., Getimis, P., & Blotevogel, H. (2014). Spatial planning systems and practices in Europe. Taylor and Francis.

Repp, A., & Weith, Th. (2015). Building bridges across sectors and scales: Exploring systemic solutions towards a sustainable management of land. *Land, 4,* 325–336. https://doi.org/10.3390/land4020325

Salet, W. (2014). The authenticity of spatial planning knowledge. *European Planning Studies, 22,* 293–305.

Schwilch, G., Bachmann, F., Valente, S., Coelho, C., Moreira, J., Laouina, A., et al. (2012). A structured multi-stakeholder learning process for sustainable land management. *Journal of Environmental Management, 107,* 52–63. https://doi.org/10.1016/j.jenvman.2012.04.023

Siedentop, S., Junesch, R., Straßer, M., Zakrzewski, P., Samaniego, L., & Weinert, J. (2009). *Einflussfaktoren der Neuinanspruchnahme von Flächen.* Bonn: Bundesamt für Bauwesen und Raumordnung.

Siedentop, S., & Fina, S. (2010). Monitoring urban sprawl in Germany: Towards a GIS-based measurement and assessment approach. *Journal of Land Use Science, 5,* 73–104.

Sikor, T., Auld, G., Bebbington, A. J., Benjaminsen, T. A., Gentry, B. S., Hunsberger, C., et al. (2013). Global land governance: From territory to flow? *Current Opinion in Environmental Sustainability, 5,* 522–527.

Storper, M., & Scott, A. J. (2009). Rethinking human capital, creativity and urban growth. *Journal of Economic Geography., 9*(2), 147–167. https://doi.org/10.1093/jeg/lbn052

Tschakert, P., Tuana, N., Westskog, H., Koelle, B., & Afrika, A. (2016). TCHANGE: the role of values and visioning in transformation science. *Current Opinion in Environmental Sustainability, 20,* 21–25.

UN. (2014). World urbanisation prospects—United Nations, Department of Economic and Social Affairs, Population Division (2014).

Weber, G., & Höferl, K.-M. (2009). Schrumpfung als Aufgabe der Raumplanung – eine Annäherung aus Österreichischer Sicht. In Th. Weith (Ed.), *Alles Metropole?* (pp. 121–129). Berlin-Brandenburg zwischen Hauptstadt: Hinterland und Europa. Planungsrundschau. Kassel.

Weith, Th., Warner, B., & Susman, R. (2019). Implementation of international land use objectives—Discussions in Germany. *Planning Practice and Research, 34*(4), 454–474.

Young, O. R. (2002). The institutional dimensions of environmental change. Fit, Interplay, and Scale. Cambridge: The MIT Press.

Zasada, I. (2011). Multifunctional peri-urban agriculture—A review of societal demands and the provision of goods and services by farming. *Land Use Policy, 28,* 639–648.

Zscheischler, J., & Rogga, S. (2015). Transdisciplinarity in land use science—a review of concepts, empirical findings and current practices. *Futures, 65,* 28–44.

Part I
Land-Use: State and Drivers in Europe

Chapter 2
Landscape Change in Europe

María García-Martín, Cristina Quintas-Soriano, Mario Torralba,
Franziska Wolpert, and Tobias Plieninger

Abstract The study of the evolution and change of landscapes' ecological conditions through history has fascinated professional and amateur scientists for centuries.
However, the understanding of why these changes happen and what these changes
fully entail is still an emerging field of research, which nowadays broadly covers
the study of the evolution of landscapes as complex social-ecological systems. This
field has become particularly relevant in the current context of rapid global change,
widespread environmental degradation and increasing land use conflicts, as an important source of information to facilitate sustainable landscape management. In this
chapter, we provide an overview of the current state of landscape change research
in Europe and of the main findings and methodological challenges therein. These
methodological challenges are bound up with the complex, dynamic and interlinked
nature of landscapes, which require co-designed approaches that combine different
perspectives, such as quantitative analysis with participatory approaches, and that
capture diverse spatial and temporal scales. Together, these make it possible to
achieve a comprehensive understanding of past changes and future trajectories.

Keywords Land use and land cover change · Intensification · Abandonment ·
Drivers of change · Transdisciplinary approaches

M. García-Martín · T. Plieninger (✉)
Department of Agricultural Economics and Rural Development, University of Göttingen, Platz
der Göttinger Sieben 5, 37073 Göttingen, Germany
e-mail: plieninger@uni-kassel.de

M. García-Martín
e-mail: maria.garcia-martin@uni-goettingen.de

C. Quintas-Soriano · M. Torralba · F. Wolpert · T. Plieninger
Faculty of Organic Agricultural Sciences, University of Kassel, Steinstr. 19, 37213 Witzenhausen,
Germany
e-mail: cristina.quintas@uni-kassel.de

M. Torralba
e-mail: mario.torralba@uni-kassel.de

F. Wolpert
e-mail: franziska.wolpert@uni-kassel.de

© The Author(s) 2021   17
T. Weith et al. (eds.), *Sustainable Land Management in a European Context*,
Human-Environment Interactions 8, https://doi.org/10.1007/978-3-030-50841-8_2

2.1 Introduction

Landscapes are dynamic; the use of the land is in constant change as societal aspirations and natural conditions evolve. The ever-growing human capacity to modify landscapes raises concerns about the consequences these changes will bring for humans' and nature's well-being. Knowing more about land use patterns, rates of land use change and the drivers behind them are all important in preventing and reducing tensions between conflicting land uses, predicting future scenarios, developing strategies to achieve more desirable futures and designing adequate policies. In Europe, the study of landscape changes is currently developing rather vividly, in part driven by initiatives by the Council of Europe and the European Union. Here, we refer to "landscape change" as an umbrella concept for the different forms of land use change and land cover change.

In this chapter, we present an overview of landscape change research developed in Europe, focusing on meta-studies, case-study research and cross-site comparison studies of European landscapes that identify processes and trajectories in land use changes and the driving forces behind them. This chapter is organised into five sections: Sect. 2.2 an introduction to the current state of landscape change research in Europe, where we also present the studies included in this overview; Sects. 2.3, 2.4 and 2.5 a review of the main findings of these studies on the trends, drivers and future scenarios of landscape change; and Sect. 2.6 a summary of research gaps and possible ways forward for landscape research towards sustainable land management.

2.2 Landscape Change Research in Europe

In Europe, landscape research dates back to the nineteenth century, when Alexander von Humboldt (1769–1859) and Carl Ritter (1779–1859) introduced the term "landscape" (*Landschaft* in German) as a scientific concept (Kirchhoff et al. 2013). Since then it has developed as a vibrant field of study, particularly after the beginning of the twenty-first century, when the German geographer Carl Troll (1899–1975) coined the term "landscape ecology" to examine the reciprocal interactions between social and ecological processes (Turner 2005). At a political level, landscape research has received increasing attention since the 1970s. It was at this time that environmental sustainability concerns gained momentum and became a target of United Nations development programmes, and scholars and policymakers realised the need to overcome the shortcomings of single-sector policies and management strategies, which prompted the adoption of the "landscape approach" (Sayer et al. 2013). One of the milestones in the inclusion of the landscape approach in the political agenda was the European Landscape Convention, in which landscapes were defined as the result of the interaction of natural and human factors (Council of Europe 2000).

Overall, there have been concerns that landscape changes are of such magnitude that societies can no longer accept them without putting landscape sustainability at risk (Antrop 2005; Bürgi et al. 2017; Plieninger et al. 2016). According to this perspective, understanding the causes, processes and outcomes of landscape change is becoming absolutely crucial (Plieninger et al. 2016). For this reason, landscape change research has become an emerging field, in which quantitative approaches to measure the expansion or decline of different land covers and changes in land use intensity have coevolved with more qualitative studies that have tried to understand what is driving such changes. More specifically, landscape researchers argue that being aware of and understanding the rates and patterns of change, as well as the driving forces behind them is necessary for context-specific and effective policy-making and for taking action towards more sustainable land management (Jepsen et al. 2015; Kuemmerle et al. 2016; Levers et al. 2015). Here, we understand "sustainable land management" as a multidimensional and evolving concept, in which a diverse range of stakeholders are involved and where ecological, economic and social aspects need to be integrated; this means that no single definition is explanatory enough (Weith et al. 2013). Such knowledge on the trends and drivers of landscape change contributes to exploring and mitigating trade-offs and impacts on biodiversity and ecosystem services. These trade-offs result from land management practices (Kuemmerle et al. 2016; Levers et al. 2015) in the context of increasing competition for land for multiple and sometimes incompatible uses (Levers et al. 2016; Pérez-Soba et al. 2015; Verkerk et al. 2018). Finally, a comprehensive understanding of how and why landscapes have changed facilitates the anticipation and projection of possible future scenarios, which in turn may be used to develop land management strategies and policy decisions to avoid undesirable futures (van Vliet et al. 2015; Verkerk et al. 2018).

The study of landscape change can be broadly divided into: (1) land cover change and (2) land use and management intensity change (Erb 2012; Levers et al. 2015). Land cover change is understood as the "alterations of biophysical characteristics of the Earth's surface", such as the spreading of forests or the reduction of agricultural land; while land use and management intensity change are defined as the "changes in the levels of socioeconomic inputs (e.g., labour, resources, water, energy or capital) and/or altered outputs (value or quantity) per unit area and time" (Erb 2012, p. 8). Erb (2012) and Verburg et al. (2013a) suggested that the former has been more commonly studied due to the availability of land cover datasets and the methodological challenge of quantifying and understanding land use intensity and intensification processes. However, in the past decade, the study of intensification processes has become a significant topic (e.g. Levers et al. 2014, 2016). In addition to that, substantial research has focused on identifying drivers of change to understand why landscapes change or remain unchanged, why they evolve faster or slower and to identify the causal mechanisms of regime shifts (Kizos et al. 2018). Studies typically differentiate between proximate drivers—i.e. the human actions that have a direct effect on the landscape changes; and underlying drivers—i.e. the social and ecological factors that trigger those human actions (van Vliet et al. 2015). In recent years, the study of landscape change has further broadened to identify patterns of landscape stability (i.e. land cover types that remain unchanged over a certain period; Lieskovský and Bürgi 2018) and their related stabilising factors.

This knowledge can help to protect and manage valuable landscapes in a rapidly changing world and regulate undesirable land use changes (Lieskovský and Bürgi 2018). Finally, there has been a focus on analysing how current trends would evolve in the future under different political, environmental and socioeconomic scenarios (Verburg et al. 2013a; Stürck et al. 2018; Verkerk et al. 2018). In this sense, the implementation of transdisciplinary projects has been discussed as a promising approach in land use science in order to address complex multifaceted "real-world problems" and to design strategies and solutions for sustainable development (Zscheischler et al. 2017). Transdisciplinary approaches are defined as a collaborative process of knowledge production that involves scientists from different disciplines and societal actors to address highly complex, real-world problems (e.g. Pohl 2008; Wickson et al. 2006). Transdisciplinary research has become a widespread research approach in sustainability science and is increasingly promoted by research programmes and agencies (e.g. Future Earth). It can develop collaborative research approaches in land use science to bring promising means of initiating change in the current course of action (Zscheischler et al. 2017).

The study of agricultural and forestry changes has become particularly exhaustive, not only because agriculture and forestry are the most extensive land cover types in Europe, but also because they have significant impacts on the provision of a wide range of services (regulating, provisioning and cultural) (van Vliet et al. 2015) and on the environment at large (Levers et al. 2016). The study of agricultural and forestry intensification has gained attention (Levers et al. 2014, 2016; van der Sluis et al. 2016) under the assumption that knowledge on the processes and trade-offs of intensification is necessary to mitigate its negative impacts in a context of rapidly changing resource demand (Erb 2012; van der Sluis et al. 2016). While some areas face processes of intensification, land use disintensification and abandonment are also important landscape change processes occurring in Europe (Plieninger et al. 2016). "Disintensification" refers to changes to reduce the intensity of land management and the contraction of agricultural land, including abandonment (see van Vliet et al. 2015); while "abandonment" describes the abandonment of any area previously used for agricultural purposes, including croplands and grazing areas (see Benayas et al. 2007). Both intensification and its counterpart have important social and ecological consequences, and have become equally important in sustainable landscape management research (Alcantara et al. 2012; Levers et al. 2016; Plieninger et al. 2014).

Considering the different objectives and questions addressed, Bürgi et al. (2017) identified three main approaches to landscape change research:

- *Local case studies* to grasp the specifics of a place and its development, which is essential for the systematic understanding of local changes. Research usually takes place via oral history interviews, local texts and historical maps (e.g. Bürgi et al. 2017; Lieskovský and Bürgi 2018).
- *Large-scale analyses* to search for the main trends and processes over large areas using European spatial and statistical data analysis (e.g. Kuemmerle et al. 2016) and broad-scale narratives (e.g. Jepsen et al. 2015).

- *Meta-analyses* and overview publications to identify general patterns within the case studies on how specific landscape characteristics and socioeconomic circumstances in combination with potential driving forces can lead to predictable change (e.g. van Vliet et al. 2015).

In the next section, we present an overview of the findings of 18 studies on recent major landscape change trends and driving forces at the European level (from a few decades to 200 years). These studies have been carried out taking a pan-European approach, including both European large-scale analysis studies and meta-analysis of case studies (see Table 2.1). These studies were performed within two collaborative projects sponsored by the EU—HERCULES (Sustainable Futures for Europe's Heritage in Cultural Landscapes) and VOLANTE (Visions of land use transitions in Europe). Both projects share many features with the BMBF's Programme on Sustainable Land Management, such as in taking a systems perspective on landscapes, in investigating the links between land management and ecosystem services and in creating actionable knowledge for land use policy and practice.

2.3 Trends in Land Use and Land Cover Change

A recent large-scale analysis of land use changes in Europe between 1990 and 2006 undertaken by Kuemmerle et al. (2016) serves as an appropriate starting point to introduce this section. The current composition of land cover types in Europe is dominated by agricultural land (representing 41.1% of the territory, mainly dedicated to arable land, permanent crops and grassland), followed by forest land (32.6%) and unused or abandoned land (15.8%), leaving the remaining land to urban and industrial areas and infrastructure (Eurostat 2017). The main land cover change in recent decades identified in this study is a decrease in croplands, followed by an expansion of areas covered by pastures and forests (partly due to the aforementioned decrease in crop production) and, to a lesser extent, by an increase in urban land (Fig. 2.1; Kuemmerle et al. 2016). The most dramatic declines in croplands were found in the east of Europe and the Mediterranean, while some hotspots for their expansion were found in areas of the Netherlands, Germany, France and Ireland (Kuemmerle et al. 2016). The expansion of urban land happened mainly around capital cities and along the Mediterranean coast (Kuemmerle et al. 2016; Levers et al. 2015).

Interestingly, in contrast to the widespread perception that landscapes have been undergoing intense transformation in Europe, Levers et al. (2015) and Kuemmerle et al. (2016) identified stability in land cover as one of the most common recent trajectories, particularly in Central, Western and Northern Europe. Lieskovský and Bürgi (2018) carried out a study of the persistence of the land cover across Europe since 1900 (Fig. 2.2) and found that the most persistent land covers were forests and settlements (about 80% of the 1900-era forest cover and settlement areas were persistent), while grasslands and croplands were the most dynamic and least persistent ones.

Table 2.1 Studies synthesised in this chapter

Study	Approach	Major contribution to landscape change research in Europe
Alcantara et al. (2012)	Large-scale analysis (Eastern Europe)	Advanced research methods for mapping abandoned agricultural land at broad scales by using coarse-resolution satellite imagery and plant phenology data. It also provided some insights into agricultural abandonment trends in Eastern Europe
Prishchepov et al. (2012)	Large-scale analysis (Eastern Europe)	Used the case of the collapse of socialism in Eastern Europe and the former Soviet Union as a natural experiment to investigate whether the rates of agricultural land abandonment responded to different types of institutional changes, based on multi-seasonal Landsat TM/ETMC satellite images
Griffith et al. (2013)	Large-scale analysis	Assessed agricultural land change in the Carpathian ecoregion from 1985 to 2010, using Landsat imagery
Verburg et al. (2013b)	Review	Provided an overview of current research practices in landscape assessments and advocated a land use change research approach that would not only focus on dominant land covers, but also on the landscape structure and composition, and its importance for the functioning of the landscape
Levers et al. (2014)	Large-scale analysis	Compiled time series of sub-national forest harvesting intensity patterns in Europe from 2000 to 2010 and quantified the influence of a wide set of biophysical, infrastructure-related and socioeconomic variables in shaping these patterns

(continued)

Table 2.1 (continued)

Study	Approach	Major contribution to landscape change research in Europe
Munteanu et al. (2014)	Meta-analysis	Analysed broad landscape change configurations and processes over the past 250 years as well as the underlying drivers. This meta-analysis covered the Carpathian region, using 102 case studies from 66 publications
Plieninger et al. (2014)	Meta-analysis (Mediterranean Basin)	Examined the consequences of land use abandonment on biodiversity in the Mediterranean Basin and found that the directions and intensities of response in species richness and abundance to land abandonment were heterogeneous and context-dependent throughout the Mediterranean region
Estel et al. (2015)	Large-scale analysis (Europe)	Developed a new methodology to map the extent and spatial patterns of active and fallow farmland annually at a continental scale based on MODIS satellite data
Jepsen et al. (2015)	Large-scale analysis (Europe)	Went back 200 years to identify broad management regimes and the institutional, social and technical forces within each regime that drove land use changes in Europe. The study combined narratives of change compiled by land use experts with quantitative data
Levers et al. (2015)	Large-scale analysis	Identified and mapped landscape archetypes as well as archetypical change trajectories of landscapes between 1990 and 2006 using a clustering approach based on self-organising maps and 12 land use indicators

(continued)

Table 2.1 (continued)

Study	Approach	Major contribution to landscape change research in Europe
van Vliet et al. (2015)	Meta-analysis	Systematically analysed case studies on land use change to provide a review of the manifestations and underlying drivers of agricultural land change in Europe in recent decades
Kuemmerle et al. (2016)	Large-scale analysis	Studied spatial patterns in the distribution of hotspots and cold spots of land cover and land use intensity changes across Europe between 1990 and 2006. The study made use of European statistical data to compile a database of high-resolution land use change indicators
Levers et al. (2016)	Large-scale analysis	Provided insights into broad-scale agricultural intensity patterns in Europe between 1990 and 2007 by focusing on yields and fertiliser application for six major crop-type groups
Plieninger et al. (2016)	Meta-analysis	Systematically reviewed 144 studies to provide insights into the driving forces of landscape change in Europe
van der Sluis et al. (2016)	Cross-site comparison	Analysed 437 landowner interviews in relation to changes in land use intensity and agricultural production in six case studies in Europe between 2001 and 2011
Bürgi et al. (2017)	Cross-site comparison	Analysed landscape changes in the last 150 years, their drivers and the perception of these changes by locals in six European municipalities. The study combined land use and land cover analysis based on historical maps with oral history interviews

(continued)

Table 2.1 (continued)

Study	Approach	Major contribution to landscape change research in Europe
Lieskovský and Bürgi (2018)	Cross-site comparison and large-scale analysis	Presented an innovative approach to studying patterns of landscape stability and the corresponding stabilising factors. The paper developed a persistence index, and combined different scales of analysis using historical land cover and topographic maps
Verkerk et al. (2018)	Large-scale analysis	Built on existing participatory scenarios of desired land use configuration in Europe and plausible future projections based on current conditions to identify potential policy pathways to link the two

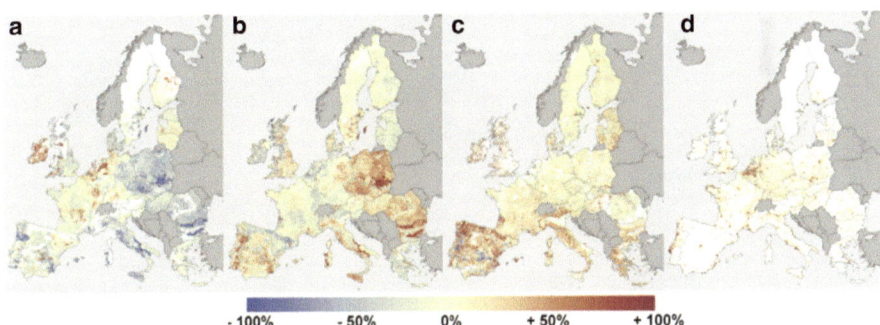

Fig. 2.1 Relative area changes in the extent of: **a** cropland, **b** pasture, **e** forestland and **f** urban land in Europe between 1990 and 2006. *Source* Kuemmerle et al. (2016) (images C and D in the original figure have not been included here)

Some crops, such as vineyards, agriculture mosaics and orchards, displayed a larger degree of persistence than others. For example, olive groves persist in a landscape over long timelines even after being abandoned. Hotspots of persistence were found in remote areas where conversion into intensive agriculture would not be viable, but also in areas particularly suitable for agriculture that have remained stable. Hotspots of change were also found in areas of major political instability, such as in the Baltic area or in areas in the south of Europe.

While land cover in Europe has remained relatively stable, the level of intensity in which the land has been used and managed has not. In fact, in the past few decades, European land use has predominantly changed along intensification gradients (Kuemmerle et al. 2016). In light of these changes, land use intensity as

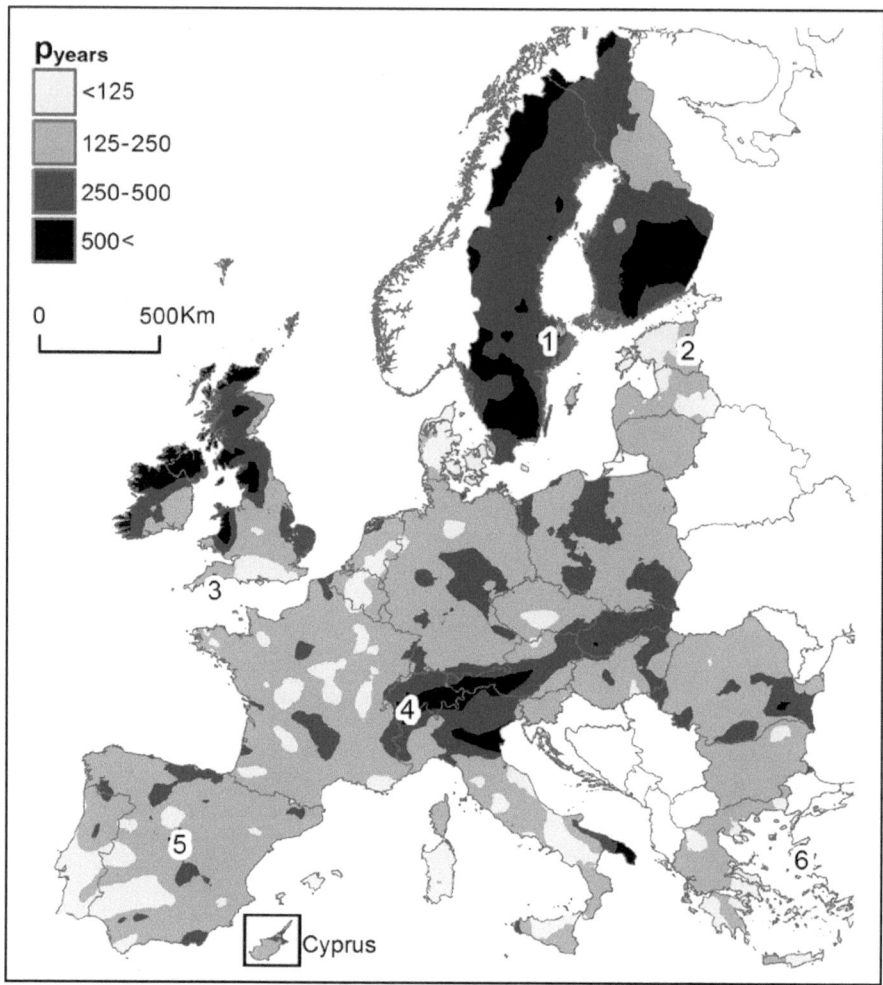

Fig. 2.2 Persistence index for Europe (years needed for the transformation of land cover if it would occur at the same speed as it occurred in the time period 1900–2010). *Source* Lieskovský and Bürgi (2018)

such has become a key area of study. Comparing these levels of intensity provides a comprehensive picture of how European landscapes have evolved (e.g. croplands may present very different characteristics depending on the intensity of how agricultural production is carried out).

There are two main processes of intensity changes: disintensification and intensification (Plieninger et al. 2016) of agriculture and of forestry. van Vliet et al. (2015, p. 28) reviewed 218 case study research articles on agricultural land use change, and defined these processes in the following way:

- "Intensification of agricultural land primarily manifests itself as an increase in land management intensity, for example through increase in livestock density or mechanization. In addition, intensification was observed as an expansion of agricultural land, a decrease in landscape elements, changes toward more intensive agricultural activities and specialization of land use activities"
- "Disintensification of agricultural land is primarily manifested as contraction, partly caused by farmers abandoning their land, but also partly caused by conversion to urban land and natural areas. To a lesser extent, disintensification is manifested as a decrease in land management intensity, as a change to a less intensive agricultural activity, as on-farm diversification and as an increase in landscape elements."

The most prominent process, as highlighted by several authors, is the disintensification of land use (Kuemmerle et al. 2016; Plieninger et al. 2016). Kuemmerle et al. (2016) found this process in approximately 30% of Europe's coverage, while intensification processes were present only in about 11%. Expansion and contraction of the agricultural land area and agricultural abandonment has always occurred (Alcantara et al. 2012); however, agricultural land abandonment in Europe became the most prominent change after 1990 (Estel et al. 2015; Levers et al. 2016; Plieninger et al. 2016). This is partly due to the drastic institutional and socioeconomic reorganisation that occurred after 1990 in former socialist countries, where land abandonment has been particularly prominent (Griffiths et al. 2013; Levers et al. 2015; Prishchepov et al. 2012). For instance, cropland abandonment was the most common land use change in the Carpathians (Griffiths et al. 2013; Munteanu et al. 2014). Land abandonment has also been particularly intense in remote mountain areas, less productive soils and areas where urbanisation processes have been intense, such as on the Mediterranean coast (Levers et al. 2015). Notwithstanding the magnitude of this process, some authors have claimed that land abandonment has not yet received enough attention, in part due to general attention to the worldwide expansion of land management activities and accelerating competition for land (Estel et al. 2015; Plieninger et al. 2016).

Land use intensification was most pronounced between 1960 and 1980, and since that decade, intensity levels have remained stable (Fig. 2.3; Kuemmerle et al. 2016; van der Sluis et al. 2016). When looking at crop yields and the amounts of nitrogen and pesticide application as indicators of land use intensity in recent decades, crop yields have stabilised, while nitrogen and pesticide use has generally decreased thanks to increased farming efficiency, greater environmental awareness and more restrictive regulations (van der Sluis et al. 2016). These increasing yields were most pronounced in Western Europe (Kuemmerle et al. 2016). It is also in Western Europe where nitrogen application rates were higher, although the use of nitrogen has decreased since the 1990s. As for the use of fertilisers, declines were observed in Southeastern Europe (e.g. in Romania) and in some countries from Central and Western Europe (e.g. in Germany and France), while there were increases in some Eastern countries (e.g. in Poland) (Kuemmerle et al. 2016).

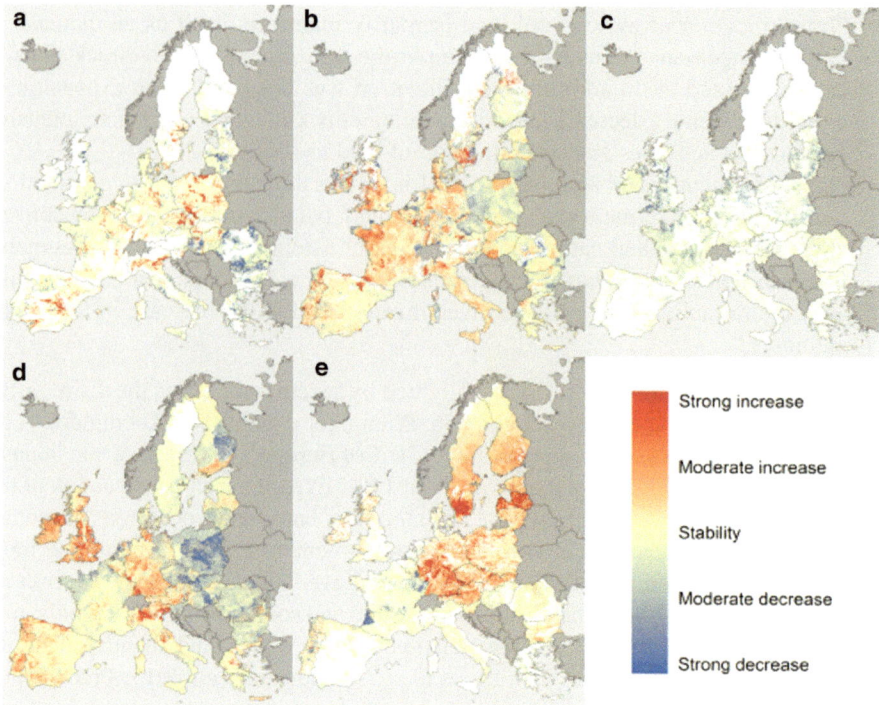

Fig. 2.3 Spatial patterns of changes in intensity: **a** fertiliser use on cropland, **b** crop yields, **c** livestock density, **d** biomass removal from grazing land and **e** roundwood production within broad land use classes in Europe between 1990 and 2006. *Source* Kuemmerle et al. (2016)

Summing up, in line with the geographic patterns of land cover change, since the 1990s intensification of agriculture has mainly taken place in Northern and Western Europe, while land abandonment and disintensification has prevailed in Eastern and Southern Europe (Kuemmerle et al. 2016; Levers et al. 2015; Plieninger et al. 2016).

However, when narrowing down the scale of analysis, land use intensification and abandonment often appear together within the same landscape (Plieninger et al. 2016). van Vliet et al. (2015) identified this mutual occurrence as a major trajectory of land use change in Europe as a result of the globalisation of agricultural markets. This polarising trend leads to a homogenisation of the landscape, where more productive areas are immersed in specialisation and intensification processes, concentrating most of the production, while marginal areas are abandoned as its use as agricultural land turns unprofitable (Levers et al. 2015), with the loss of traditional agricultural landscapes as one of the consequences (Kuemmerle et al. 2016). In the same vein, due to differences in the productivity of regions, the intensification of forestry did not necessarily happen in areas where forest land cover was expanding (Levers et al. 2015). In fact, forest cover expanded in areas that had been abandoned (Kuemmerle et al. 2016; Levers et al. 2015).

2.4 Drivers of Landscape Change

Plieninger et al. (2016) identified five groups of underlying drivers of landscape change: political/institutional (e.g. agricultural and forest policy, spatial development policy and property rights); economic (e.g. structural changes in agriculture, prices for agricultural products, market growth and commercialisation); technological (e.g. modernisation of society and land management, such as introduction of mineral fertiliser and tractors; Jepsen et al. 2015); cultural (e.g. demography, attitudes and behaviour); and natural/spatial (e.g. climate, topography and spatial configuration). It is important to note that these different types of drivers are usually combined (Plieninger et al. 2016). Here, we describe some of the most frequently mentioned ones.

2.4.1 Political and Institutional Drivers

Political and institutional factors, such as policies that regulate agriculture, forestry and spatial development, land reforms and property rights, appear as the dominant drivers of change in various studies (e.g. Jepsen et al. 2015; Munteanu et al. 2014; Prishchepov et al. 2012).

On a broad political and institutional scale, Jepsen et al. (2015) reviewed narratives on the drivers of land use change to provide a broad storyline of the succession of the main land management regimes in Europe in the last two centuries. They described two main regimes after World War II that were to have an important impact on land use across Europe. These two regimes help scholars understand the frequently mentioned East–West dichotomy (Kuemmerle et al. 2016; Levers et al. 2015; van Vliet et al. 2015): in Western Europe, the industrialisation regime; in Eastern Europe, the establishment and subsequent collapse of the collectivisation regime (Jepsen et al. 2015). The industrialisation regime in Western Europe was characterised by the introduction of new technologies and the adoption of commercial farming specialised in crop or livestock production oriented towards the global market (Jepsen et al. 2015). In the Eastern European countries, there were two distinct phases. The first phase was the collectivisation regime between 1945 and 1991, characterised by major land reform and the establishment of large collective and state farms and centrally planned intensification of agriculture (Jepsen et al. 2015). The second phase was characterised by the collapse of these collectivisation regimes and the consequent dismantling of the collective farm structure, as well as the state-supported, capital-intensive socialist farming model. Here, the intensity of farming practices persisted, but according to commercial premises. Since then, two trends have been observed in Central and Eastern Europe: on the one hand, the acquisition of former state and collective farms by large agro-businesses and, on the other hand, the orientation towards subsistence farming or the abandonment of the most marginal land (Jepsen et al. 2015).

This East–West dichotomy is particularly recognisable in changes to cropland systems, with fairly constant cropland area but stable or increasing land management intensity in the West, partly due to the strategic support from the EU's Common Agricultural Policy (CAP) (Kuemmerle et al. 2016); and abandonment in the East (Kuemmerle et al. 2016; Plieninger et al. 2016; Prishchepov et al. 2012). However, there are also regional differences on how institutional and political drivers have affected landscape change patterns. For example, national policies during Soviet times led to agricultural expansion in some areas (e.g. Hungary) and to abandonment of agriculture in others (e.g. Romania) (Munteanu et al. 2014). Prishchepov et al. (2012) observed higher abandonment rates in countries with changing or inadequately established institutions designed to regulate land use changes (e.g. Latvia, Lithuania and Russia).

Since the entry of Eastern countries into the EU starting in 2004, the situation has changed. Jepsen et al. (2015) identified a new regime beginning in the 1990s all over Europe driven by environmental awareness of the impact of agricultural production, triggering agro-environmental policies that have subsequently had an impact on land use and land cover across Europe.

Political and institutional drivers are also important explanatory factors of the prevalent land cover stability observed especially in Western Europe. Land use policies frequently hinder drastic changes in landscapes across the EU, for example by providing economic support to farmers in less favoured areas (Levers et al. 2015). The CAP plays a crucial role in this stability, with the decoupling of CAP payments regarded to be an important element for preserving extensive grazing systems that otherwise would be abandoned (Levers et al. 2015). However, the CAP has also had the opposite effect in some places, leading to land abandonment (Bürgi et al. 2017) and changes in land use when the EU implemented eligibility criteria for payments based on a quota system.

2.4.2 Economic, Technological and Cultural Drivers of Change

Of the driving factors that shape society as a whole, urbanisation in particular appears to be a prominent factor triggering landscape change (Bürgi et al. 2017). In Europe, almost 75% of the population nowadays live in urban areas (Eurostat 2016), which typically have expanded at the expense of agricultural land. The rural exodus due to diminishing income opportunities in marginal areas in contrast to increasing opportunities in urban areas is one of the most recurrent sociocultural drivers explaining land abandonment (Levers et al. 2015). Moreover, with an increasingly urban society, new uses such as recreation activities have emerged around urban areas that can compete with agriculture and forestry activities (van Vliet et al. 2015). Urbanisation is typically accompanied by an increased purchasing power and higher demand for commodities. This in turn translates into increasing pressure on ecosystems and a specialisation of the service supply of many landscapes (Verburg et al. 2013b).

Globalisation and other related economic drivers have also brought about important changes in land use. The global agriculture market has pushed farmers to intensify their production methods in order to remain competitive, while those that do not succeed have had to find their livelihood outside agriculture. This has often led to the abandonment of areas with less favourable conditions (van Vliet et al. 2015). Another consequence of globalisation is the specialisation of farm production, with the expansion of monocultures and the "outsourcing" of, for example, fodder production not only outside of livestock farms, but even outside Europe. This decreases the need to intensification of production in Europe and contributes to the abandonment of those areas that are less suitable for agriculture (Levers et al. 2015). The technological innovations associated with globalisation have also brought about important changes in farming systems (Bürgi et al. 2017). Mineral nitrogen application is one of the most mentioned technological innovations when studying the intensification of agriculture. This has important effects on landscape configuration, with the disappearance of traditional features and a loss of biodiversity (van Vliet et al. 2015).

Apart from these commonly mentioned drivers of change, growing environmental, social and political awareness is reflected in land use policies and subsidies, with an increasing focus on environmental management, nature preservation and landscape restoration, rather than on agricultural production (van Vliet et al. 2015). But beyond these institutional levels, farmers' decisions are also an important factor that can ultimately lead to diverging land use trajectories (van Vliet et al. 2015). Most of these decisions are the result of the behaviour of land owners and land managers responding to market prices and policy incentives in varying ways (Verburg et al. 2013b). While farmers' attitudes (e.g. productivist or environmentalist) may not be an important driver of land cover changes, they do influence the intensity of management practices (van Vliet et al. 2015).

2.4.3   Spatial and Natural Factors

In the context of urbanisation and globalisation, accessibility is another important factor when explaining the geographic distribution of more intensively or less intensively managed areas. In areas with good accessibility, land use management is often more intensive; in more remote areas, land abandonment is more frequent (Levers et al. 2015). Bürgi et al. (2017) analysed narratives on the driving forces of landscape change, using oral history interviews with local residents in six case studies. They found that access and infrastructure (e.g. railways and highways) were important drivers in most of the cases (Bürgi et al. 2017). With respect to agricultural and forestry production, this means that local products can be exported and therefore, production patterns might change. However, at the same time farmers reported that they needed to become more competitive against products that came from the outside. The flow of people also increases with better accessibility and can bring about significant changes in the landscape if, for instance, the surrounding areas of a big city turn into commuter cities or second-home areas.

Finally, as Levers et al. (2015) observed, notwithstanding technical improvements and the increasing capacity of humans to modify the land, agro-climatic conditions still constitute an important factor to take into consideration. Intensified crop production prevails in areas with favourable conditions; forest and grasslands dominate in areas with disadvantageous edaphic and climatic conditions, although institutional and socioeconomic factors can alter this pattern to a certain extent. Nevertheless, climate change as a driver of landscape change has still not played a very evident role in the studies considered in this chapter, except for the melting of the glaciers in the Alps (Bürgi et al. 2017).

2.5 Operationalising Current Trends and Drivers of Change Towards Developing Future Scenarios of Landscape Change

In an effort to hinder the tendency towards polarisation of the landscape (intensification or abandonment) and the landscape homogenisation and loss of multifunctionality that results from it, some authors have worked together with stakeholders in the visualisation of future landscape change trajectories. Pérez-Soba et al. (2015) worked with stakeholders representing the main land use sectors in Europe to develop three overarching visions of the desired futures envisaged for Europe. These three visions shared a common ambition: to generate multifunctional land uses in Europe that would integrate multiple social, ecological, economic and cultural demands. Recently, Verkerk et al. (2018) studied potential pathways and policies required to achieve these visions of multifunctionality. These pathways would vary in the specific mechanisms involved, but all of them would entail major interventions across Europe, depending on the environmental and socioeconomic context. In general, the space dedicated to agricultural land would need to be severely restricted in favour of larger, interconnected natural areas. These studies highlighted the challenge of identifying a pathway toward reaching landscape multifunctionality at the local level. This was mostly due to methodological constraints; models still cannot capture many of the complexities associated with multifunctionality at a local-scale resolution. However, Verkerk et al. (2018) pointed out that the policy interventions that were needed to navigate that pathway would necessarily require the strengthening and maintenance of Europe's existing traditional multifunctional landscapes, most of which have been subject to long-term trends of decline.

2.6   Research Gaps and Ways Forward Towards Landscape Sustainability

Without fundamental social and economic changes, the demand for natural resources will multiply in the next few decades. The effects of this growing demand will intensify current land use trends, increasing the impacts described in the previous sections. Therefore, one of the greatest sustainability challenges in landscape change research is to identify strategies that will meet society's demands without further threatening and degrading European landscapes and their functioning. In light of this, a broad array of literature has identified pathways towards tackling the challenges of landscape change research in Europe and beyond (e.g. Bürgi et al. 2004; Plieninger et al. 2016). Here, we use our review to summarise the major contributions and derive key opportunities to make advancements in landscape change science:

- **Research for more context-specific, regionalised policymaking** (Kuemmerle et al. 2016). A deeper understanding of the outcomes of land use change for ecosystem service flows and biodiversity can contribute to mitigating the trade-offs among different land uses. In that regard, one interesting arena for future research would be to derive typologies of typical land use changes and the effects of policy interventions that characterise Europe (e.g. Levers et al. 2015; Kuemmerle et al. 2016). There is also a need for research that advances the study of landscape stability patterns and the drivers behind these patterns (Plieninger et al. 2016) in the local context, which could be very relevant for landscape management and policy.
- **Uncovering the complexity behind the drivers**. Comprehension of the underlying drivers of landscape change remains partial (Jepsen et al. 2015). Landscape research has traditionally considered the spatial determinants of land use changes (e.g. topography, soil quality, market access) and land use decisions as separate items. Future research should incorporate joint analysis that favours an understanding of complex behaviour and the linkages behind the various drivers. In addition, future studies should acknowledge that the landscape change effects of different underlying drivers reveal themselves at different time intervals, making attribution difficult if only short time spans are considered in the analysis (Jepsen et al. 2015).
- **Avoiding oversimplification of the complex realities of the land**. Verburg et al. (2013a) criticised global and supra-regional assessments for oversimplifying the complex reality of landscapes. In that regard, performing cross-site comparison studies based on place-based research (e.g. Bürgi et al. 2017) could help scholars grasp these complex realities behind landscapes. Beyond the spatial scale, another important gap in landscape change studies is the identification and comprehension of the diversity of actors and their role in landscape changes (Plieninger et al. 2016; Kizos et al. 2018). Understanding the inherent complexities of landscape change imply the incorporation of a plurality of research approaches and of underlying

conceptualisations of human–environment interactions that can encompass the full complexity of land use developments.

- **Jointly analysing changes in the extent and intensity of land use and disparate linkages**. In an increasingly connected world, the already ongoing trend of spatial disconnection between production and consumption landscapes will only increase. This entails considerable challenges to sustainability understood as imbalances in environmental degradation. Therefore, it is important to analyse how spatial patters in changes in the extent and intensity of land use relate to changes in distant places (Kuemmerle et al. 2016).
- **Incorporating innovative approaches to evaluating landscape change, and promoting the co-design of research to address societal problems**. The importance of producing actionable knowledge in collaboration with stakeholders is gaining significance across landscape research (Verburg et al. 2013a). For this, a more participatory approach, in which local knowledge and perceptions are taken into account, is needed. In that regard, previous authors have discussed the importance of combining information on land use changes derived from field and GIS procedures with perceptions of the local population. This facilitated the development of mixed-method approaches and takes advantage of the complementarity and the specific strengths of the inclusion of a variety of types of data sources (e.g. Bürgi et al. 2017).

2.7 Conclusions

Land cover in Europe in the past few decades has remained relatively stable. In those areas where it changed, it has predominately been towards the diminishing of cropland area in favour of grasslands, and a general increase in urban areas. A more nuanced perspective emerges when looking at how the intensity of landscape management has changed. Here, two opposing but co-occurring processes can be seen: the intensification of agriculture on the one hand, and the abandonment of farming activities on the other. Both of these processes generally involve the loss of biodiversity, an erosion of cultural heritage and a diminishing of landscape multi-functionality. These trends have pushed European social-ecological systems beyond the boundaries of environmental and societal well-being. In order to change these trajectories, it is crucial to identify and understand the factors that drive them. Political and institutional drivers seem to be the most prominent ones, but economic, technological, cultural and natural aspects also play a very important role, and need to be considered as well. In this sense, although European-scale studies provide extremely relevant information to identify broad trends and drivers of land cover and land use change, more place-based analyses are needed where different sources of information are combined (by engaging with local stakeholders), various approaches and disciplines are brought together and several temporal and spatial scales are taken into account. This is necessary for a comprehensive understanding of how and why landscapes are changing, and what the consequences of these changes will be.

References

Alcantara, C., Kuemmerle, T., Prishchepov, A. V., & Radeloff, V. C. (2012). Mapping abandoned agriculture with multi-temporal MODIS satellite data. *Remote Sensing of Environment, 124*, 334–347. https://doi.org/10.1016/j.rse.2012.05.019

Antrop, M. (2005). Why landscapes of the past are important for the future. *Landscape and Urban Planning, 70*(1–2), 21–34. https://doi.org/10.1016/j.landurbplan.2003.10.002

Benayas, J. M. R., Martins, A., Nicolau, J. M., & Address:, J. J. S. (2007). Abandonment of agricultural land: an overview of drivers and consequences. *CAB Reviews: Perspectives in Agriculture, Veterinary Science, Nutrition and Natural Resources, 2*(57). https://doi.org/10.1079/PAVSNNR20072057.

Bürgi, M., Hersperger, A. M., & Schneeberger, N. (2004). Driving forces of landscape change—Current and new directions. *Landscape Ecology, 19*(8), 857–868. https://doi.org/10.1007/s10980-005-0245-3

Bürgi, M., Bieling, C., von Hackwitz, K., Kizos, T., Lieskovský, J., Martín, M. G., & Printsmann, A. (2017). Processes and driving forces in changing cultural landscapes across Europe. *Landscape Ecology, 32*(11), 2097–2112. https://doi.org/10.1007/s10980-017-0513-z

Council of Europe. (2000). The European Landscape Convention. October 2000, Florence.

Erb, K. H. (2012). How a socio-ecological metabolism approach can help to advance our understanding of changes in land-use intensity. *Ecological Economics, 76*, 8–14. https://doi.org/10.1002/hbm.22248

Estel, S., Kuemmerle, T., Alcántara, C., Levers, C., Prishchepov, A., & Hostert, P. (2015). Mapping farmland abandonment and recultivation across Europe using MODIS NDVI time series. *Remote Sensing of Environment, 163*, 312–325. https://doi.org/10.1016/j.rse.2015.03.028

Eurostat. (2016). *Urban Europe—Statistics on cities, towns and suburbs—Working in cities—Statistics Explained*. Retrieved from https://ec.europa.eu/eurostat/statistics-explained/index.php/Urban_Europe_-_statistics_on_cities,_towns_and_suburbs_-_working_in_cities.

Eurostat. (2017). *Land use statistics* (Vol. 2015). Retrieved from https://ec.europa.eu/eurostat/statistics-explained/index.php?title=Land_use_statistics.

Griffiths, P., Müller, D., Kuemmerle, T., & Hostert, P. (2013). Agricultural land change in the Carpathian ecoregion after the breakdown of socialism and expansion of the European Union. *Environmental Research Letters, 8*(4). https://doi.org/10.1088/1748-9326/8/4/045024.

Jepsen, M. R., Kuemmerle, T., Müller, D., Erb, K., Verburg, P. H., Haberl, H., & Reenberg, A. (2015). Transitions in European land-management regimes between 1800 and 2010. *Land Use Policy, 49*, 53–64. https://doi.org/10.1016/j.landusepol.2015.07.003

Kirchhoff, T., Trepl, L., & Vicenzotti, V. (2013). What is landscape ecology? An analysis and evaluation of six different conceptions, landscape research. https://doi.org/10.1080/01426397.2011.640751.

Kizos, T., Verburg, P. H., Bürgi, M., Gounaridis, D., Plieninger, T., Bieling, C., & Balatsos, T. (2018). From concepts to practice: Combining different approaches to understand drivers of landscape change. *Ecology and Society, 23*(1). https://doi.org/10.5751/ES-09910-230125.

Kuemmerle, T., Levers, C., Erb, K., Estel, S., Jepsen, M. R., Müller, D., & Reenberg, A. (2016). Hotspots of land use change in Europe. *Environmental Research Letters, 11*(6), 1–14. https://doi.org/10.1088/1748-9326/11/6/064020

Levers, C., Verkerk, P. J., Müller, D., Verburg, P. H., Butsic, V., Leitão, P. J., & Kuemmerle, T. (2014). Drivers of forest harvesting intensity patterns in Europe. *Forest Ecology and Management, 315*, 160–172. https://doi.org/10.1016/j.foreco.2013.12.030

Levers, C., Müller, D., Erb, K., Haberl, H., Jepsen, M. R., Metzger, M. J., & Kuemmerle, T. (2015). Archetypical patterns and trajectories of land systems in Europe. *Regional Environmental Change, 18*(3), 715–732. https://doi.org/10.1007/s10113-015-0907-x

Levers, C., Butsic, V., Verburg, P. H., Müller, D., & Kuemmerle, T. (2016). Drivers of changes in agricultural intensity in Europe. *Land Use Policy, 58*, 380–393. https://doi.org/10.1016/j.landusepol.2016.08.013

Lieskovský, J., & Bürgi, M. (2018). Persistence in cultural landscapes: A pan-European analysis. *Regional Environmental Change, 18*(1), 175–187. https://doi.org/10.1007/s10113-017-1192-7

Munteanu, C., Kuemmerle, T., Boltiziar, M., Butsic, V., Gimmi, U., Halada, L., & Radeloff, V. C. (2014). Forest and agricultural land change in the Carpathian region—A meta-analysis of long-term patterns and drivers of change. *Land Use Policy, 38,* 685–697. https://doi.org/10.1016/j.landusepol.2014.01.012

Pérez-Soba, M., Paterson, J., & Metzger, M. (2015). *Visions of future land use in Europe Stakeholder visions for 2040.*

Plieninger, T., Hui, C., Gaertner, M., & Huntsinger, L. (2014). The impact of land abandonment on species richness and abundance in the Mediterranean Basin: A meta-analysis. *PLoS ONE, 9*(5). https://doi.org/10.1371/journal.pone.0098355.

Plieninger, T., Draux, H., Fagerholm, N., Bieling, C., Bürgi, M., Kizos, T., & Verburg, P. H. (2016). The driving forces of landscape change in Europe: A systematic review of the evidence. *Land Use Policy, 57,* 204–214. https://doi.org/10.1016/j.landusepol.2016.04.040

Pohl, C. (2008). From science to policy through transdisciplinary research. *Environmental Science & Policy 11.1,* 46–53. https://doi.org/10.1016/j.envsci.2007.06.001.

Prishchepov, A. V., Radeloff, V. C., Baumann, M., Kuemmerle, T., & Müller, D. (2012). Effects of institutional changes on land use: Agricultural land abandonment during the transition from state-command to market-driven economies in post-Soviet Eastern Europe. *Environmental Research Letters, 7*(2). https://doi.org/10.1088/1748-9326/7/2/024021.

Sayer, J., Sunderland, T., Ghazoul, J., Pfund, J. L., Sheil, D., Meijaard, E., et al. (2013). Ten principles for a landscape approach to reconciling agriculture, conservation, and other competing land uses. *Proceedings of the National Academy of Sciences, 110*(21), 8349–8356. https://doi.org/10.1073/pnas.1210595110

Stürck, J., Levers, C., van der Zanden, E. H., Schulp, C. J. E., Verkerk, P. J., Kuemmerle, T., et al. (2018). Simulating and delineating future land change trajectories across Europe. *Regional Environmental Change, 18,* 733–749. https://doi.org/10.1007/s10113-015-0876-0.

Turner, M. G. (2005). Landscape ecology: What Is the state of the science? *Annual Review of Ecology Evolution and Systematics, 36,* 319–344. https://doi.org/10.1146/annurev.ecolsys.36.102003.152614

van der Sluis, T., Pedroli, B., Kristensen, S. B. P., Lavinia Cosor, G., & Pavlis, E. (2016). Changing land use intensity in Europe—Recent processes in selected case studies. *Land Use Policy, 57,* 777–785. https://doi.org/10.1016/j.landusepol.2014.12.005

van Vliet, J., de Groot, H. L. F., Rietveld, P., & Verburg, P. H. (2015). Manifestations and underlying drivers of agricultural land use change in Europe. *Landscape and Urban Planning, 133,* 24–36. https://doi.org/10.1016/j.landurbplan.2014.09.001

Verburg, P. H., Mertz, O., Erb, K. H., Haberl, H., & Wu, W. (2013a). Land system change and food security: Towards multi-scale land system solutions. *Current Opinion in Environmental Sustainability, 5*(5), 494–502. https://doi.org/10.1016/j.cosust.2013.07.003.

Verburg, P. H., van Asselen, S., van der Zanden, E. H., & Stehfest, E. (2013b). The representation of landscapes in global scale assessments of environmental change. *Landscape Ecology, 28*(6), 1067–1080. https://doi.org/10.1007/s10980-012-9745-0.

Verkerk, P. J., Lindner, M., Pérez-Soba, M., Paterson, J. S., Helming, J., Verburg, P. H., & van der Zanden, E. H. (2018). Identifying pathways to visions of future land use in Europe. *Regional Environmental Change, 18*(3), 817–830. https://doi.org/10.1007/s10113-016-1055-7

Weith, T., Besendörfer, C., Gaasch, N., Kaiser, D. B., Müller, K., Repp, A., et al. (2013). Nachhaltiges land management: Was ist das? Diskussionspapier Nr. 7, April 2013, Leibniz-Zentrum für Agrarlandschaftsforschung (ZALF) e.V.

Wickson, F., Carew, A. L., & Russell, A. W. (2006). Transdisciplinary research: Characteristics, quandaries and quality. *Futures, 38*(9), 1046–1059.

Zscheischler, J., Rogga, S., & Busse, M. (2017). The adoption and implementation of transdisciplinary research in the field of land use science—A comparative case study. *Sustainability, 9*(11), 1926. https://doi.org/10.3390/su9111926

Chapter 3
New Trends and Drivers for Agricultural Land Use in Germany

Dieter Kirschke, Astrid Häger, and Julia Christiane Schmid

Abstract Agricultural land use in Germany is faced with new drivers and conflicts. There has been a continuous downward trend in agricultural land use since reunification, and agriculture seems to be increasingly squeezed by various land use demands. Whereas land prices and land rents have stayed rather stable during the 90ies and at the beginning of the new millennium, they have started to go up considerably during the last ten years. At the same time the agricultural sector is faced with deteriorating environmental indicators and a changing land use structure and concentration. International agricultural prices have become a key determinant for land prices in Germany contributing to increasing land prices. Equally, new demands for nature conservation and natural resource protection under the Common Agricultural Policy have contributed to make agricultural land scarcer, and bioenergy production under the Renewable Energy Act has considerably affected land demand and prices in various regions. In East Germany, some land market specialties relate to the farm structure and the land privatization process after reunification. In view of these developments, there is a new policy debate on agricultural land market interventions.

Keywords Agricultural land use · Land markets · Common Agricultural Policy

3.1 Introduction

Land is a limited and scarce natural resource faced with competing and rising demands. These demands, basically, comprise land use for agriculture, nature and natural resource protection, industrial areas and human settlements, and infras-

---

D. Kirschke (✉) · A. Häger · J. C. Schmid
Faculty of Life Sciences—Thaer-Institute—Department of Agricultural Economics,
Humboldt-Universität zu Berlin, Unter den Linden 6, 10099 Berlin, Germany
e-mail: dieter.kirschke@hu-berlin.de

A. Häger
e-mail: astrid.haeger@agrar.hu-berlin.de

J. C. Schmid
e-mail: j.c.schmid@agrar.hu-berlin.de

T. Weith et al. (eds.), *Sustainable Land Management in a European Context*,
Human-Environment Interactions 8, https://doi.org/10.1007/978-3-030-50841-8_3

tructure. Consequently, land use conflicts evolve, and these conflicts may change over time and are driven by various factors. A study by Kirschke et al. (2013) gives a comprehensive overview on land use and land use change in Europe and the main economic drivers behind, bringing together theoretical approaches from regional economics, economic geography, agricultural economics, environmental and resource economics, and infrastructure planning.

Our paper is based on these general findings and specifically looks into new developments in German agricultural land use. Are there new trends for agricultural land use change in the new millennium and, if so, what are the new conflicts and drivers?

In Sect. 3.2, we briefly sketch out the background for our discussion, addressing general land use trends in Germany after reunification. We will then discuss ongoing and new conflicts in agricultural land use in recent years in Sect. 3.3. The section focusses on rising land prices, deteriorating environmental indicators, and land use structure. In Sect. 3.4, we will look at major drivers behind this development which, basically, reflect market forces and a new policy framework. We will discuss international agricultural price developments, technological developments, Common Agricultural Policy (CAP) reforms, the Renewable Energy Act (REA), and particularities related to German reunification. Some concluding remarks are given in Sect. 3.5.

3.2   General Land Use Trends

Land use in Germany is characterized by industrial development and high population density, but also by agriculture. In the "Land use and cover area frame survey (LUCAS)", Eurostat (2012) has compared land use in EU member states. They find a higher share of human settlements and transport infrastructure in Germany as compared to the EU average. Also, the forest share in Germany (33%) is lower than in the EU (37%). However, the share of permanent grassland is somewhat higher (22% as compared to 19%), and the share of arable land is considerably higher (32% as compared to 23%).

Overall trends in land use have been similar in Germany and the EU. According to the "Coordination of information on the environment (CORINE)-Project" there has been a continuous decline of agricultural areas and a continuous increase of areas for human settlements and transport infrastructure. Table 3.1 shows a more detailed picture of land use change in Germany from 1992 to 2015/16.

According to Table 3.1, areas for human settlements and transport infrastructure have considerably increased since reunification, but this trend has come down in the recent period. This picture holds for both human settlements and transport infrastructure individually. For industrial areas and recreation areas there has been an increasing demand over the whole period whereas changing trends can rarely be identified.

Table 3.1 Absolute and relative land use change in Germany, 1992–2015/16

	1992–2000		2000–2008		2008–2015		Land use 2015 (ha)
	Hectares	%	Hectares	%	Hectares	%	
Human settlements and transport infrastructure	363,373	9.0	319,832	7.3	192,914	4.1	4,906,641
of which							
Human settlements	234,745	11.3	133,541	5.8	66,047	2.7	2,507,666
Industry	18,268	33.2	5500	7.5	26,222	33.3	104,961
Recreation	40,379	17.9	112,867	42.5	66,764	17.6	445,484
Transport infrastructure	67,680	4.1	67,229	3.9	31,812	1.8	1,810,805
Cemeteries	2,301	7.0	696	2.0	2,069	5.8	37,725
Extraction land	−8,180	−4.4	−12,696	−7.1	−11,012	−6.6	155,871
Total agricultural area	−408,408	−2.1	−338,197	−1.8	−331,346	−1.8	18,433,248
of which							
Agricultural area used	117,263	0.7	−141,633	−0.8	−266,800[a]	−1.6	16,658,900[a]
of which							
Arable land	336,019	2.9	128,991	1.1	−169,500[a]	−1.4	11,763,000[a]
Permanent grassland	−195,030	−3.7	−258,943	−5.1	−94,200[a]	−2.0	4,694,500[a]
Forest	77,858	0.7	203,477	1.9	216,569	2.0	10,951,461
Water	24,761	3.2	39,688	4.9	7,063	0.8	855,213

Source Own compilation according to Statistisches Bundesamt (2017a); [a]Data from 2016

For agriculture, there has been a continuous and considerable downward trend regarding total agricultural area. For the agricultural area used, there has been an increase in the 90ies which may be explained by adjustment processes in East German agriculture after reunification. Since 2000, however, agricultural area used is continually going down with an increasing trend. In contrast to agriculture, forest areas have increased over the whole period, with an increasing trend since 2000.

Within agriculture, interesting changes of trends for arable land and permanent grassland can be noted. Whereas arable land use has increased during the first periods considered—though with a downward trend in the second period—arable land use has considerably declined in the last period. For permanent grassland, land use has continuously declined over the whole period, though with a less dramatic trend in recent years. Indeed, a more detailed, yearly breakdown of the figures for permanent

Table 3.2 Permanent grassland in Germany, 2013–2018

Permanent grassland	2013	2014	2015	2016	2017	2018
1,000 ha	4,621.0	4,650.7	4,677.1	4,694.5	4,710.2	4,713.4
% of agricultural area used	27.7	27.8	28.0	28.2	28.2	28.3

Source Own compilation according to Statistisches Bundesamt (various years), Fachserie 3, Reihe 3.1.2

grassland shows that from 2013 onwards a slight but visible positive trend can be observed (Table 3.2).

In sum, agricultural land use seems to be increasingly squeezed between various non-agricultural land use demands. Whether or not these developments should be restricted, and if so in which way, have become popular policy topics. The German government (Bundesregierung) has claimed that the increase of areas for human settlements and for transport infrastructure should not surpass 30 ha per day until 2030[1] (it has been 63 ha per day in 2014) (Die Bundesregierung 2016, pp. 158–159). The German Farmers Association (Deutscher Bauernverband—DBV) argues that the loss of agricultural land is still about 70 ha per day and demands a legal framework for the conservation of agricultural areas (DBV et al. 2016, p. 2). Within agriculture, the increase of arable land, at the cost of permanent grassland, has equally been blamed by various actors (e.g. DAFA 2015, p. 11).

3.3 Ongoing and New Conflicts in Agriculture Land Use

The trends for agricultural land use in Germany in recent years indicate that there are both ongoing as well as potentially new land use conflicts. These conflicts will probably result in changes on land markets. In what follows we will look into the development of land market prices first, and then turn to ongoing major conflicts between agricultural land use and nature and natural resource protection. Furthermore, we will focus on land use structure and concentration in German agriculture.

3.3.1 Rising Land Prices

Despite high demand for land and conflicts between different actors, land prices have remained quite stable during the 90ies and at the beginning of the new millennium, both in Western and Eastern Germany. Since 2007/08, however, land prices have gone up considerably.

---

[1]The Bundesregierung recently had to shift the objective year from 2020 to 2030 (Die Bundesregierung 2016, p. 159).

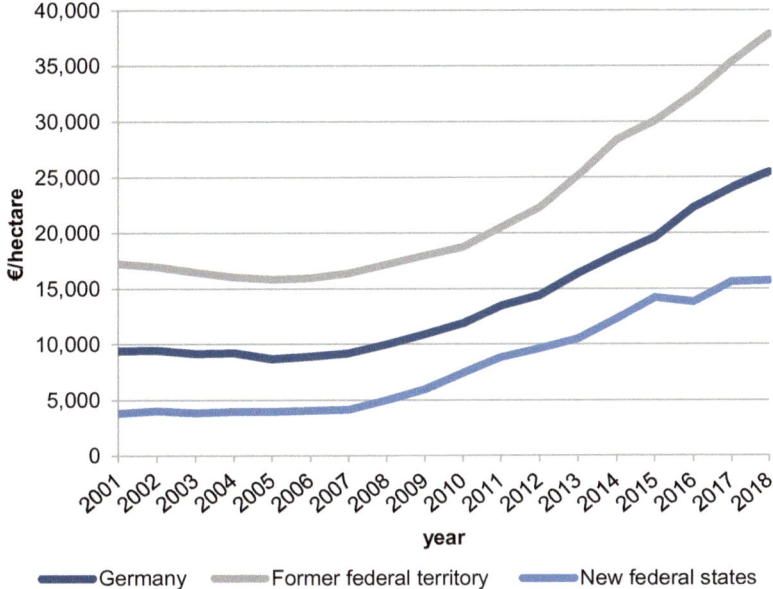

Fig. 3.1 Agricultural land prices in Germany, 2001–2018, €/ha (*Source* Own compilation according to Statistisches Bundesamt (various years), Fachserie 3, Reihe 2.4, Tab. 1.4.)

Figure 3.1 shows the development of agricultural land prices in the period 2001–2018. The figure shows average land purchase values of around 10,000 €/ha until 2007, with around 4,000 €/ha in East Germany and around 17,000 €/ha in West Germany. Since then, land prices almost doubled in West Germany and more than quadrupled in East Germany, which is a remarkable increase in a few years. Though there has been some convergence of land prices in East and West Germany in relative terms, the absolute difference has remained at around 15,000 €/ha for many years, and increased to 22,000 €/ha in 2018.

The rise in land prices is only partly accompanied by more active land markets in terms of the number of sales and the area sold. The number of sales has been around 45,000 between 2010 and 2014, with a slightly increasing trend since then. Half of this increased land market activity takes place in East Germany. The area sold has been around 40,000 ha per year in West Germany over the whole period whereas it has been around 65,000 ha per year in East Germany for many years, decreasing from 2015 to 45,000 ha in 2018.

The overall share of land market sales in total agricultural land used is rather small, with 0.7% for Germany in total and 0.3% for West Germany. However, the picture is different for East Germany: Here, the land mobility has evolved from 1% in 2004 to 1.2% in 2007, to 1.4% in 2015 and 0.8% in 2018 (Statistisches Bundesamt, various years, Fachserie 3, Reihe 2.4).

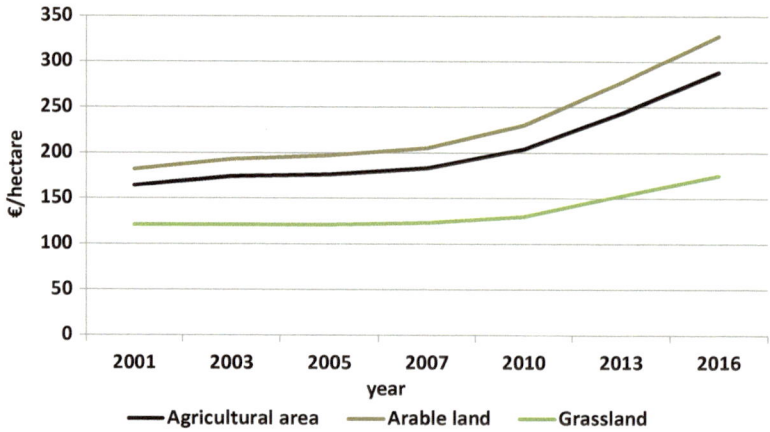

Fig. 3.2 Average land rent in Germany, 2001–2016, €/hectare [*Source* Own compilation according to Statistisches Bundesamt (2016a)]

Interestingly, the development on land markets is not correlated to the quality of land sold. The yield index, a German indicator for the yield potential of agricultural land, for areas sold has, in fact, slightly decreased over the period both for East and West Germany. There is, however, a remarkable difference in the area sold per sale. Whereas this area continuous to stay rather low in West Germany with 1.4 ha per sale, the figure has been around 4–5 ha per sale in East Germany, with increasing numbers of sales regarding areas above 50 ha (Statistisches Bundesamt, various years, Fachserie 3, Reihe 2.4).

The picture for land rents in Germany in the last decade is similar to the picture of land sales. Figure 3.2 shows that average land rents have gone up slightly until 2007, with a sharp increase in the last years. Whereas the average land rent for agricultural areas was 164 €/ha in 2001, it increased to 288 €/ha in 2016. The average rent for arable land was 328 €/ha and 175 €/ha for grassland in this year.

Figure 3.3 shows the land rent for new and renegotiated rental contracts for the German federal states (Bundesländer) during 2015–2016. The increase of rental prices is remarkable for all Bundesländer, both in East and West Germany. Whereas rental prices are of comparable size among the East German Bundesländer there is a considerable differentiation in the West German Bundesländer. Land rents are particularly high in Lower Saxony, Schleswig-Holstein, North Rhine-Westphalia and Bavaria, all of which are important agricultural regions. For Germany as a whole, in 2015 and 2016, the new and renegotiated land rents amounted to 430 €/ha for arable land, as compared to the overall average land rent of 328 €/ha in 2016. The respective figures for grassland rentals are 234 and 175 €/ha.

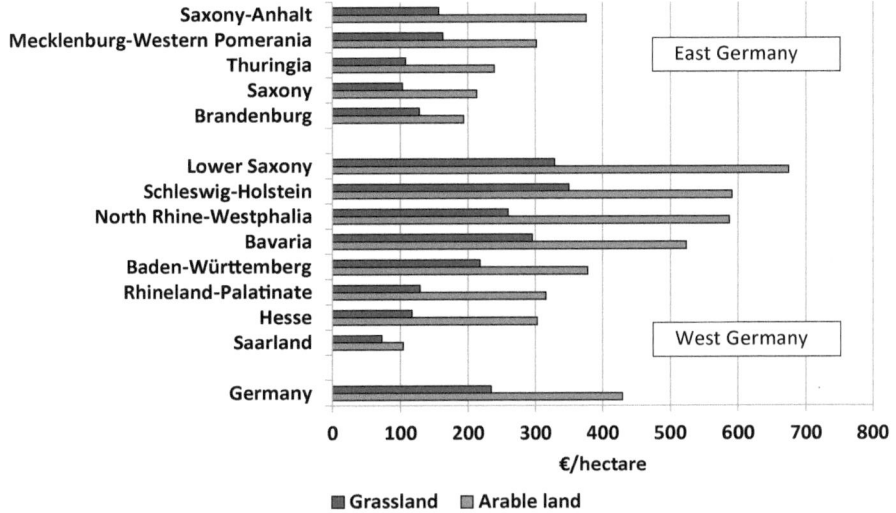

Fig. 3.3 Land rent for new and renegotiated rental contracts in German Bundesländer, 2015 and 2016, €/hectare [*Source* Own compilation according to Statistisches Bundesamt (2016a)]

3.3.2 Deteriorating Environmental Indicators

In a densely populated country like Germany with a highly intensive agriculture that accounts for a high share in agricultural land use, the conflict between agricultural land use and nature and natural resource use protection is obvious. Various environmental indicators are in place to describe the state of nature in the EU and in Germany. These indicators show that there are several problems in Germany and that the situation is not getting better in recent years but worse.

Biodiversity or rather the loss of biodiversity is among the key environmental problems. As one mayor indicator for biodiversity, the bird index was developed (Statistisches Bundesamt 2018, pp. 100–101). Using the bird index, the German government has defined a biodiversity and landscape quality indicator within the context of its "Sustainability Strategy" (Die Bundesregierung 2016). Figure 3.4 shows the development of this indicator since 2000, with a reference to reconstructed former index values in the 70ies and the target value of 100 in 2030. Based on these figures, there has been a drastic decline of biodiversity until 2000 and no improvement since then. However, a more detailed picture emerges when we look at the different land use activities individually: Whereas biodiversity in forests has improved, particularly the biodiversity on agricultural land has more or less steadily decreased. While biodiversity has come under threat from various human land use activities, the picture is particularly bad and even deteriorating for agricultural land use.

The problem of declining biodiversity on agricultural land increasingly enters the public and political debate and there is a multitude of research analyzing various aspects associated with the problem. A recent study from the German Council for

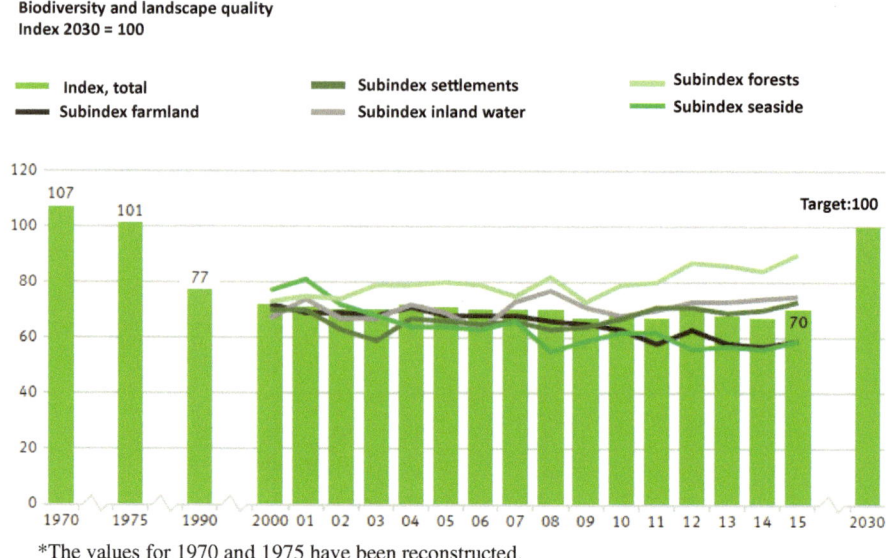

Fig. 3.4 Biodiversity and landscape quality in Germany, 1970–2015*, 2030 = 100 (*Source* Edited according to Statistisches Bundesamt (2018), p. 100.)

Landscape Conservation, for example, analyzes the declining common hamster population (Deutscher Rat für Landschaftspflege 2014). Further endangered species are grey partridges and hares, which are hardly seen anymore in the countryside (Adelmann et al. 2017). These are visible signs of declining biodiversity which—like the declining bird population—only mark the tip of the iceberg. It is well known that the main determinants for the partridge population decline are the lack of insects in the fields caused by insecticides and the rising pressure of predators (Gottschalk and Beeke 2014). The common hamster population is threatened by structural change and modern cultivation methods: large field plots, narrow crop rotation, use of pesticides, deep soil tillage, and unfavorable crops like maize (Adelmann et al. 2017). A recent survey on biodiversity on agricultural land is presented by Bundesamt für Naturschutz (BfN) (2017).

The High Nature Value (HNV) farmland indicator describes the quality of farmland for nature and is, thus, used to evaluate the impacts of farming on biodiversity (Peppiette 2011; Oppermann 2011). Figure 3.5 looks at the development of this indicator for Germany. In this figure, HNV I marks the highest quality of land, followed by HNV II and III. The picture shows that the aggregated indicator has gone down from 13.1% of total German farmland in 2009 to 11.5% in 2017. Interestingly, the decrease in the HNV value occurs in the HNV III class. Hence, one may conclude, that highly valued areas (HNV I and II) stay small, but are not threatened whereas less valued HNV III areas are increasingly endangered by intensive agricultural production.

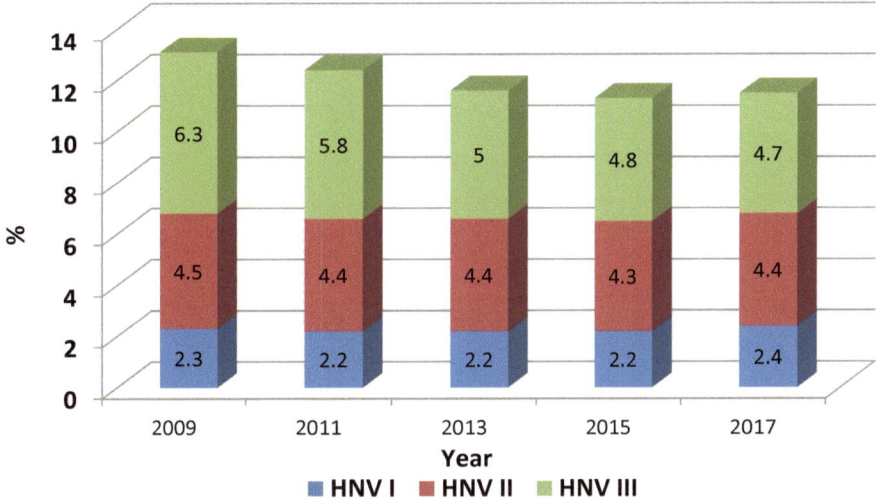

Fig. 3.5 High Nature Value (HNV) farmland in Germany in percent, 2009–2017 [*Source* Own compilation according to BfN (2018).]

The high nitrogen fertilizer use in German agriculture has caused a severe environmental problem: nitrate pollution of groundwater, surface water, and the sea. The problem is well known for years, and the total amount of nitrogen surplus on agricultural land has gone down from 110 kg/ha in 2000 to 84 kg/ha in 2014, but increased again to 102 kg/ha in 2016 (Statistisches Bundesamt 2018). Hence, the nitrate problem persists and has even deteriorated. All EU member countries are obliged to implement the Water Framework Directive (Directive W. F. 2000/60/EC) and the Nitrate Directive (Council Directive 91/676 EEC). According to the EU Commission's nitrate report from 2013 (European Commission 2013) the groundwater nitrate content has surpassed the 50 mg per liter limit at many German groundwater stations in intensive agricultural regions, and the Commission has asked for policy changes. In view of Germany's unsatisfactory reaction the EU Commission opened an infringement procedure at the European Court (Europäische Kommission 2016), and in his judgement from 21 June 2018 the Court, basically, followed the Commission's argumentation (Kirschke et al. 2019, p. 10).

Figure 3.6 shows groundwater nitrate pollution in Germany at groundwater stations of the European Environmental Agency (EEA) monitoring network. The figure exemplifies that there is no uniformly distributed nitrate pollution problem in Germany, but that some regions, typically characterized by intensive agriculture, are heavily polluted. This holds for regions in North-West Germany, the intensive arable farming regions in East Germany and the Rhine-Main area. In fact, the situation has deteriorated in many regions since 2008–2011, and this is true both for the non-polluted and the polluted areas (BMUB and BMEL 2017).

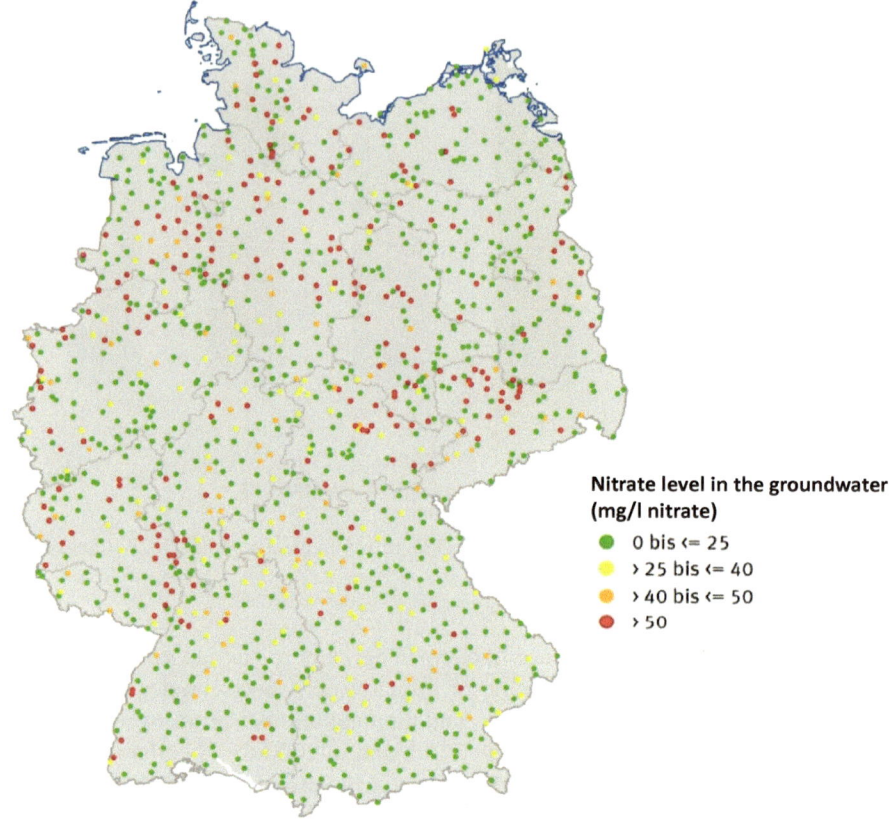

Fig. 3.6 Average nitrate levels at the measuring points (EEA monitoring network), 2012–2014 (*Source* Edited according to BMUB and BMEL: (2017), p. 45)

Hence, the nitrate problem has developed into a serious threat to society and nature. After an intensive debate the German Bundestag has finally decided upon tightening the Act on Fertilizers (Düngegesetz) (Bundesgesetzblatt 2017a) and a reform of good practice in fertilization (Düngemittelverordnung) (Bundesgesetzblatt 2017b) that govern the use of fertilizer and nitrogen fertilizer in Germany. Major new elements are e.g. fertilizer requirement statements for each plot, modified blocking periods for N-fertilizers, a limitation of nitrogen surplus on agricultural land to 60 kg/ha until 2020 and 50 kg/ha thereafter. The European Commission is not convinced that the revisions are sufficient and has requested the German government to provide proposals for a further revision; a new infringement procedure with threatening financial penalties is becoming more likely (Agra-Europe 2019). In any case, the nitrate problem will remain a key problem for agricultural land use in Germany since the revitalization of groundwater resources takes a considerable amount of time.

3.3.3 Land Use Structure and Concentration

There has been a continuous structural change in German agriculture resulting in a declining number of farms and an increasing average farm size. In 2001 the average farm size has been 28 ha in West Germany and 182 ha in East Germany (BMVEL 2002), increasing to 46 and 223 ha, respectively, in 2019 (Statistisches Bundesamt 2019). There is a new political debate in Germany restricting the access to land to avoid land concentration and non-farm land investments. In Lower Saxony the "Niedersächsisches Agrarstruktursicherungsgesetz" (Landtag Niedersachsen 2017) is to conserve structures and family-based farming and to reduce land price increases. This is an interesting new field of land use policy, however, we will not further follow up the topic since the relevance to land use seems to be limited. It may be that bigger farms and, thus, the growth of farms have an impact on the size of field plots and crop rotation, but talking about biodiversity the problem is not "big farms", but "big plots" and intensive agriculture (BfN 2015). It is widely argued that environmental problems of farming can hardly be linked to the size of a farm and, thus, the farm structure, but to the kind of farming on any farm.

Looking at agricultural land use in Germany specifically, there have been remarkable structural changes in recent years. Permanent grassland is a specific point in case. There has been a continuous decline of permanent grassland since 2000 until 2013, but since then the trend is stopped and permanent grassland is increasing. This overall picture is dominated by the situation in West Germany with a share of permanent grassland of 76% in 2016 whereas the area of permanent grassland has stayed roughly the same in East Germany from 2004 onwards. The West German picture is reflected by Bavaria and Lower Saxony, which are the German Bundesländer with the largest area of permanent grassland. In Bavaria the downward trend, with 1.15 million hectares in 2002, seems to be stopped in recent years, with 1.06 million hectares in 2016. In Lower Saxony, the figures are 0.78 million hectares in 2002 and 0.69 million hectares in 2016, respectively (Statistisches Bundesamt, various years).

Some remarkable land use changes can also be stated for arable land in Germany since 2000. Figure 3.7 shows land use changes between 2004 and 2019 for the main crops farmed in Germany. With a share of 55% in 2019 cereals cover most of the arable land in Germany. In recent years there has been a downward trend in cereal land use contrasted by a stable trend for wheat. A downward trend can also be noted for root crops (potatoes and sugar beets) with an increase to 680,000 ha in 2019 whereas there has been a remarkable extension of the silage maize area since 2010. Additionally, some trends for minor crops are worth mentioning. The total area farmed with pulses decreased to 92,000 ha in 2014 but doubled within two years with 196,000 ha in 2019. Set-aside land has been coming down from 939,000 ha in 2003 to 147,000 ha in 2019.

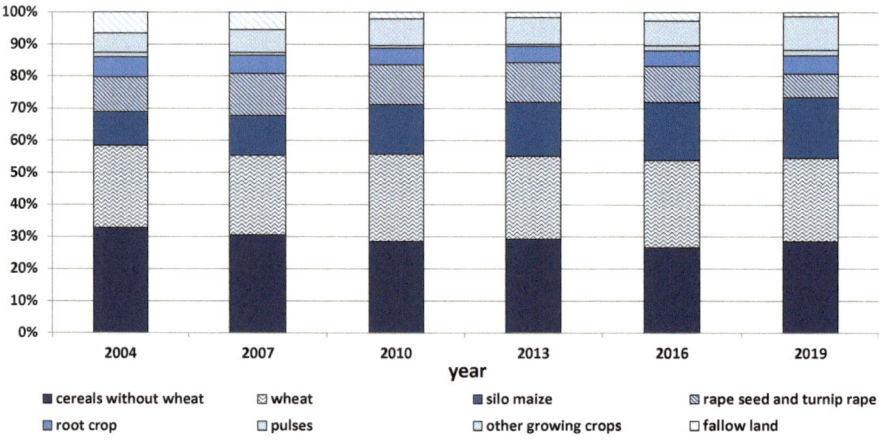

Fig. 3.7 Use of arable land in Germany, 2004–2019, in % (*Source* Own compilation according to Statistisches Bundesamt (various years), Fachserie 3)

3.4 New Drivers

Recent agricultural land use trends in Germany indicate that land use conflicts are increasing. An obvious indication for this development is rising land prices pointing to increased demand for land both within and outside agriculture. The traditional conflict between agricultural production and nature and natural resource protection did not diminish, but rather deteriorate. New trends in the agricultural land use structure both reflect changing demand for land within the agricultural sector and seem to underline the conflict between agriculture and environment. Interestingly, some of these trends (permanent grassland, pulses) seem to have been reversed in recent years. What are the new drivers behind these developments?

3.4.1 Market Forces

In a market economy agricultural production and agricultural land use basically depend on prices. The relationship between commodity and factor prices, and land prices in particular, has been a well-known and analyzed topic in agricultural economics. Since land supply is fairly inelastic, changes in land demand have a major impact on land prices. High commodity prices translate into higher demand for land and other factors and, thus, result in higher land prices and intensification.

Figure 3.8 illustrates the theoretical aspects showing how an increased commodity price e.g. for wheat changes land prices and affects the nitrogen market. On both factor markets, the higher commodity price shifts the marginal value product function to the right showing the increased demand for production factors. For simplicity, land

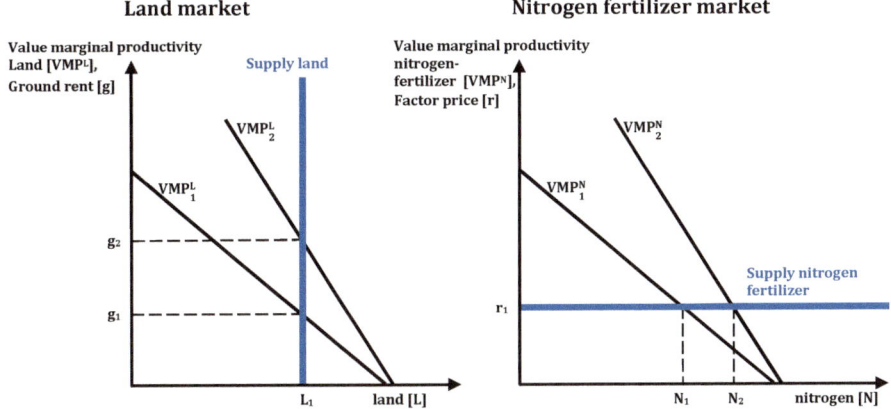

Fig. 3.8 Impact of rising commodity prices on factor markets (*Source* Own compilation.)

supply is shown as a totally inelastic supply curve in the figure whereas nitrogen supply is supposed to be totally elastic. Hence, the commodity price increase results in a high increase of the land price with no impact on land use whereas nitrogen use increases with a constant nitrogen price. Consequently, nitrogen intensity, defined as nitrogen use per hectare, increases.

The importance of commodity prices for land prices (and agricultural intensification) has been broadly discussed in the literature. The EU's protectionist agricultural policy of the past has certainly contributed to higher land prices as compared to free trade conditions whereas the liberalization of this policy during the last two decades would have had the opposite effect under the historical framework conditions. However, this framework has changed and the major factors for a new increased demand for land relate to the introduction of direct payments, tied to land, and increasing world market prices. Today, with liberalized EU markets and domestic prices driven by world markets and no longer by a protectionist policy framework, international agricultural prices and their changes have a crucial impact on land prices in the EU and Germany.

International agricultural prices have, in fact, been going up in recent years with a well-known price boom in 2007–2008. Whereas a long-term decreasing price trend could be observed on international agricultural markets in the past (Tyers and Anderson 1992, p. 49), this trend seems to be reversed for the future (European Commission 2011, p. 49). The Food and Agriculture Organization's (FAO) food prices index shows a clearly increasing trend in the new millennium and the Organization for Economic Co-operation and Development (OECD) and FAO expect stable and even increasing price trends for agricultural commodities in the future (OECD/FAO 2016).

Figure 3.9 shows the average producer price index for German farmers from 2000 to 2018. After the liberalization of the EU's CAP agricultural prices in the EU are widely determined by international prices. This is visualized by the price trend in

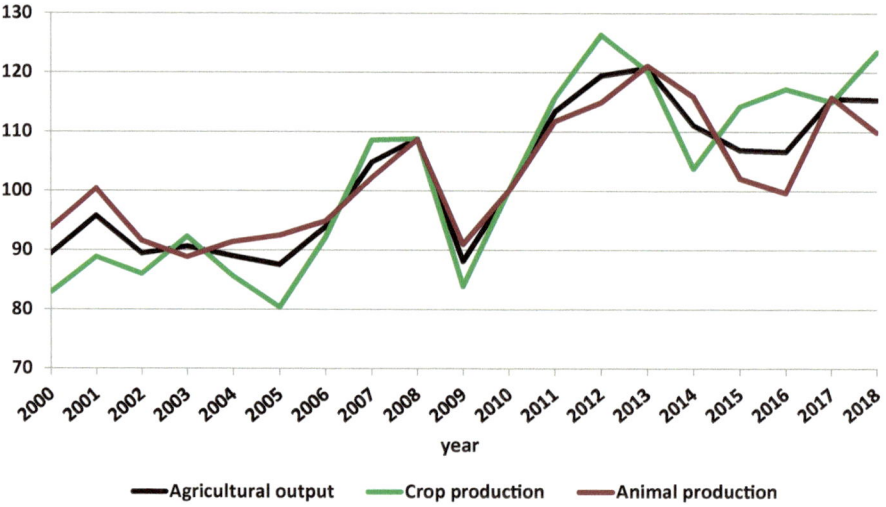

Fig. 3.9 Agricultural producer price index in Germany, 2000–2018 (2010 = 100) (*Source* Own compilation according to Statistisches Bundesamt (various years), Fachserie 17, Reihe 1.)

Fig. 3.9. Hence, international agricultural prices have become a major determinant for land prices in Germany and will be so in the future contributing to stable and, possibly, increasing land prices.

The demand for agricultural land certainly reflects the development and the competitiveness of the agricultural sector, but increasing land prices also reflect the demand from non-agricultural sectors. We have argued that the demand for land for human settlement and transport infrastructure remains to be strong, and this is certainly a "pull factor" for land prices. Some interesting new "pull factors" have become the demand for land for nature conservation and natural resource protection and bioenergy production, but since these demands are clearly policy-driven, they are discussed in the following chapter.

There may be some additional factors contributing to rising land prices, but relationships are less obvious. Farm succession has, generally, been considered a problem in times of decreasing agricultural prices, but times seem to change, at least in some regions. There is some indication that the taking over of farms by young farmers is connected with farm growth and increasing land demand, e.g. in Bavaria (Statistische Ämter des Bundes und der Länder 2011). Additionally, investment into agricultural land becomes attractive for investors outside the agricultural sector. Such a development has been observed for East German agriculture, even though this development cannot yet be considered as a major new trend (Tietz 2015). Low interest rates caused by the European Central Bank's monetary policy might actually support this development.

It is a well-known debate in agricultural economics that technological progress and increased productivity contribute to a higher demand for land and, thus, rising land prices. These "pull factors" for land markets continue to be relevant since new technologies are often linked to economies of scale and larger plots, e.g. in the case of precision farming and efficient fertilizer use. An interesting point in this regard is that a reconciliation of farming and nature protection might require appropriate large scale and expensive technologies, thus, contributing to structural change (Schaft and Balmann 2010). It is an open question whether such incentives for large-scale farming might have an effect on land market and prices.

New technological developments might actually reverse such trends. Interesting ideas relate to the miniaturization of agricultural technologies like small harvesters and to ongoing digitalization of farming e.g. drone technology (ATB 2017). In fact, such technological options might allow for competitive farming on smaller agricultural plots, thus, counteracting the need for large-scale farming and structural change.

3.4.2 Policy Framework

There is no question that demand for agricultural land is highly policy-induced in Germany and the European Union. The former protectionist CAP has contributed to rising farm prices in the past whereas the liberalization of this policy had the opposite effect. Furthermore, direct payments as introduced with the MacSharry reform certainly contributed to rising land prices. Such impacts of price policies and direct payments have been well analyzed and documented in the agricultural economics literature (Offermann et al. 2012; Hennig et al. 2014).

In addition, a new feature of the CAP may have had an impact on land use and on land markets and prices. For the financial period 2014–2020, the EU has decided upon a new CAP framework and a major feature of this is the "greening" of direct payments. Farmers can receive 30% of direct payments if, and only if, they fulfill certain land use requirements: Arable land use must comprise a minimum number of crops, permanent grassland is to be maintained, and 5% of arable land has to be Ecological Focus Areas (EFA) (European Parliament and Council 2013, Article 43). Despite such obligations, the general implications for agricultural land use seem to be limited. Due to major exemptions (small farms, grassland farms) only 52% of the EU agricultural land is affected by the EFA obligation (Pe'er et al. 2014), and the minimum number of crops required on arable land hardly impacts land use (BfN 2017).

However, the requirement to maintain grassland use certainly will have an impact. In Germany, the downward trend of permanent grassland use in the past seems to have stopped and this is most probably due to the restrictive CAP policy framework already existing in the preceding financial period and the restrictive greening regulation since 2014. Hence, if regulations stay the same we expect the area under permanent grassland to persist in the future.

The EFA framework will also have an impact on arable land use. Figure 3.10 shows the use of EFA in the German Bundesländer in 2015. According to the figure most of this area has been used for cash crops. This may have beneficial effects for

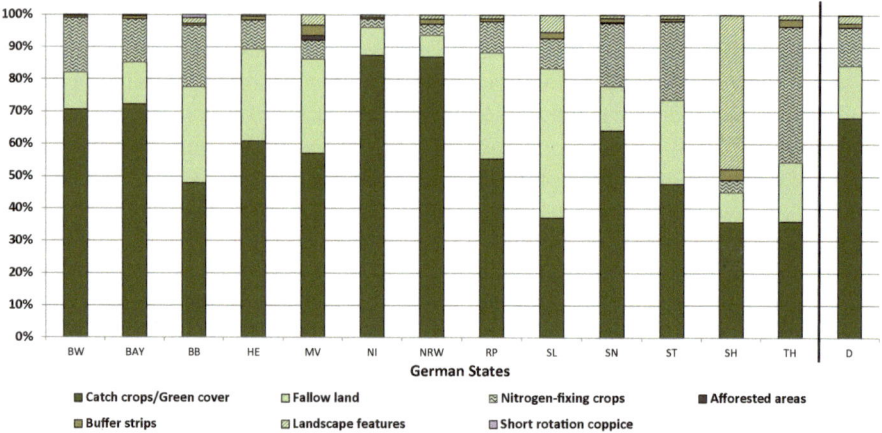

Fig. 3.10 Use of Ecological Focus Areas in Germany, 2015 (*Source* Own compilation according to Lakner et al. (2016).)

soil and groundwater, but biodiversity is hardly affected. Set-aside land, on the other hand, is certainly positive for biodiversity, though the area is limited. An interesting new development is to use pulses on EFA (Fig. 3.7), which explains the revival of pulses production in Germany in recent years. This example shows how land use may more be affected by the policy framework than by comparative advantage.

Summing up, the new greening regulation under the CAP certainly has an impact with respect to permanent grassland and pulses, but the overall land use implications are limited. In particular, the greening regulation will hardly contribute to alleviating the conflict between agricultural land use and nature and natural resource protection. Consequently, the greening regulation is a rather inefficient instrument to enhance biodiversity and ecological objectives (Pe'er et al. 2014; BfN 2017). On the other hand, greening contributes to making agricultural land scarcer, thus, impacting on land markets and prices.

The specific feature of the German REA is to support various types of renewable energies like wind and water power, solar energy, and bioenergy by guaranteeing sale and prices of the electricity produced. The production of biogas has become an important activity field for German farmers: From 2000 the installed electric output of biogas plants in Germany increased from 65 megawatt to 1,377 MW in 2008; it jumped to 3,905 MW in 2014 and 5,228 MW in 2019, respectively. In recent years installation somewhat stagnates on this high level due to corrections of the REA and reduced incentives (Statista 2019).

This development has led to a remarkable land use change in German agriculture. Since silage maize has proven to be the most profitable crop for biogas production, the area of maize production has increased considerably. It amounted to 1.2 million hectares in 2003, increased to 1.6 million hectares in 2008 and was about 2.2 million hectares in 2019. This increase in maize production area is not evenly distributed all over Germany, but maize production is concentrated in some regions like North-West Germany (Lower Saxony 24%, North Rhine-Westphalia 10%) and Bavaria (21%) (Statistisches Bundesamt 2016b). In some regions the maize production share amounts to more than 45% of agricultural area used. The reasons for this local concentration relate to animal production (and, thus, to liquid manure use), technological synergy effects, and opportunity costs in agricultural production (Scholz 2015; Breustedt and Habermann 2011). Hence, this new development in agricultural land use is clearly (energy) policy-induced.

The increasing "Vermaisung" (increasing maize cultivation) of the countryside has been analyzed and criticized from various sides. Maize cultivation has led to reduced crop rotation and to monoculture tendencies, which is counterproductive from an agronomic point of view and a threat to biodiversity from an ecological point of view (Linhart and Dhungel 2013). Within the agricultural sector, the biogas boom has led to new income opportunities for some farmers (and for non-farm investors), but it has also influenced the competitiveness between farms and production orientation. E.g., the opportunity costs of milk production have increased in some regions due to competition for the same production factor: maize. More generally, the increased land use for biogas production has contributed to higher land prices, thus, increasing cost of production for agricultural production. Various studies have shown this land market effect of maize cultivation and biogas production (Hennig and Latacz-Lohmann 2017; Bund-Länder-Arbeitsgruppe 2015; Breustedt and Habermann 2011). An interesting side effect of the increased maize cultivation area is the potential effect on the wild boar population (Hahn 2014).

Additional competition on land markets may also be caused by demand for wind power installations. The profit of wind power installations has been calculated to be around 9,000 €/ha/year (Ritter et al. 2015). Direct land requirement for wind power installations is in the range of 200–400 m^2 per unit, supplemented by additional indirect land requirements for distance and compensatory areas. In 2018, the total number of onshore wind power units in Germany amounts to 29,213 (Bundesverband Windenergie 2018). The effect of wind power installations on land price increases in Brandenburg has been calculated to be 5% (Ritter et al. 2015).

A German land market specialty is the privatization of land in the East German Bundesländer after reunification. In the former German Democratic Republic the state owned 2.1 million hectares (almost a third) of agricultural land that was to be privatized after reunification. The Bodenverwertungs- und -verwaltungsgesellschaft (BVVG) was created to guide and to carry out the land privatization process. At the beginning the BVVG favored long-term leasing of land; land sales started in 1995. A part of the land privatization took place in accordance with the rules of the EALG (Entschädigungs- und Ausgleichsleistungsgesetz). Until 2009, entitled persons (tenant farmers and previous owners) could buy land at special conditions (35% under market value) (Jochimsen 2010). In 2007, the BVVG fundamentally modified the guidelines of privatization: The principles now are call for tender and sale to highest bidder. Since 1992 a total of 869,100 ha of agricultural land have been privatized, of which 80% were sold to local tenant farmers, and 51% were sold under the rules of the EALG (16% to previous owners, 84% to tenant farmers) (BVVG 2018). The still possible EALG-sales to previous owners will end in the foreseeable future. It can be argued that land prices in East Germany would have increased faster in the past without the former EALG privatization rules.

There are now concerns that the recent BVVG land selling prices are higher than the non-BVVG land sales, probably for all new Bundesländer and particularly for Saxony-Anhalt. There is an increasing debate whether the new BVVG privatization guidelines contribute to the rising land prices in East Germany. The changed land privatization process with public tender has led to increased market transparency and, possibly, to a new interest of non-agricultural investors in land by reducing market entry barriers (Hüttel et al. 2015). In a recent study for Brandenburg, increasing activities of non-agricultural and non-regional investors on land markets have been identified (Agra-Europa 2017), but there is no comprehensive overview on such new developments on agricultural land markets in East Germany so far. Balmann (2015) argues that the BVVG land activities basically enhance market transparency and contribute to market efficiency. Hence, the fragmented land market in East Germany rather reflects "market prices" for BVVG land sales whereas non-BVVG land sales might be influenced by inadequate information, non-economic interests, or market power of local big farms. Hence, the change of the BVVG land privatization guidelines has certainly an impact on land markets, but a specific price increasing effect in recent years can hardly be identified. Generally, Tietz and Forstner (2014) argue that the rising agricultural land prices in recent years basically reflect market forces and do not point to a speculative bubble.

It is also true that BVVG land sales mark a considerable share of overall land sales, e.g. amounting in 2015 to 35% for the East German average and to even 57% for Mecklenburg-Western Pomerania (own calculation according to BVVG, various years, and Agra-Europe 2016). The BVVG privatization system is scheduled to end in 2030.

There is no doubt that land mobility has increased considerably in East Germany in recent years (Sect. 3.3). However, this is not entirely reflected in official land statistics. The official figures will rather be lower than real activities on land markets since shareholder changes in legal entity farms affect land allocation, but not official statistics. Tietz (2015) estimates that such shareholder deals surpass the statistically documented land mobility by about 20%. Hence, one may argue that non-agricultural and non-regional land investors contribute to higher land mobility (Emmann et al. 2015).

Summarizing this chapter the political framework certainly had and will continue to have an impact on land markets and land price developments. New demands under the CAP have contributed to make agricultural land scarcer and the REA has considerably affected land use in various regions. In East Germany, there are some specialties related to the farm structure and the land privatization process under the BVVG.

3.5 Conclusions

Recent developments regarding agricultural land use and land markets in Germany show that land has become an increasingly scarce natural resource faced with competing and rising demands. The general picture is that agricultural land prices have gone up considerably and there are ongoing and accentuated as well as new conflicts. Basic conflicts refer to agricultural land use in competition to nature and natural resource protection as well as to new demands on agricultural land use related to bioenergy production. Key drivers are international agricultural markets and prices as well as changing policy frameworks related to the CAP and to the REA.

What kind of policy conclusions can be drawn? First, the scarcity of a limited natural resource like land has to be recognized and can hardly be changed. This requires proper policy-making between market efficiency and policy intervention. Second, policy-making should generally avoid contributing to land scarcity and enhancing land use conflicts. The support of bioenergy production under the REA is a point in case: There may be better ways to achieve renewable energy objectives instead of directly enhancing land use conflicts. Also, indirect and international land use effects should not be neglected in national and EU policy-making in order to contribute to a proper global land use between food and energy production and nature and natural resource protection. Third, interventions on land markets will become a key policy issue. There is an increasing debate in Germany and notably in East Germany on land market interventions. Various political claims have been put forward such as the transparency on agricultural land ownerships, authorization for share deals, and the establishment of a public fund for land (Agra-Europe 2017). It needs to be questioned whether such political ideas are worth to be pursued, but finding the right answers to the real problems on land markets is a key challenge for the future.

References

Adelmann, W., Sturm, P., Stettmer, C., Burkart-Aicher, B., & Hoiss, B. (2017). Kommentar: Faktencheck zu den "neuen Bauernregeln". *Anliegen Natur,, 39*(1), 136–143.
Agra-Europe. (2016). Preisanstieg für Agrarland etwas gedämpft. 34/16, Markt + Meinung 10–12.
Agra-Europe. (2017). Investoren weiter auf dem ostdeutschen Bodenmarkt aktiv. 26/17, Länderberichte 13.
Agra-Europe. (2019). Noch keine Entscheidung im Düngestreit. 36/19, EU-Nachrichten 8.
ATB. (2017): 22. Workshop Computer-Bildanalyse und Unbemannte autonom fliegende Systeme in der Landwirtschaft. 23. Workshop Computer-Bildanalyse in der Landwirtschaft. *Bornimer Agrartechnische Berichte*, Heft 93.
Balmann, A. (2015). Braucht der ostdeutsche Bodenmarkt eine stärkere Regulierung? *Agra-Europe 13/15, Länderberichte Sonderbeilage.*
BfN. (2015). Artenschutz-Report 2015. Tiere und Pflanzen in Deutschland. https://www.bfn.de/fileadmin/BfN/presse/2015/Dokumente/Artenschutzreport_Download.pdf (22 November 2019).
BfN. (2017). Agrar-Report 2017. Biologische Vielfalt in der Agrarlandschaft. https://www.bfn.de/fileadmin/BfN/landwirtschaft/Dokumente/BfN-Agrar-Report_2017.pdf (22 November 2019).

BfN. (2018). Fachtagung "10 Jahre HNV-Farmland-Monitoring" am 13. September 2018 in Erfurt. Tagungsbericht. https://www.bfn.de/fileadmin/BfN/monitoring/Dokumente/Zusammenfassung_HNVJub_Tagung_fuer_HP.pdf (22 November 2019).

BMVEL. (2002). Statistisches Jahrbuch über Ernährung, Landwirtschaft und Forsten 2002. Tab. 31 Landwirtschaftliche Betriebe nach Rechtsformen (pp. 32–33).

BMUB; BMEL. (2017). Nitratbericht 2016. https://www.bmub.bund.de/fileadmin/Daten_BMU/Download_PDF/Binnengewaesser/nitratbericht_2016_bf.pdf (22 November 2019).

Breustedt, G., & Habermann, H. (2011). Einfluss der Biogaserzeugung auf landwirtschaftliche Pachtpreise in Deutschland. *GJAE, 60*(2), 85–100.

Bundesgesetzblatt. (2017a). Erstes Gesetz zur Änderung des Düngegesetzes und anderer Vorschriften. Teil I Nr. 26 vom 5. Mai 2017, pp.1068–1073.

Bundesgesetzblatt. (2017b). Verordnung zur Neuordnung der guten fachlichen Praxis beim Düngen. Teil I Nr. 32 vom 25.Mai 2017, pp.1305–1348.

Bundesverband Windenergie. (2018). Anzahl der Windenergieanlagen in Deutschland. https://www.wind-energie.de/themen/zahlen-und-fakten/deutschland/ (22 November 2019).

Bund-Länder-Arbeitsgruppe (2015). Landwirtschaftliche Bodenmarktpolitik: Allgemeine Situation und Handlungsoptionen. Bericht der Bund-Länder-Arbeitsgruppe "Bodenmarktpolitik". https://www.bmel.de/SharedDocs/Downloads/Landwirtschaft/LaendlicheRaeume/Bodenmarkt-Abschlussbericht-Bund-Laender-Arbeitsgruppe.pdf?__blob=publicationFile (22 November 2019).

BVVG. (2018). Zahlen und Fakten 2018. https://www.bvvg.de/INTERNET/internet.nsf/vSysDok/dPDFZahlenundFakten/$File/BVVG%20Zahlen%20und%20Fakten%202019%20Web.pdf (22 November 2019).

BVVG (various years). Informationen für Presse, Funk und Fernsehen. https://www.bvvg.de/INTERNET/internet.nsf/HTMLST/ARCHIV (22 November 2019).

DAFA. (2015). Fachforum Grünland. Forschungsstrategie der DAFA. https://www.dafa.de/wp-content/uploads/FF_Gruenland.pdf (22 November 2019).

DBV; EDL; KLB; dlv. (2016). Erntedank 2016. Erklärung von EDL—KLB—DBV—dlv. https://media.repro-mayr.de/69/661469.pdf (22 November 2019).

Deutscher Rat für Landschaftspflege. (2014). Bericht zum Status des Feldhamsters. BfN-Skripten 385. https://www.bfn.de/fileadmin/BfN/service/Dokumente/skripten/skript385.pdf (22 November 2019).

Die Bundesregierung. (2016). Deutsche Nachhaltigkeitsstrategie. Neuauflage 2016. https://www.bundesregierung.de/Content/Infomaterial/BPA/Bestellservice/Deutsche_Nachhaltigkeitsstrategie_Neuauflage_2016.pdf?__blob=publicationFile&v=18 (22 November 2019).

Directive, W. F. (2000). Directive 2000/60. EC of the European Parliament and of the Council of, 23(1).

Emmann, C. H., Surmann, D., & Theuvsen, L. (2015). Charakterisierung und Bedeutung außerlandwirtschaftlicher Investoren: Empirische Ergebnisse aus der Sicht des landwirtschaftlichen Berufsstandes. Diskussionspapier 1504: Department für Agrarökonomie und Rurale Entwicklung der Georg-August-Universität Göttingen.

European Commission. (2011). Prospects for agricultural markets and income in the EU 2011–2020. https://ec.europa.eu/agriculture/publi/caprep/prospects2011/fullrep_en.pdf (22 November 2019).

European Commission. (2013). Report from the Commission to the Council and the European Parliament on the implementation of Council Directive 91/676/EEC concerning the protection of waters against pollution caused by nitrates from agricultural sources based on Member State reports for the period 2008–2011. COM (2013) 683 final.

European Parliament and Council. (2013). Regulation (EU) No 1307/2013 of The European Parliament and of The Council of 17 December 2013 establishing rules for direct payments to farmers under support schemes within the framework of the common agricultural policy and repealing Council Regulation (EC) No 637/2008 and Council Regulation (EC) No 73/2009.

Europäische Kommission. (2016). Wasser: Kommission verklagt Deutschland vor dem Gerichtshof der EU wegen Gewässerbelastung durch Nitrat. Pressemitteilung IP/16/1453.

Eurostat. (2012). LUCAS Bodenbedeckungs-/Bodennutzungsstatistik. Datenbank Bodenbedeckung [lan_lcv_ovw]. https://ec.europa.eu/eurostat/de/web/lucas/data/database (22 November 2019).

Gottschalk, E., & Beeke, W. (2014). Wie ist der Rückgang des Rebhuhns (Perdix Perdix) aufzuhalten? Erfahrungen aus 10 Jahren mit dem Rebhuhnschutzprojekt im Landkreis Göttingen. *Vogelschutz, 51*, 95–116.

Hahn, N. (2014). Brennpunkt Schwarzwild. Abschlussbericht. Studie im Auftrag der bayrischen Landesanstalt für Wald und Forstwirtschaft. In: Wissenschaftlicher Beirat für Agrarpolitik, Ernährung und gesundheitlichen Verbraucherschutz (WBAE) beim BMEL.

Hennig, S., & Latacz-Lohmann, U. (2017). The incidence of biogas feed-in tariffs on farmland rental rates—evidence from northern Germany. *European Review of Agricultural Economics, 44* (2), 231–254. https://doi.org/10.1093/erae/jbw023 (22 November 2019).

Hennig, S., Breustedt, G., & Latacz-Lohmann, U. (2014). Zum Einfluss mitgehandelter Zahlungsansprüche auf die Kauf- und Pachtpreise von Ackerland in Schleswig-Holstein. *GJAE, 63*(4), 219–239.

Hüttel, S., Odening, M., & Schlippenbach, Vv. (2015). Steigende landwirtschaftliche Bodenpreise: Anzeichen für eine Spekulationsblase? *DIW-Wochenbericht, 82*(3), 37–42.

Jochimsen, H. (2010). 20 Jahre Grüner Aufbau Ost. *Berichte über Landwirtschaft, 88*(2), 203–246.

Kirschke, D., Häger, A., & Noleppa, S. (2013). Ökonomische Aspekte eines nachhaltigen Landmanagements. Nachhaltiges Landmanagement Diskussionspapier Nr. 5. https://d-nb.info/103310 4450/34 (12 June 2017).

Kirschke, S., Häger, A., Kirschke, D., & Völker, J. (2019). Agricultural nitrogen pollution of freshwater in Germany. The governance of sustaining a complex problem. *Water, 11*, 2450. https://doi. org/10.3390/w11122450 (23 November 2019).

Lakner, S., Schmitt, J., Schüler, S., & Zinngrebe, Y. (2016). Naturschutzpolitik in der Landwirtschaft: Erfahrungen aus der Umsetzung von Greening und der ökologischen Vorrangfläche 2015. Vortrag anlässlich der 56. Jahrestagung der GEWISOLA. https://ageconsearch.umn.edu/ bitstream/244768/2/Lakner.pdf (22 November 2019).

Landtag Niedersachsen. (2017). Drucksachen der 17. Wahlperiode. Drs. 17/8003. https://www.lan dtag-niedersachsen.de/drucksachen/ (22 November 2019).

Linhart, E., & Dhungel, A.-K. (2013). Das Thema Vermaisung im öffentlichen Diskurs. *Berichte über Landwirtschaft*, Vol. 91(2).

OECD/FAO. (2016). OECD-FAO Agricultural outlook 2016–2025. Paris: OECD Publishing. https://doi.org/10.1787/agr_outlook-2016-en (22 Nov. 2019).

Offermann, F., Banse, M., Ehrmann, M., Gocht, A., Gömann, H., Haenel, H.-D., Kleinhanß, W., Kreins, P., Ledebur, O., Osterburg, B. von Pelikan, J., Rösemann, C., Salamon, P., & Sanders, J. (2012). vTI-Baseline 2011–2021: agrarökonomische Projektionen für Deutschland. Braunschweig: vTI, 82 p, Landbauforsch SH 3.

Oppermann, R. (2011). High nature value farming. Durch Landbewirtschaftung einen hohen Naturwert schaffen und erhalten.*Der kritische Agrarbericht , 96.*

Pe'er, G., Dicks, L. V., Visconti, P., Arlettaz, R., Báldi, A., Benton, T. G., Collins, S., Dieterich, M., Gregory, R. D., Hartig, F., Henle, K., Hobson, P. R., Kleijn, D., Neumann, R. K., Robijns, T., Schmidt, J. A., Shwartz, A., Sutherland, W. J., Turbé, A., Wulf, F., & Scott, A. V. (2014). Eu agricultural reform fails on biodiversity. *Science, 344*, 1090–1092. Supplement Materials Table S1.

Peppiette, Z. (2011). Bewertungsmodelle der so genannten HNV-Flächen in der EU. *Newsletter des Europäischen Evaluierungsnetzwerkes für ländliche Entwicklung, 6*,1–6.

Ritter, M., Hüttel, S., Walter, M., & Odening, M. (2015). Der Einfluss von Windkraftanlagen auf landwirtschaftliche Bodenpreise. *Berichte über Landwirtschaft, 93*(3), 1–15.

Schaft, F., & Balmann, A. (2010). Determinanten des Erfolgs der Umstrukturierung der ostdeutschen Landwirtschaft. In: Institut für Wirtschaftsforschung Halle - IWH (Hrsg.): 20 Jahre Deutsche

Einheit - Von der Transformation zur europäischen Integration, Tagungsband, IWH-Sonderheft 3, Halle (Saale), pp. 217–233.

Scholz, L. (2015). Bestimmungsfaktoren der Verteilung und Konzentration der Biogasproduktion in Deutschland – Eine räumlich-ökonometrische Analyse. Berliner Schriften zur Agrar- und Umweltökonomik 22.

Statista. (2019). Installierte elektrische Leistung der Biogasanlagen in Deutschland in den Jahren 1999 bis 2019 (in Megawatt). https://de.statista.com/statistik/daten/studie/167673/umfrage/instal lierte-elektrische-leistung-von-biogasanlagen-seit-1999/ (22 November 2019).

Statistische Ämter des Bundes und der Länder. (2011). Agrarstrukturen in Deutschland Einheit in Vielfalt. Regionale Ergebnisse der Landwirtschaftszählung 2010. https://www.destatis.de/ DE/Themen/Branchen-Unternehmen/Landwirtschaft-Forstwirtschaft-Fischerei/Landwirtschaftl iche-Betriebe/Publikationen/Downloads-Landwirtschaftliche-Betriebe/agrarstrukturen-in-deu tschland-5411203109004.pdf?__blob=publicationFile (22 November 2019).

Statistisches Bundesamt. (2016a). Land-, Forstwirtschaft und Fischerei. Eigentums- und Pachtver-hältnisse. Fachserie 3, Reihe 2.1.6.

Statistisches Bundesamt. (2016b). Land-, Forstwirtschaft und Fischerei. Landwirtschaftliche Bodennutzung. Anbau auf dem Ackerland. Fachserie 3, Reihe 3.1.2, Vorbericht.

Statistisches Bundesamt. (2017a). GENESIS-Online Datenbank. Flächennutzung. Bodenfläche Code 33111-0001; Siedlungs- und Verkehrsfläche Code 33111-0003; Landwirtschaftlich genutzte Fläche Code 41241–0007. https://www-genesis.destatis.de/genesis/online (22 November 2019).

Statistisches Bundesamt. (2018). Nachhaltige Entwicklung in Deutschland. Indikatorenbericht 2018. https://www.destatis.de/DE/Themen/Gesellschaft-Umwelt/Nachhaltigkeitsindikatoren/ Publikationen/Downloads-Nachhaltigkeit/indikatoren-0230001189004.pdf?__blob=publicati onFile (22 November 2019).

Statistisches Bundesamt. (2019). Landwirtschaftliche Betriebe. https://www.destatis.de/DE/The men/Branchen-Unternehmen/Landwirtschaft-Forstwirtschaft-Fischerei/Landwirtschaftliche-Betriebe/Tabellen/betriebsgroessenstruktur-landwirtschaftliche-betriebe.html (22 November 2019).

Statistisches Bundesamt. (various years). Preise. Preisindizes für die Land- und Forstwirtschaft. Fachserie 17, Reihe 1. https://www.destatis.de/GPStatistik/receive/DESerie_serie_00000006?lis t=all (22 November 2019).

Statistisches Bundesamt (various years): Land-, Forstwirtschaft und Fischerei. Kaufwerte für Landwirtschaftliche Grundstücke. Fachserie 3, Reihe 2.4., Tabelle 1.4.

Statistisches Bundesamt. (various years). Fachserie 3, Reihe 3: Landwirtschaftliche Boden-nutzung und pflanzliche Erzeugung. Tabelle: Grund- und Verhältniszahlen der landwirtschaftlich genutzten Fläche. Wiesbaden.

Tietz, A. (2015). Überregional aktive Kapiteleigentümer in ostdeutschen Agrarunternehmen: Bestandsaufnahme und Entwicklung. Thünen Report 35. https://www.thuenen.de/media/publik ationen/thuenen-report/Thuenen-Report_35.pdf (22 November 2019).

Tietz, A., & Forstner, B. (2014). Spekulative Blasen auf dem Markt für landwirtschaftlichen Boden. *Berichte über Landwirtschaft* (3)3, 1–17. https://buel.bmel.de/index.php/buel (22 November 2019).

Tyers, R., & Anderson, K. (1992). Disarray in World food markets. A quantitative assessment. Cambridge: University Press.

Chapter 4
Demographic Change and Land Use

Jens Hoffmann

Abstract Demographic change is increasingly being described as one of the central factors of human influence on land-use change. But do changes in population size and composition directly effect changes in land use? These reflections prompted the author and two colleagues to conduct a study aimed at (a) finding answers to the questions of whether and to what degree clear correlations between demographic change and observable land-use changes could be found in the existing literature, and (b) establishing what this means for regional studies and regional development policies. After presenting the methodological approach, the current state of knowledge regarding correlations between demographic change and land-use change is reflected and will finally be followed by conclusions and the need for further research.

Keywords Land use · Demographic change · Impact

4.1 Does Demographic Change Cause Changes in Land Use?

Aside from natural processes, it is the many and varied demands that people make on land that lead to changes in land cover and land use. The drivers of change in land cover and land use are wide-ranging, complex, and overlapping, as are the effects of those changes. Drivers vary depending on the location, time frame, and institutional architecture of the human–environment system under observation. Some drivers have long-term impacts that unfold slowly while others trigger rapid and visible changes. (McNeill et al. 1994; Lambin et al. 2001; Lambin and Geist 2006).

According to Lambin et al. (2001), the debate on drivers of land-use change is dominated by simplifications and even myths that affect concepts, values, and decisions in politics and policies shaping land use.

J. Hoffmann (✉)
Naturschutz und Landnutzungsplanung (FB LG), University of Applied Sciences
Neubrandenburg, Brodaer Straße 2, 17033 Neubrandenburg, Germany
e-mail: jenshoffmann@hs-nb.de

© The Author(s) 2021
T. Weith et al. (eds.), *Sustainable Land Management in a European Context*,
Human-Environment Interactions 8, https://doi.org/10.1007/978-3-030-50841-8_4

A distinction is made between biophysical drivers, i.e., natural influences, and socio-economic drivers, i.e., drivers of anthropogenic use (Schinninger 2008; Cortesi and Hepperle 2011). The latter category is sometimes divided or broadened to include socio-economic, political, technological, and cultural factors. (Bürgi et al. 2004; Schneeberger et al. 2007).

Demographic change is increasingly being described as one of the central factors of human influence on land-use change (Hersperger and Bürgi 2009; EEA 2010; Moorfeld 2011). But does demographic change really contribute directly to land-use change? Do changes in population size and composition directly effect changes in land use (see remarks made in studies such as EEA 2010; BfN and BBSR 2011, 8; Milbert 2013, 45)? These reflections prompted the author and two colleagues to conduct a study aimed at (a) finding answers to the questions of whether and to what degree clear correlations between demographic change and observable land-use changes could be found in the existing literature, and (b) establishing what this means for regional studies and regional development policies.

4.2   Methodological Approach

The study was, at its core, an extensive literature review. No original empirical research or other inquiries were conducted. The findings were rigorously debated at two expert workshops and subsequently summarized in a review paper (Behrens et al. 2012[1]).

One of the first steps was to define the two terms "demographic change" and "land use":

Demographic change refers to changes in the composition of a population (Hoßmann and Münz, n.d.). This includes not just changes in population size, and in age and sex structures, but also changes in ethnic composition, regional distribution, and lifestyle (BMI 2011, 11). Demographic change cannot be characterized as a single, coherent phenomenon. It is shaped by preexisting social processes and takes very different regional and local forms (Demuth et al. 2010, 26).

The concept of land use was based largely on Spitzer's definition. He uses the category "direct land use" (*direkte Landnutzung*) to group together types of land use that visibly cover the surface of the land (Spitzer 1991, 161). Then there are various forms of "indirect land use" (*indirekte Landnutzung*), for example nature conservation, which may be included in multiple-use scenarios alongside the principal land-use types (Spitzer 1991, 172). Data on direct land use is recorded in Germany's official land-use statistics.

The catalog of land-use types used by the Working Committee of the Surveying Authorities of the Länder of the Federal Republic of Germany (AdV and AK LK

[1]The study was funded by the Sustainable Land Management program, established by the German Federal Ministry of Education and Research.

2011) was adopted in order to establish a connection to official statistics. The land-use classifications contained in the catalog were tied to activity sectors with the goal of obtaining a problem-and-action-oriented overview of the issues.

The following activity sectors were established: agriculture, forestry, tourism, settlement expansion,[2] transportation, and water management. In addition, although it is not included in the AdV's catalog of use types, the sector of nature conservation was defined within the category of indirect land use. In addition to research aimed at specific sectors, the review looked at studies into overlapping sectors, in particular those with a focus on rural areas.

The search for sources focused on the Federal Republic of Germany, that is, the review looked for findings regarding the impacts of demographic change on land use that could be observed in Germany or that were held to be relevant for Germany. A time frame of 2005–2012 was chosen.[3]

The search for sources turned up a total of 222 publications (research reports, monographs, multi-author volumes, and periodical articles) whose titles or subtitles indicated that they addressed the issues under review. The sources were then examined to determine if they contained findings on the relationship between demographic change and the selected sectors that were tangible and susceptible of analysis. This resulted in the number of sources subjected to detailed examination being reduced to 133 (for detailed information on these sources, see Behrens et al. 2012, 99–110). With respect to the sources as a whole, it can be said that although there is now an abundance of sources on the topics of demographics and population trends, those numbers are greatly reduced when the field is narrowed down to the impacts of demographic change on land use and even smaller when we look at specific land-use types. Upon closer examination of the sources, moreover, it was found that the studies frequently promised more in their titles than they actually delivered in substance.

The 133 sources that constituted the final selection were then assessed with respect to the following three points: (1) general effects of demographic change per sector, (2) effects of demographic change on land use per sector, and (3) ways to control or respond to these effects.

---

[2]Settlement expansion refers to the expansion of human settlements, including residential, commercial, industrial and recreational areas.

[3]The decision to restrict the review to Germany was made in order to be able to analyze the findings within a single legal and administrative framework. The time frame was selected so that the review could address current topics of interest, i.e., debates going back approximately ten years.

A second survey undertaken in early 2017 revealed that virtually no new studies on the issues under review had been published since 2012. Isolated studies on settlement expansion were found. Most noteworthy among them was the Environmental Report 2016 of the German Advisory Council on the Environment, in particular the chapter titled "Land consumption and demographic change" (SRU 2016).

The findings of the review can therefore still be seen as reflecting the current state of knowledge.

4.3 Current State of Knowledge Regarding Correlations Between Demographic Change and Land-Use Change

4.3.1 Impacts of Demographic Change on Agriculture, Forestry, and Nature Conservation

Shrinking populations in rural areas do not automatically result in reductions in the amount of land used for agriculture and forestry. The dominance of agriculture and forestry among land-use types is primarily a matter of agricultural/energy policy and economics, and is not determined by demographic factors. Unless this changes, there is no reason to expect agriculture's share of land use to decrease (Demuth et al. 2010; for an opposing opinion, see Wolf and Appel-Kummer 2005). What is more, agriculture and forestry trends are affected to a great extent by global interdependencies in the production, distribution, and consumption of resources. Thus in forestry, too, demographic change is not seen as a significant factor in trends regarding forest use. The possible consequences of demographic change at a national level that are identified in the studies—*a decrease in the demand for wood, forests reverting to a state of relative wilderness, a reduction in the use of local natural areas for recreation* (Wurz 2007)—lose their significance when global realities are factored into the equation.

One of the consequences of generally increasing competition for land use is that larger-scale areas will not be available for nature conservation in the future and that wilderness areas will therefore not be a large-scale option for nature conservation. This is not to say that there may not be small-scale, positive developments in nature conservation that can be traced back directly or indirectly to demographic change. In certain regions, abandoned city neighborhoods and districts may present expanded opportunities for nature conservation. Especially where protected areas and landscapes are on the rural–urban fringe and difficult to access, there are (limited) chances for a reduction in the intensity of use. It is expected that these areas will experience less recreational use by regional and local populations.

4.3.2 Impacts of Demographic Change on Transportation, Tourism, and Settlement Expansion

Parallel processes of growth and decline are expected with regard to trends in transportation and tourism infrastructure. In growing regions, steady or growing demand will result in greater pressure on infrastructure, eventually to the point that supply is expanded. In declining regions, pressure on infrastructure will decrease, leading to a reduction in supply (Grimm et al. 2009a; Canzler 2007; among others). In particular with respect to transportation, the sources conclude that, at least in the short term, declining areas cannot do without existing infrastructure. This means that as demand stagnates or even shrinks over the long term, infrastructure will tend to continue to grow (FGSV 2006). Congestion and bottlenecks will be seen in growth regions and highly traveled transport corridors. In these cases, investments aimed at expanding existing infrastructure and efforts to increase its efficiency should be made early on. Additional investments in transportation infrastructure should only be made if sufficient demand is expected in the medium and long term (Knie and Canzler 2009, 177 f.).

With regard to the tourism sector, demographic change—*and in particular, the aging of the population*—will cause far-reaching changes on the supply side. The shift in age structure—*at least in the short and medium term*—is considered to be more serious than the potential decrease in demand as a whole (Petermann et al. 2006; Schröder et al. 2007). In terms of demand volume, a decrease in domestic travel with overnight accommodation is anticipated. This will result in poorer utilization not only of tourism infrastructure but also of associated regional infrastructure. In terms of demand structure, a reversal in market share is expected as demand shifts to older age groups. Tourism among older people will be an engine for growth in the tourism sector, at least in the medium term (until 2030) (Grimm et al. 2009a). Concrete, empirically substantiated predictions of whether and to what degree a change in tourism demand will impact land consumption for tourism infrastructure cannot be found in the literature.

In order to explain developments in the consumption of land for settlement (i.e., residential, commercial, industrial, and recreational areas) and transportation, a complex collection of variables must be brought to bear. While interdependencies between demographic change and land consumption for settlement and transportation were reasonably clear in the 1990s, cause and effect are no longer easy to identify today. Trends in land use for settlement and transportation are not tied to economic or even demographic factors in the way that they once were. Though demographic trends continue to be relevant to land consumption, the nature of that relationship is no longer clear and other factors are gaining in significance. Planning approval for new developments of single-family homes was long considered to be one of the principal drivers of new land consumption while the construction of high-rise buildings was not expected to contribute greatly to overall demand for housing land (BMVBS and BBSR 2009). This is starting to change, however: thanks to positive economic growth and continuing in-migration into urban areas, the number of approvals granted to high-rise apartments is considerably higher than those granted to single-family homes (SRU 2016). Other important factors will probably continue to contribute to land consumption, however. These include exaggerated growth expectations on the part of municipal administrations and resultant supply planning, low building density, and what Germans call the Remanenzeffekt (the "staying-put" effect), i.e., the phenomenon of elderly people staying in their apartments or houses, resulting in a rise in living space per person in regions with shrinking populations (BMVBS and BBSR 2009).

4.3.3 Impacts of Demographic Change in Multicausal Models

In addition to sources that dealt with specific sectors individually, the review identified and analyzed a number of sources that looked at multiple (parallel or overlapping) sectors and examined possible impacts of demographic change in qualitative terms (BMVBS and BBSR 2009; Rudolph et al. 2007; Siedentop et al. 2011). They did not reveal any findings beyond those offered by studies on single sectors, however.

There are only a few quantitative studies on the relationship between demographic change and land-use patterns (including Schaldach and Priess 2008; Murray-Rust et al. 2013; Wyman and Stein 2010). The quantitative study undertaken by Kroll and Haase looked at this relationship in counties in Germany. The results of the study, for the purpose of the review at any rate, can be summed up in the following key points (Kroll and Haase 2010, 730–737):

- With respect to regions in eastern Germany, the data show that land-use variables do not correlate with demographic variables. Economic trends tend to play a greater role in land-use change in these regions.
- In contrast, western Germany was found to exhibit high correlations between population trends and migration on the one hand and changes in settlement and transportation areas on the other. The increase in settlement and transportation areas in turn affects the amount of agricultural land, which consequently decreases.

4.3.4 Recommendations on Guiding Land-Use Impacts

Recommendations for guiding impacts on land-use sectors are just as unsubstantiated and imprecise as the findings regarding the impacts on land use. Neither for those sectors in which use is directly linked to the land, i.e., agriculture and forestry, nor for nature conservation, do the sources provide suggestions for new or improved approaches to guiding land-use patterns with respect to either the type or intensity of land use.

For sectors such as tourism and transportation, which are tied to land use indirectly through infrastructure, the principal challenge is managing simultaneous growth and decline. In view of this new situation, it is necessary to reexamine conventional models, legal regulations and standards created as part of infrastructure policies, and societal objectives such as creating equal living conditions. Investments in infrastructure should be made only in regions with sufficient demand over the medium and long term. Better coordination among sectors and more strongly integrated approaches—in terms of both perspective and action—are needed (Knie and Canzler 2009; FGSV 2006; among others).

The settlement expansion sector boasts a relatively sophisticated range of control mechanisms, created in part with the aim of achieving the German federal government's target of 30 ha per day (Die Bundesregierung 2002). A reduction in the amount of new land zoned for settlement areas is recommended, alongside measures to increase the availability of unused urban spaces. The sources contain numerous recommendations for sustainable land management (at local and regional levels) and for management tools and incentives (Dosch and Bergmann 2005; Bürkner et al. 2007; BMVBS and BBSR 2009; Weith 2009; BBSR 2011; Bock et al. 2011).

4.3.5 General Effects of Demographic Change in the Activity Sectors

In addition to the specific impacts of demographic change on land use in the activity sectors targeted by the review, the sources also describe the general effects of demographic change in these sectors. The following key effects were observed (Heiland et al. 2004; Wiener et al. 2004; Heiland et al. 2005; FGSV 2006; Petermann et al. 2006; TLL 2006; Oeltze et al. 2007; Rudolph et al. 2007; Schröder et al. 2007; Wurz 2007; Grimm et al. 2009a; Knie and Canzler 2009; Thoroe 2009; IG BAU 2014):

- changes in demand for agricultural, forestry, and tourism products (extent, structure, and quality) and changes in transportation-related demand (extent, structure, and distribution over space and time) as a result of decreasing population sizes and changing population structures
- labor and skills shortages as a result of decreasing population sizes and changing population structures
- regionally differentiated changes in core concepts and values (for example, guiding principles in forest management and nature conservation, or principles of spatial planning such as equal living conditions)
- diminishing public funds, associated with a diminishing scope for control and guidance.

A common feature of all the findings and recommendations is that they provide no—or only very vague—indications of whether the type and intensity of land use will change at some further, future stage, and if so, what form that change might take.

The following approaches regarding the management of or response to general effects can be found in the literature (Wiener et al. 2004; TLL 2006; TLL 2007; Grimm et al. 2009b; BMELV 2011; IG BAU 2014):

- rationalization and improvement of labor productivity
- marketing and PR work aimed at developing new markets and positioning products and services in the context of new challenges
- training and retraining programs as a counter-strategy to labor and skills shortages
- cooperation and cross-sector approaches to action.

4.4   Conclusions and the Need for Further Research

An analysis of the literature revealed that there are virtually no sources that contained findings or allowed conclusions to be derived regarding the direct impacts of demographic change on land-use change. This is especially true of the agriculture, forestry, tourism, and nature conservation sectors. The one exception was settlement and transportation areas, since research (including some empirical studies) has been conducted on the correlation between population trends and settlement expansion. Otherwise, findings regarding the impacts of demographic change on land use were based on more or less substantiated analytical arguments. It is possible that this finding is due to a time lag, i.e., that demographic changes already in progress have not yet made a sufficiently visible or tangible impact on certain land-use groups or types for society (or the literature, in this case) to have become aware of them.

In view of the complexity of spatial dynamics, it is difficult to isolate and examine individually the components of demographic change that are causal factors in land-use change. Clear cause-and-effect relationships cannot be established with any great certainty. Demographic factors have nonspecific effects and overlap with other influences on land use, such as economic trends, climate change, scarcity of resources, or institutional influences such as policy objectives. These influences may strengthen, neutralize, or even reverse the impacts of demographic change. In many cases, it would appear that other factors have—and will continue to have—a considerably stronger influence on developments in land-use patterns than will demographic factors.

In light of this fact, the findings of the literature under review do not support, or only marginally support, oversimplified conclusions such as those below (see remarks made in studies including EEA 2010; BfN and BBSR 2011, 8; Milbert 2013, 45):

- Shrinking populations lead to lower land-use intensities and/or to reduced land use in certain areas, which in turn opens up options for new or different land-use types.
- The combination of an aging population and greater cultural diversity causes changes in demand structures that lead to different land-use types and/or changes in land-use intensity.

The literature review and the subsequent discussion of its findings at two expert workshops came to the following conclusions regarding the need for further research:

Lack of empirical data: the absence of empirical evidence for assertions regarding the impacts of demographic change on land-use sectors proved to be a major shortcoming, as became clear over the course of the review. Above all, there is a lack of empirically substantiated, spatially differentiated studies on the correlation between demographic change and land use.

Comparative, regional case studies: the effects of demographic change vary from region to region. When combined with other trends and processes, a highly differentiated spatial structure emerges which is characterized by the coexistence, on regional and local levels, of both sprawl and decline. Thus research is needed that takes regional variation into account and that is conducted over time periods which permit the effects of a region-specific combination of causes to be identified in their entirety.

Inclusion of demographic indicators in land-use statistics: the collection of land-use data should be improved so that it factors in demographic information, thus expanding the knowledge base for impacts of demographic change on land use. Correlations between population trends and land-use dynamics should be examined in long-term, comparative studies that focus on relatively small areas; these studies would have to tie data collected for land-use statistics to population data. An appropriate methodology for such studies has yet to be developed.

Correlation between changes in lifestyle and land use: one topic of research that urgently needs to be investigated, and where much is currently unknown, is the correlation of changes in the lifestyles, leisure time activities, motivations, and attitudes of present and future generations to changes in land-use demands.

References

AdV, & AK LK/Arbeitsgemeinschaft der Vermessungsverwaltungen der Länder der Bundesrepublik Deutschland; Arbeitskreis Liegenschaftskataster. (2011). Katalog der tatsächlichen Nutzungsarten im Liegenschaftskataster und ihrer Begriffsbestimmungen (AdV-Nutzungsartenkatalog). Stand: November 2011.

BBSR/Bundesinstitut für Bau-, Stadt- und Raumforschung. (2011). Auf dem Weg, aber noch nicht am Ziel – Trends der Siedlungsflächenentwicklung. Bonn. = BBSR-Berichte KOMPAKT Nr. 10.

Behrens, H., Dehne, P., & Hoffmann, J. (2012). Demografische Entwicklung und Landnutzung. Müncheberg. = Leibniz-Zentrum für Agrarlandschaftsforschung (ZALF) e.V., Nachhaltiges Landmanagement, Diskussionspapier Nr. 3.

BfN and BBSR/Bundesamt für Naturschutz & Bundesinstitut für Bau-, Stadt- und Raumforschung (Hrsg.). (2011). Kulturlandschaften gestalten! Zum zukünftigen Umgang mit Transformationsprozessen in der Raum- und Landschaftsplanung. Bonn.

BMELV/Bundesministerium für Ernährung, Landwirtschaft und Verbraucherschutz (Hrsg.). (2011). Waldstrategie 2020. Nachhaltige Waldbewirtschaftung – eine gesellschaftliche Chance und Herausforderung. Berlin.

BMI/Bundesministerium des Innern. (2011). Demografiebericht. Bericht der Bundesregierung zur demografischen Lage und künftigen Entwicklung des Landes. Berlin.

BMVBS, & BBSR/Bundesministerium für Verkehr, Bau und Stadtentwicklung; Bundesinstitut für Bau-, Stadt- und Raumforschung (Hrsg.). (2009). Einflussfaktoren der Neuinanspruchnahme von Flächen. Bonn. = Schriftenreihe Forschung Heft 139.

Bock, S., Hinzen, A., & Libbe, J. (Eds.). (2011). *Nachhaltiges Flächenmanagement – Ein Handbuch für die Praxis*. Berlin: Ergebnisse aus der REFINA-Forschung.

Bürgi, M., Hersperger, A. M., & Schneeberger, N. (2004). Driving forces of landscape change—current and new directions. *Landscape Ecology, 19,* 857–868.

Bürkner, H.-J., Berger, O., Luchmann, C., & Tenz, E. (2007). Der demografische Wandel und seine Konsequenzen für Wohnungsnachfrage, Städtebau und Flächennutzung. Erkner. = IRS Working Paper 36.

Canzler, W. (2007). Verkehrsinfrastrukturpolitik in der schrumpfenden Gesellschaft. In O. Schöller, W. Canzler, A. Knie (Hrsg.). Handbuch Verkehrspolitik, 510–532. Wiesbaden.

Cortesi, F., & Hepperle, E. (2011). Impacts of megatrends on soils. A new approach to sustainable resource management. In E. Hepperle, R. W. Dixon-Gough, T. Kalbro, R. Mansberger, K. Meyer-Cech (Eds.), Core themes of land use politics: Sustainability and balance of interests. Zürich: 55–69.

Demuth, B., Moorfeld, M., & Heiland, S. (2010). Demografischer Wandel und Naturschutz. Bonn-Bad Godesberg. = Bundesamt für Naturschutz,. Naturschutz und Biologische Vielfalt 88.

Die Bundesregierung. (2002). Perspektiven für Deutschland. Unsere Strategie für eine nachhaltige Entwicklung. Berlin.

Dosch, F., & Bergmann, E. (2005). Schwerpunkt Flächeninanspruchnahme in der Nachhaltigkeitsstrategie – Trends, Strategien und Initiativen auf Bundesebene. In: Forum Stadt- und Regionalplanung (Hrsg.): Das Flächensparbuch. Diskussion zu Fläschenverbrauch und lokalem Bewusstsein, 65–75. Berlin.

EEA/European Environment Agency. (2010). The European environment—State and outlook 2010. Land use. Synthesis, Copenhagen. https://www.eea.europa.eu/soer/europe/land-use (Zugriff am April 11, 2014).

FGSV/Forschungsgesellschaft für Straßen- und Verkehrswesen, Arbeitsgruppe Verkehrsplanung. (2006). Hinweise zu verkehrlichen Konsequenzen des demografischen Wandels. Köln.

Grimm, B., Lohmann, M., Heinsohn, K., Richter, C., & Metzler, D. (2009a). Auswirkungen des demografischen Wandels auf den Tourismus und Schlussfolgerungen für die Tourismuspolitik. AP 2, Teil 1: Trend- und Folgenabschätzung für Deutschland. Berlin, Kiel, München.

Grimm, B., Lohmann, M., Heinsohn, K., Richter, C., & Metzler, D. (2009b). Auswirkungen des demografischen Wandels auf den Tourismus und Schlussfolgerungen für die Tourismuspolitik. AP 4: Folgerungen. Berlin, Kiel, München.

Heiland, S., Regener, M., & Stutzriemer, S. (2004). Endbericht zum Forschungs- und Entwicklungsvorhaben Folgewirkungen der demografischen Entwicklung in Sachsen im Geschäftsbereich des SMUL. Im Auftrag des Sächsischen Staatsministeriums für Umwelt und Landwirtschaft, vertreten durch das Sächsische Landesamt für Umwelt und Geologie. Dresden.

Heiland, S., Regener, M., & Stutzriemer, S. (2005). Auswirkungen des demografischen Wandels auf Umwelt- und Naturschutz. Blinder Fleck in Wissenschaft und Praxis. *Raumforschung und Raumordnung, 63* (3), 189–198.

Hersperger, A. M., & Bürgi, M. (2009). Going beyond landscape change description: quantifying the importance of driving forces of landscape change in a Central Europe case study. *Land Use Policy, 26,* 640–648.

Hoßmann, I., & Münz, R. (o. J.). Glossar. In: *Berlin-Institut für Bevölkerung und Entwicklung: Online-Handbuch Demografie.* Berlin. https://www.berlin-institut.org/online-handbuchdemografie/glossar.html#c1400 (Zugriff am August 25, 2014)

IG BAU/Industriegewerkschaft Bau-Agrar-Umwelt. (2014). Handlungsbedarf. Demographischer Wandel in der Landwirtschaft. Berlin, Frankfurt am Main.

Knie, A., & Canzler, W. (2009). Forschungsvorhaben de.wi.mob.i.n. Demografische und wirtschaftsstrukturelle Auswirkungen auf die Mobilität in der Gesellschaft in den nächsten Jahrzehnten. Konsequenzen für die Verkehrsträger und die Zukunft staatlicher Daseinsvorsorge, Endbericht. Berlin.

Kroll, F., & Haase, D. (2010). Does demographic change affect land use patterns? A case study from Germany.

Lambin, E. F., Turner, B. L., Geist, H. J., et al. (2001). The causes of land-use and land-cover change: moving beyond the myths. *Global Environmental Change, 11*(4), 261–269.

Lambin, E. F., & Geist, H.J. (Eds.). (2006). Land-use and land-cover change. In *Local processes and global impacts,* Berlin, Heidelberg.

McNeill, J. et al. (1994). Toward a typology and regionalization of land-cover and land-use change: Report of working group B. In W. B Meyer, B. L. Turner II (Eds.), *Changes in land use and land cover: A global perspective* (pp. 55–65), Cambridge.

Milbert, A. (2013). Vom Konzept der Nachhaltigkeitsindikatoren zum System der regionalen Nachhaltigkeit. *Informationen zur Raumentwicklung, 1,* 37–50.

Moorfeld, M. (2011). Landscapes in Eastern Germany at a turning point—linkages between population decline, ageing and land consumption.

Murray-Rust, D., Rieser, V., Robinson, D. T., Milicic, V., & Rounsevell, M. (2013). Agent-based modelling of land use dynamics and residential quality of life for future scenarios. *Environmental Modelling & Software, 46,* 75–89.

Oeltze, S., Bracher, T., Eichmann, V., Dreger, C., Ludwig, U., Lohse, D., Zimmermann, F., & Heller, J. (2007). Mobilität 2050. Szenarien der Mobilitätsentwicklung unter Berücksichtigung von Siedlungsstrukturen bis 2050. Berlin. = Deutsches Institut für Urbanistik, Edition Difu – Stadt Forschung Praxis, Band 1.

Petermann, T., Revermann, C., & Scherz, C. (2006). Zukunftstrends im Tourismus. Studien des Büros für Technikfolgen-Abschätzung beim Deutschen Bundestag. Berlin.

Rudolph, A., Regener, M., Müller, B., & Meyer-Künzel, M. (2007). Soziodemografischer Wandel in Städten und Regionen – Entwicklungsstrategien aus Umweltsicht. Dessau. = Umweltbundesamt. UBA-Texte 18/07.

Schaldach, R., & Priess, J. A. (2008). Integrated models of the land system: A review of modelling approaches on the regional to global scale. *Living Reviews on Landscape Research, 2,* 1.

Schinninger, I. (2008). Globale Landnutzung. Externe Expertise für das WBGU-Hauptgutachten "Welt im Wandel: Zukunftsfähige Bioenergie und nachhaltige Landnutzung", Wissenschaftlicher Beirat der Bundesregierung Globale Umweltveränderungen, Materialien, Zürich, Berlin.

Schneeberger, N., Bürgi, M., Hersperger, A. M., & Ewald, K. C. (2007). Driving forces and rates of landscape change as a promising combination for landscape change research—An Application on the northern fringe of the Swiss Alps. *Land Use Policy, 24,* 349–361.

Schröder, A., Widmann, T., & Brittner-Widmann, A. (2007). Wer soll in Zukunft eigentlich noch reisen? Tourismus in Deutschland zwischen Geburtenrückgang und Überalterung. In: C. Haehling von Lanzenauer, K. Klemm (Hrsg.). Demografischer Wandel und Tourismus. Zukünftige Grundlagen und Chancen für touristische Märkte, 57–89. Berlin. = Deutsche Gesellschaft für Tourismuswissenschaft e.V., Schriften zu Tourismus und Freizeit Band 7.

Siedentop, S., Gornig, M., & Weis, M. (2011). Integrierte Szenarien der Raumentwicklung in Deutschland.Berlin = DIW Berlin, Politikberatung kompakt 60.

Spitzer, H. (1991). Raumnutzungslehre, Stuttgart.

SRU/Sachverständigenrat für Umweltfragen. (2016). Umweltgutachten 2016 Impulse für eine integrative Umweltpolitik. Berlin.

Thoroe, C. (2009): Wald im Wandel – gesellschaftliche Herausforderungen. In: B. Seintsch, M. Dieter (Hrsg.). Waldstrategie 2020. Tagungsband zum Symposium des BMELV, 10.-11. Dez. 2008, 5–9, Berlin. Braunschweig. = Johann Heinrich von Thünen-Institut, Landbauforschung – vTI Agriculture and Forestry Research Sonderheft 327.

TLL/Thüringer Landesanstalt für Landwirtschaft. (2006). 1. Teilbericht Auswirkungen des demografischen Wandels auf die Thüringer Landwirtschaft. Jena.

TLL/Thüringer Landesanstalt für Landwirtschaft. (2007). Auswirkungen des demografischen Wandels auf die Thüringer Landwirtschaft. Teilbericht Betriebsbefragung zur Personalentwicklung und Weiterbildung. Jena.

Weith, T. (2009). Bausteine zur Nachhaltigkeit in Flächenpolitik und Flächenmanagement. In T. Weith (Hrsg.), Flächenmanagement im Wandel, 9–19. Berlin. = Zeitschrift für angewandte Umweltforschung, Sonderheft 16.

Wiener, B., Richter, T., & Teichert, H. (2004). Abschätzung des Bedarfs landwirtschaftlicher Fachkräfte unter Berücksichtigung der demografischen Entwicklung (Schwerpunkt neue Bundesländer). Halle/Saale. = Forschungsberichte aus dem Zentrum für Sozialforschung Halle e. V.

Wolf, A., & Appel-Kummer, E. (2005). Demografische Entwicklung und Naturschutz – Perspektiven bis 2015. Essen. = Bundesamt für Naturschutz, BfN-Skripten 196.

Wurz, A. (2007): Waldzukünfte – Basispapier (Kurzfassung). Zukunftsfeld „Demografische Entwicklung" im Rahmen des Projektes Zukünfte und Visionen Wald 2100. Freiburg.

Wyman, M. S., & Stein, T. V. (2010). Modelling social and land-use/land-cover change data to assess drivers of smallholder deforestation in Belize. *Applied Geography, 30*(3), 329–342.

Chapter 5
Urbanisation and Land Use Change

Henning Nuissl and Stefan Siedentop

Abstract Urbanisation is one of the major driving forces behind the formation of today's land use systems. It almost always involves the conversion of land use from non-urban to urban uses. A great deal of contemporary urbanisation has been characterised as urban sprawl, i.e. a highly extensive form of land take for urban uses having environmentally detrimental effects. However, urban land use change can occur in relatively diverse forms in terms of layout, building density and speed of change, to name but a few aspects. In recent decades, researchers have made substantial progress in empirically addressing the various forms of urban land use and its change over time. As a consequence, the global dimension of urbanisation-related land use change is now on the agenda of policymakers and researchers worldwide. In order to provide an overview of the many geographical, environmental, sociological and political aspects that are relevant with respect to urban land use change, this contribution strives to make (1) some conceptual clarification regarding the notions associated with urban land use change, before (2) highlighting its (economic, social and political) drivers, as well as its (3) impacts. The text then moves on to (4) briefly systematising the instruments and strategies that have been put in place to cope with urban land use change. Finally, (5), we reflect on the current state of the art regarding research and policies on urban land use change.

Keywords Urban land use change · Measurement of urban sprawl · Drivers and impacts · Planning policy

H. Nuissl (✉)
Geography Department, Humboldt-Universität zu Berlin, Unter den Linden 6, 10099 Berlin, Germany
e-mail: henning.nuissl@geo.hu-berlin.de

S. Siedentop
Research Institute for Regional and Urban Development, ILS, Brüderweg 22 – 24, 44135 Dortmund, Germany
e-mail: stefan.siedentop@ils-forschung.de

© The Author(s) 2021
T. Weith et al. (eds.), *Sustainable Land Management in a European Context*, Human-Environment Interactions 8, https://doi.org/10.1007/978-3-030-50841-8_5

5.1   What is Urban Land Use Change and Why It is a Relevant Issue?

Apart from cultivation, the use of land for residential and related purposes has been part of the encroachment of human civilisation upon natural ecosystems since the beginning. However, the amount of land covered by settlements was largely negligible until the advent of industrialisation and the processes of massive urbanisation it brought along. Urbanisation—understood as an increase in the urban (as compared to rural) population and an "urban' workforce—i.e. manufacturing as compared to agricultural workforce—has almost always involved the conversion of land use from non-urban to urban uses, because it requires an increased need for space in (existing) settlement areas. The visible outcome of land use change in the wake of urbanisation is the spatial expansion of built-up areas (which implies a significant alteration of land cover features), accompanied by changes in the urban spatial structure and the urban form.

The rapid conversion of open, mostly agricultural land into settlement areas has been accompanied by pronounced criticism since the heyday of industrialisation in the nineteenth century. Even at that time, the rapid growth of industrial urban centres raised great suspicion and was blamed not only for the accumulation of human disorder, vice and despair (i.e. the "traditional" anti-urban concerns), but also for the destruction of the traditional (pastoral) landscape due to its greed for land. In a 1937 speech to US urban planners, Earl Draper, the Director of the Tennessee Valley Authority, became the first to use the term "sprawl" to indicate a specific pattern of urban growth that makes the countryside—from his point of view—"ugly, uneconomic [in terms] of services and doubtful social value" (cited in Wassmer 2002). Since then and up to today, the scientific and political discussion about the negative impacts and drawbacks of urban land use change in both developed *and* developing countries has largely been linked to the notion of urban sprawl (e.g. Whyte 1958; Clawson 1962; Harvey 1965; Benfield et al. 1999; Burchell et al. 1998; Burchell et al. 2002; Peiser 2001; Gillham 2002; Squires 2002; Nechyba and Walsh 2004).

More recently, and linked to a growing concern for ecological issues and the finiteness of natural resources (Meadows et al. 1972), urban land use change has often been labelled "land consumption" (Frenkel 2004; Köck et al. 2007; Nuissl et al. 2009), which is a somewhat imprecise notion because the (amount of) land does not diminish on account of altering its use. However, concepts such as urban sprawl, land consumption and land take clearly indicate the association of urban land use change with negative side effects.

For the period from 1990 to 2000, Angel et al. (2005: 56) estimated that the annual increase in built-up areas in developing countries was around 3.6%, whereas it amounted to only 2.9% on average in industrialised countries. Among world regions, East Asia, including the Pacific, and Southeast Asia witnessed the most intensive land consumption, with growth rates of 7.2% and 6.4%, respectively. In Europe, the annual growth of urban land is expected to range between a maximum of 2% in rapidly growing areas and nearly zero in remote rural regions (EEA 2006). Focusing on the European Union, Kuemmerle et al. (2016) put this observation in relation to other kinds of land use change: "The most widespread changes in the extent of land-use categories in the EU between 1990 and 2006 were cropland decline ($\sim$136,660 km^2), followed by expansion of grazing land ($\sim$75,670 km^2), and expansion of forest areas ($\sim$70,630 km^2). The least common conversion among broad land-use categories was urban expansion ($\sim$16,820 km^2). … At the European scale, these area changes translate into moderate land-conversion rates in the agricultural sector between 1990 and 2006, ranging from -13.4% for permanent crops to $+6.5\%$ for meadows and pastures, while urban areas expanded by approximately 21%" (Kuemmerle et al. 2016: 5). In addition, various other studies have provided empirical evidence that the spreading of urban land uses has clearly exceeded population growth, resulting in declining overall densities (e.g. Fulton et al. 2001; Glaeser and Kahn 2003; Lopez and Hynes 2003; Angel et al. 2005; Theobald 2005). At the same time, urban density gradients have significantly levelled off over time in metropolitan areas. Urban densities decreased between 1990 and 2000 worldwide, in East Asia by as much as 4.9% per year and in Europe by a relatively moderate 1.9%.

However, looking merely at the size or the growth of urban areas would provide only poor insights into the dynamics of urban land use change (even if related to population growth), because there are different kinds of urban land use change that have rather diverse impacts (McGranahan and Marcotullio 2005). For instance, residential development on former agricultural land usually damages considerably fewer and other wildlife habitats than industrial development on a drained wetland site; likewise, new development in the vicinity of existing settlements infringes on the landscape matrix to a lesser degree than the development of many small and unconnected patches of urban land. Hence, it is not only the quantity of land converted to urban uses that needs to be considered, but also:

- the previous land use and land cover (agricultural, forest and natural);
- the dominant purpose of the new urban use (residential, commercial, industrial, recreational or other) and the corresponding land cover features (such as the imperviousness of surfaces and the emission of pollutants);
- the location and pattern of new urban land; and
- the efficiency of land use.

Recognition of these aspects is key not only to a comprehensive understanding of land use change dynamics and their knock-on effects on environmental qualities, but also as a basis for urban planning and management.

Significant improvements in the resolution and quality of digital land use and land cover data have opened up new possibilities for a more complex monitoring of land use change dynamics that captures differences in aspects such as urban form, land uses, development and location patterns as well as efficiency of land use (e.g. Schneider and Woodcock 2008). Drawing on the availability of such data, numerous methodological approaches have been introduced to provide a quantitative assessment of urban land use change (for a brief overview see, e.g. Chin 2002; Frenkel and Ashkenazi 2008, or Siedentop and Fina 2010). Table 5.1 lists prominent measure-

Table 5.1 Frequently proposed land consumption and urban sprawl measures

Indicator	Description	Sources
Size or share of urban land	Size of urban land (km^2 or sqm); percentage of urban land (%)	Galster et al. (2001), Kolankiewicz and Bleck (2001), Angel et al. (2005), Schneider and Woodcock (2008), Siedentop and Fina (2010), Wolff et al. (2018)
New land consumption	Converted urban land (in hectares or acres)	Anthony (2004), Siedentop and Fina (2010)
Urban density	Number of people, jobs or housing units per hectare of urban land (gross or net)	Razin and Rosentraub (2000), Torrens and Alberti (2000), Chin (2002), Ewing et al. (2002), Glaeser and Kahn (2003), Angel et al. (2005), Siedentop and Fina (2010), Wolff et al. (2018)
Change in urban density	Change in urban density between two base years (percentage)	Emison (2001), Anthony (2004), Angel et al. (2005), Siedentop and Fina (2010), Wolff et al. (2018)
Density gradient	Regression of density against distance by ordinary least squares (OLS)	Torrens and Alberti (2000)
Land use mix/land use separation	Degree to which different urban land uses exist in close vicinity to each other	Galster et al. (2001), Chin (2002), Ewing et al. (2002), Song and Knaap (2004), Torrens (2008)

(continued)

Table 5.1   (continued)

Indicator	Description	Sources
Concentration/decentralisation	Degree to which urban development is located near to the CBD (e.g. measured by the percentage of population and employment within concentric rings around the CBD or the median person's/worker's distance in distance units from CBD)	Galster et al. (2001), Glaeser and Kahn (2003), Lopez and Hynes (2003), Weber and Sultana (2005), Huang et al. (2007), Torrens (2008)
Continuity/dispersion/fragmentation/complexity	The degree to which developable land is built up continuously; the degree of irregularity of built-up patches (measured using certain indices such as patch density or more complex statistical measures of spatial regularity)	Galster et al. (2001), Chin (2002), Huang et al. (2007), Schneider and Woodcock (2008), Siedentop and Fina (2010), Salvati and Carlucci (2016)

ments (indicators) of urban land use change along with their function (description) and sources in the literature.

The final indicator listed in Table 5.1 ("Continuity/dispersion/fragmentation/complexity") concerns the spatial pattern of urban land use change in relation to the existing settlement area, and has been widely used to describe the general shape of urban land use change in a particular urban region. Over time, most metropolitan areas have changed their urban form from a highly concentrated compact structure to a more irregular, discontinuous or dispersed urban land use pattern (e.g. Nelson 1992; Theobald 2001; Carruthers and Ulfarsson 2002; Lang and LeFurgy 2003; Salvati and Carlucci 2016). This trend has often been discussed using terms such as "leapfrogging" or "ribbon" development.

5.2   Drivers of Urbanisation and Urban Land Use Change

In order to fully comprehend the phenomenon of urban land use change, simply observing it and measuring it—even using sophisticated methods of geoinformatics and statistics—is not sufficient. What is also needed is an account of its driving factors. This helps to explain, for instance, why urban land use change occurs at a given rate and in a given pattern, and why these significantly differ among countries and even regions (e.g. Kolankiewicz and Beck 2001; Lambin et al. 2001; Huang et al. 2007; Creutzig et al. 2019). However, first of all, it is crucial to note that there is no grand theory of urbanisation or comprehensive explanatory model of urban land use change that would make it possible to interpret and explain actual observations. Instead, social sciences and economics have offered various theories that hint at important drivers of land-consuming human activities.

Neoclassical economic theory in particular has provided a closed framework for the explanation of urban growth. Basically, it postulates an unregulated land market, where land rents near the urban core are highest (because of maximum accessibility to urban services and correspondingly negligible transport costs), and argues that location decisions by both private households and firms reflect the goal of achieving maximum utility by balancing space needs, location preferences and financial budget constraints. Based on these primary considerations, it might be plausible to assume that high-income groups (with more land-demanding aspirations) would prefer to live at a greater distance from the city centre where large building lots are available, while low-income households would choose a location near the urban core which would incur lower transport costs. Starting from these assumptions, the neoclassical monocentric model of urban spatial structure (the "Alonso-Muth-Mills model") explains the spatial expansion of cities and density gradients with just a few variables, specifically the demand for new housing and commercial land, rising incomes, innovations in intra-urban transport systems, and decreasing transport costs (Alonso 1964; Muth 1969; Mills 1972; Mieszkowski and Mills 1993). This meant that the growing physical footprint of cities and their declining density was the combined effect of a growing population, rising affluence and enhanced individual mobility due to the increasing affordability of the private motor car.

In addition to economic theorising, technical viewpoints underpin the importance of distance and transport costs with respect to the spatial diffusion of urban land uses. They associate the compactness of the pre-industrial city with the fact that most trips had to be made on foot or similarly slow modes of transport. This constraint disappeared with the availability of faster mass transport technologies and the private automobile (e.g. Antrop 2004). Following this logic, the physical growth of cities became a function of transport technology. Nelson (1992) pointed out that other improvements in technology, such as the personal computer, cellular phones and the internet, may have encouraged the spatial decentralisation of people and firms even further, setting up conditions for more land-intensive forms of urbanisation.

Empirical observations of how urban areas develop, such as the emergence of polycentric urban configurations, often defy the simple assumptions of pure neoclassical urban theory. In contrast to the fundamental assumption in the neoclassical city model that employment was concentrated in central business districts (and gradually decreased with increasing distance to the urban core), modern agglomerations in developed countries are characterised by their multi-nodal settlement system, with a complex pattern of primary and secondary centres (Garreau 1991; Champion 2001; Davoudi 2003). Accordingly, many additional factors other than land prices and commuting costs have been identified that affect the location decisions of individual households or firms (Nechyba and Walsh 2004). Examples include the quality of urban services, specific priorities and demands of different social groups in terms of urban and environmental amenities, or the desire to live in a socially homogeneous neighbourhood. The Tiebout Local Public Finance Model (Tiebout 1956) suggested that people decided to locate in a particular jurisdiction based on their preferences and taste for local amenities. Tiebout described factors that "pull" people out of the central areas of metropolitan regions on account of attractive characteristics of suburban communities (e.g. good service levels or lower taxes) and others that "push" people out of central areas as a result of inner city problems such as poor environmental quality and services, or crime. Theoretical accounts of this nature hint at the importance of particular social-cultural trends that mould the current demand for urban land, such as the proliferation of both land-consuming urban lifestyles (tourism and recreational activities, second homes) on the one hand and a (re)orientation (particularly among upper middle-class households) towards urban centres ("reurbanisation") on the other.

With urban sprawl and land consumption being a major environmental concern, recent scholarly efforts have broadened the knowledge on its causes and drivers significantly. However, while it is often possible to explain the intensity of urban land use change on a broader scale, e.g. on the European (Oueslati et al. 2015) or global scale (Creutzig et al. 2019), predicting its spatial patterns remains a challenging issue. As a consequence, spatially explicit land use models have been developed which not only explain *at what rates* urban land use change occurs in a given period of time, but also address the question *where* it will take place, i.e. its likely location (Frenkel 2004). Poelmans and van Rompaey (2010) have distinguished five groups of explanatory variables that have been frequently used in models of urban land use change:

- Biophysical factors, such as the slope or water table, have an impact on the suitability of land tracts for the construction of buildings or infrastructure facilities, and can explain why certain areas are excluded from development.
- Social factors reflect the location preferences of households (or household types). Examples include the income level or ethnic composition of nearby neighbourhoods, and the availability of open green spaces. These factors may encourage or discourage a household's choice of development site.
- Economic factors refer to accessibility features as proxy values for market access. Frequently used measurements include the distance to urban centres or main roads, and the availability of public transport services within a walkable distance. Undeveloped properties with good accessibility are more likely to become urbanised in the future.
- Neighbourhood interactions refer to an observed spatial autocorrelation between new developments and existing urbanised areas. In contrast, some potentially conflicting land uses (e.g. residential and industrial development) are unlikely to be located directly next to each other.
- Spatial policy and planning include the possibility to legally define, i.e. distinguish the usability of different land parcels. These policies can be labelled "negative planning" inasmuch as they aim to protect current land uses (habitat conservation,

prime farmland) or "positive planning" inasmuch as they define the suitability of a piece of land for a specific use (i.e. where they explicitly designate sites for urban development).

While some of these determinants of urban land use change illuminate the total pressure on the land within a region or even a nation-state, others—such as moderate land prices or above-average accessibility—are suited to identifying the local hot spots of development, but are largely unable to explain the aggregate regional growth rate of urbanised land. Table 5.2 presents a set of relevant variables with their estimated explanatory capacity.

Various studies (Ulfarsson and Carruthers 2006; Siedentop et al. 2009) have found that urban land use change is to a large extent a supply-driven process. They have argued that it is not only the result of demand driven by demographic and economic growth pressures or social preferences, but is also fuelled and facilitated by policies at national as well as local levels. For instance, the political agenda of local decision-makers in stagnating or economically declining areas often emphasises the importance of cheap land for residential or commercial uses as a means to attract people and enterprises and thus to generate tax revenue. This can explain why some regions and municipalities without demographic or economic demand pressure nevertheless show significant land consumption rates (Nuissl and Rink 2005). Government policies such as the commuter tax allowance in Germany, the financing of highway

Table 5.2 Explanatory variables used in land consumption model applications (dark grey = strong explanation, pale grey = moderate explanation, white = not relevant)

Factor	Examples	Explanation of …	
		rate of land consumption	location of land consumption
Biophysical	Slope		■ dark grey
	Hazardous land		■ dark grey
Economic	Economic growth	■ dark grey	
	Land prices	■ dark grey	
	Distance to urban centres		■ dark grey
	Distance to the main road		■ dark grey
	Fiscal motives to convert land into urban use	■ dark grey	
Demographic/social	Population growth	■ dark grey	
	Income growth and changes in lifestyle	■ dark grey	
	Motorisation	■ dark grey	
	Social preferences for housing types and locations	▨ pale grey	■ dark grey
Spatial policies	Land use regulation (positive and negative planning)	■ dark grey	▨ pale grey
	Revitalisation and regeneration policies	■ dark grey	
	Public funds for greenfield development	■ dark grey	

infrastructure in the US, or subsidies for the development of industrial or retail development by the European Union are likely to support this effect (Persky and Kurban 2001).

Some scholars have presented evidence that the institutional fragmentation of local authorities could be another important factor explaining the rate and pattern of land consumption. According to this position, decentralised land use governance with numerous local governments controlling urban land use is more likely to promote urban sprawl, as it increases the number of jurisdictions seeking extra-budgetary revenue through land conversion to urban uses (Downs 1998; Razin and Rosentraub 2000; Ulfarsson and Carruthers 2006). The size of local government units is also important in other ways—the bigger they are, the less likely they will be reliant on one particular investor or project, and the less vulnerable they will be to the influence of individual local land owners with regard to planning policies and decisions. Furthermore, smaller communities are more likely to permit exclusionary zoning policies, where local governments attempt to exclude low-income groups from their municipalities (Pendall 1999; Clingermayer 2004). These policies are driven by suburban residents' desires to protect their housing investments and to maintain their social status (Downs 1998).

5.3 Impacts of Urban Land Use Change

While urban land use change on the global scale has only become a hot topic in recent decades (e.g. Foley et al. 2005; Seto et al. 2012; Creutzig et al. 2019), urbanisation, suburbanisation and urban sprawl—i.e. urban land use change at the local and regional level—have been a subject of major concern and passionate debate for quite some time because of their obvious effect on the morphology of urban systems (increases in artificial surfaces, changes in densities, alteration of land use patterns) and consequent impacts on the environment and other amenities. While the unintended effects of urban growth have been a matter of discussion in the United States since as early as the mid-twentieth century, they are an issue of concern all over the globe today. Initially, urban growth was mainly blamed for endangering landscape beauty, weakening community life and overloading the transport and network infrastructure (e.g. Nechyba and Walsh 2004). However, the debate has clearly broadened its scope and increased in intensity over the decades, now also raising concerns regarding the loss of habitats and biodiversity, the rise in greenhouse gas emissions, and environmental justice in general, to name but a few.

Despite its various drawbacks, it would be an inappropriate simplification to simply blame urban land use change as an environmentally harmful and generally non-sustainable phenomenon. On the one hand, the use of land for urban purposes inevitably infringes on its "value" in other (mainly environmental) respects; but on the other hand, urban land use change is essentially a by-product of demographic change and economic growth, and it is difficult—yet not impossible—to conceive of prosperous and dynamic societal development without any kind of "land consumption"—particularly in economically growing nations (e.g. Deng et al. 2010). The dispute on whether urban land use change is a curse or a blessing is not only one about the prioritisation of goals, but also includes a fervent academic debate about the validity of countless empirical findings on its adverse impacts. Critics mark these findings as well as the methodologies with which they were obtained, as largely ideological, i.e. based on the normative assumption that urban land use change should be contained. This kind of criticism, put forth in defence of laissez-faire urbanisation, however, usually appears at least as "ideological" as the criticised studies.

The scholarly debate about the impacts of urbanisation and urban land use change has become almost incomprehensible. However, it is possible to distinguish a few major threads of debate each of which emphasises a particular issue of concern. First of all, there are major concerns regarding the environmental outcomes of urban land use change. These impacts are largely related to changing land cover, i.e. the sealing of land, which almost inevitably occurs when land is being developed (Johnson 2001; Pauleit et al. 2005). In other words: urban land use change leads to an increased share of artificial, impervious surfaces, including built-up land, i.e. rooftops, roads, parking lots, pavements, etc. (Arnold and Gibbons 1996; Haase and Nuissl 2007). Imperviousness physically limits the infiltration of rainfall into the ground. Rainfall and snowmelt that is unable to infiltrate instead must become surface runoff (Alberti 1999). Thus, soil sealing in highly urbanised areas is widely viewed as an important causal factor for flood risks (Frenkel 2004). Due to the fact that urban runoff water carries with it chemical pollutants (e.g. from automobile traffic or industrial land uses), imperviousness also contributes to the biochemical degradation of water resources. Based on many empirical studies, Moglen and Kim (2007) estimated that, once the rate of paved surfaces exceeds a threshold of 10–15%, various indicators of biological stream quality begin to markedly decrease. Moreover, the spatial concentration of artificial land cover with specific thermal characteristics also creates local temperature anomalies. This leads to a higher average temperature in the dense urban fabric compared to the urban periphery ("urban heat island") (Voogt 2002; Watkins et al. 2007).

In addition to the magnitude of urban land use change, its spatial pattern has also to be taken into account. Dispersed and fragmented land use patterns are a crucial contributor to landscape fragmentation, which is characterised by a process of perforation, dissection and isolation of habitat areas and natural or semi-natural ecosystems (Jaeger 2000). Thus, many scholars have regarded urban land use change as a major cause of the alarming loss of species all over the world (Theobald et al. 1997; Cieslewicz 2002).

The overall impact of urban land use change not only depends on the environmental "quality" of resulting land use patterns. Indeed, the characteristics of the land (e.g. soil quality, habitat quality, vegetation, etc.) that became urbanised within a specific period of time also have to be examined from an economic perspective. One particular concern is the loss of prime agricultural land, which has major importance for the long-term competitiveness and sustainability of agriculture generally (Hasse and Lathrop 2003). The European Environment Agency (EEA 2005: 176) has argued that the "continent's best soils" have been sealed off due to the fact that most urban centres were built on fertile valley soils and around estuaries (see also American Farmland Trust 1994; Kuemmerle et al. 2016; Creutzig et al. 2019). Hasse and Lathrop (2003) presented quantitative findings that prime farmland was more vulnerable to urbanisation than farmland of lesser quality. Urban development in increasingly fragmented agricultural landscapes can also be problematic for the production of food and fibre on the remaining farmland. For instance, conflicts between farmers and their residential neighbours "can arise over noise, chemical applications, and smells that are part of farming" (Merenlender et al. 2005: 2).

The (not only monetary) costs for providing settlements with public services have often been addressed in urban sprawl studies that focus on the economic effects of urban land use change. In 1974, the "costs of sprawl" study (Real Estate Research Corporation 1974) presented empirical evidence for a negative interdependency between the density of residential developments and the fiscal costs for providing basic urban services. The findings of this study triggered an intensive dispute not only with respect to the implications for urban development policies, but also in terms of methodological uncertainties. A couple of subsequent studies confirmed the results of the 1974 work (see Burchell et al. 1998 with many references); others disputed the relevance of urban form variables on infrastructure costs (Peiser 1989; Ladd 1991). Ultimately, researchers today widely acknowledge the idea that low-density and dispersed urban developments are more cost-intensive than more compact development patterns (see also Speir and Stephenson 2002; Carruthers and Ulfarsson 2003; Burchell et al. 2005).

Urban land use change has also been criticised for unintended social outcomes, particularly in association with the broad process of suburbanisation that has affected cities and urban agglomerations worldwide for many decades (e.g. Power 2001). The "spatial mismatch" debate, starting in the 1960s (Gordon et al. 1989; Kain 1992), addressed the extent of limits on residential choices for minority populations (especially people of colour in the US), combined with the intra-regional decentralisation of employment. Proponents of the spatial mismatch hypothesis have argued that the exclusion of low-income and non-white households from suburban communities, together with the continuous spatial dispersal of jobs, especially for low-skilled employees, is responsible for the high rates of unemployment and the low earnings of minority populations living in inner cities. More recently, studies have found evidence that low job accessibility in public transport catchments has a negative effect on the likelihood of employment among social groups that tend to lack access to cars (Matas et al. 2010). Moreover, suburban development has also been associated with gender issues, as it is usually linked to a traditional (key) family model, with the female adult being responsible for reproductive work and childrearing. In such mono-functional residential areas characteristic of suburbia in particular, women are largely unable to participate in the labour market and even have difficulties in accessing public spaces.

More recently, another social effect of dispersed urbanisation patterns has attracted major attention from researchers and environmental policymakers. Various studies have highlighted the relationship between urban form variables and physical activities with their corresponding health implications. These studies came to the conclusion that urban sprawl could have a severe impact on public health, leading to obesity and a generally insufficient level of physical activity among many people (e.g. McCann and Ewing 2003; Kelly-Schwartz et al. 2004; Committee on Physical Activity 2005).

Box: Systematisation of Systematic Accounts of Land Use Change Impacts
Numerous approaches to systematise the impacts of (urban) land use change have been introduced, each of which applies a particular dimension of categorisation (see table).

	Dimension of categorisation	Impact categories	Author
1.	Issue of concern	Environmental impacts	Chin (2002)
		Economic impacts	
		Social impacts	
2.	Causality	Direct impacts	Cooper (2004)
		Indirect or cumulative impacts	
3.	Spatial scale	Direct impacts (on the plot)	Nuissl et al. (2009)
		Cumulative impacts (on aggregated plots)	
		Contextual impacts (regional effects)	
4.	Impact pathway	Land surface-related impacts	Siedentop & Fina (2010)
		Land use pattern-related impacts	
		Density-related impacts	
5.	Appraisal (prioritisation of goals)	Negative impacts (as costs)	Burchell et al. (1998)
		Positive impacts (as benefits)	

1. Comprehensive reports on urbanisation and urban land use change issues usually classify effects and impacts according to the sphere (or policy field) in which they occur. Distinguishing between environmental (i.e. ecological), economic and social impacts, this classification often corresponds with the "classic triangle of sustainability". Sometimes additional dimensions, such as transport or politics, are considered as well.
2. At a more general level, it is possible to distinguish between single, i.e. direct, and cumulative, i.e. indirect, impacts of urban land use change. While the first denotes the direct and immediate outcome of a change of

land use on a particular plot of land, e.g. the reduction of agricultural land, significant land use-related environmental problems, such as the modification of urban climate conditions ("urban heat islands") or an increase in runoff, usually result from the cumulative effects of development activities.

3. After a closer look, it is possible to add a third kind of urban land use change impact. Contextual impacts depend on the characteristics of the larger territory (context) in which a land use change takes place. One example of this type of impact is the generation of traffic due to the development of an exurban retail facility.

4. Siedentop and Fina (2010) distinguish between three key dimensions of urban land use which they use to explain and model a broad range of land consumption impacts. These are land cover features (surface), the pattern of land use (the spatial configuration of urban and non-urban land patches) and the intensity of use (urban density).

5. Last but not least, the literature on urban sprawl in particular has often adopted a decidedly normative stance as to the impacts of urban land use change in that it distinguishes between costs and benefits. However, the sharp disagreement about the overall assessment on whether urban sprawl is "good" or "evil" illustrates that it is a matter of perspective (if not a political standpoint) if a certain issue is assessed in positive or negative terms.

Last but not least, the impact of urban land use change on motorised transport demand is probably the most frequently discussed issue in this field of research. Many scholars have argued that households in peripheral, low-density environments have long travel distances and tend to use their car extensively (Banister 1999; Naess 2003). Some critics dispute the causality between urban form and travel behaviour, pointing to the possibility that private households self-select themselves to places that are in accordance with their preferences for particular modes of transport (Handy 2005). At the same time, studies that controlled for demographic, socioeconomic and attitudinal variables (such as household income, family size or age) proved the significant effect of urban form on transport (Cervero 2003; Naess 2007; Vance and Hedel 2007; Ewing and Cervero 2010).

5.4 Policies on Urban Land Use Change

The desire to control the dynamics of land consumption was one of the earliest motivations for spatial planning. However, while this desire used to be of minor importance in comparison to the goal of mitigating land use conflicts and safeguarding the most rational form of urban growth, it has since become one of the major issues in land use policy (e.g. Gallent 2006). This issue is probably most disputed in the

US, where public concern about sprawl has grown significantly in the recent past (Bengston et al. 2005), resulting in a strong anti-sprawl movement. This in turn has prompted a number of states to adopt growth management programmes that attempt to contain urban growth and preserve open space. However, urban land consumption is also a major concern in Europe (EEA 2006; Nilsson et al. 2013), where quite a few planning strategies and instruments have been placed under scrutiny regarding their effectiveness (Hersperger et al. 2018). Such policies, however, have met with strong opposition by more liberal academics and planners who emphasise the importance of individual choice and the free market (e.g. Ewing 1994; Ewing 1997; Benfield et al. 1999; Gordon and Richardson 2001; Bruegmann 2005). Elsewhere, most notably in Western and Central Europe and Australia, the debate about the drawbacks of urban land use change has gained considerable momentum, too (e.g. Newman 1992; EEA 2006). In England, the former Labour government set a national target of delivering 60% of all new housing units on previously developed land and through conversions of the existing building stock. The government saw the reuse of urban land as a key policy in reducing development pressures on the open countryside (Downs 1999; Ganser and Williams 2007). The Chinese government, concerned about the alarming loss of prime farmland due to urbanisation, has introduced regulatory policies that have attempted to protect farmland more effectively (Lichtenberg and Ding 2006). The federal governments of Austria, Germany and Switzerland have all introduced national targets to reduce the rate of conversion of non-urban to urban land uses (Bundesregierung 2002). They argued that urban land use change and landscape fragmentation were key drivers of species loss, landscape deterioration and reductions in infrastructure efficiency.

A great variety of policy and planning instruments have been proposed to implement the goal of taming urban land use change. One common way to categorise policy and planning instruments is their classification according to whether they are concerned either with (I) regulation, or (II) spending, taxation and subsidies, or (III) advocacy (e.g. Bengston et al. 2004). Adopting and slightly modifying this approach, we propose differentiating policy and planning instruments according to where they are located on the continuum that ranges between the poles of two basic planning principles:

- *Planning*, reflecting the "traditional" regulatory approach of spatial planning to set legally binding rules for the use of land via regulatory plans, and
- *Market*, reflecting the "economic" approach of land use policy which employs "market-based instruments" that modify incentives in a way that lead actors to use the land in an intended manner. For instance, taxation schemes that put an additional cost on the development of land are clearly among the most efficient ways to minimise the total amount of urbanised land (e.g. Song and Zenou 2006).

Somewhere in the middle between these two poles, there is a wide array of instruments that are primarily managerial by character because they basically focus on influencing the decision-making processes of (either potentially land consuming or land use policy making) actors. This group of instruments can be subsumed under a third planning principle:

- *Management*; reflecting the "persuasive" approach that tries to change the behaviour of land-using actors, either by providing them with information on the consequences of their behaviour, or by involving them in a communicative process together with actors that wish to restrict land consumption.

These planning principles are of course ideal types. In reality, policy responses and planning instruments that address the problem of land consumption are frequently combinations of several instruments that entail the adoption of various principles.

Table 5.3 illustrates the three general approaches that land use policy can adopt to pursue its goals, including the goal of taming urban land use change. However, only the regulatory and the persuasive approach fall into the scope of spatial planning in a strict sense, while the modification of incentives is usually achieved in other policy fields such as taxation or social policy (Nuissl and Schröter-Schlaack 2009).

While there is a plethora of instruments that could be used in principle to interfere in the process of urban land use change, the likelihood of achieving the goal of

Table 5.3 A taxonomy of land use policy instruments

Governance principle	Planning approach	Examples of strategies and instruments
Planning	Regulation (law)	Land use planning, i.e. zoning (e.g. Hirt 2007; Köck et al. 2007)
		Urban design planning (e.g. density controls) (e.g. Acioly and Davidson 1996; Churchman 1999)
		Transit planning (Freilich 1998; Handa 1996)
		etc.
Management	Persuasion (information and communication)	Forums and roundtables (e.g. Healey 1992; Wates 2000)
		Information campaigns (e.g. Besecke et al. 2005; Haughton 1999)
		Land use change assessment and forecasting tools (e.g. Criterion Planners/Engineers 2001; EPA 2000)
		etc.
Market	Modification of incentives	Development taxes (e.g. Gihring 1999; Korthals-Altes 2009)
		Subsidies (e.g. urban regeneration) (e.g. Couch et al. 2003; Newton 2010)
		Tradable permit schemes (e.g. Nuissl and Schröter-Schlaack 2009; Pruetz 2003)
		etc.

minimising land consumption is not only dependent on the theoretical availability of such instruments, but also on the political will to use them for precisely this purpose. In this vein, the normative ideas underlying actual development policies and planning practices are of major importance. Three of the most important normative ideas to turn policy and planning towards a prudent use of land resources and minimising urban land use change are *(A)* the Green Belt and the Urban Growth Boundary, *(B)* various leitmotifs of urban development that promote a compact and mixed-use city, and *(C)* the prioritisation of urban regeneration.

1. The delineation of Green Belts or Urban Growth Boundaries are among the most famous "tools" designed by spatial planners to provide a clear orientation about where to steer new development and where to prevent urbanisation (e.g. Al-Hathloul and Mughal 2004; Bengston and Youn 2006; Abbott and Margheim 2008; Siedentop 2016). While both involve the idea of defining a ring of open land that surrounds the urban area, the latter concept usually implies a precisely defined line beyond which no building activity may take place, whereas the former is the more general concept which is often used in nonbinding regional plans and usually needs to be enforced by specific plans that prohibit development in the Green Belt areas. Urban Growth Boundaries are often difficult to implement in densely populated regions where it is hardly possible to make a sharp distinction between urban and rural areas. In addition, there is a broad debate in the US as well as in the UK on whether the definition of a rigid boundary around a settlement is indeed the most effective means for curbing urban sprawl and its associated negative impacts (e.g. Carlson and Dierwechter 2007; Gant et al. 2011). Several scholars claim to have proven this assumption (e.g. Weitz and Moore 1998), while others doubt it (e.g. Bae and Jun 2003). Likewise, Green Belt policies or Urban Growth Boundaries can prove unsuitable in a situation where informal housing is a frequent phenomenon (Wang and Scott 2008).

2. Since its beginnings, spatial planning has been heavily influenced by the predominant leitmotifs of the time regarding the "optimal" urban environment. Today, the chief guiding stars in urban planning promote, in one way or another, the economic use of land—this holds true for the ideal of the mixed use and compact city (e.g. Williams et al. 2000), which is at the heart of the New Urbanism campaign in the US (Talen 2005), for instance, as well as the discourse on the European City which has become influential in particular in Central Europe (Rietdorf 2001). These guiding stars have developed over the last several decades as a reaction to the neglect of the particular qualities of "urban" environments that was characteristic of post-war principles and trends of urban development (Jacobs 1961) and that have at their heart the idea of the "compact city" (Burgess 2000; Richardson et al. 2000; Dielemann and Wegener 2004).

3. Within the last 50 years or so, urban regeneration in many countries has become a major paradigm in spatial policies, and a variety of strategies and instruments have emerged to promote it. These include specific legislative measures that regulate urban renewal processes (e.g. Couch et al. 2003), urban regeneration schemes aiming to re-establish the attractiveness of inner urban areas (e.g. Haase

et al. 2005), congestion charges in inner urban areas (Anas and Rhee 2006), or graduated density zoning (Shoup 2008). These efforts have helped to make the existing urban area as attractive as newly developed areas as a place to invest, develop, set up a business, live and work, from the perspectives of economic return, social satisfaction and environmental quality (Couch and Karecha 2006). Moreover, the most common strategies of urban development include the densification and intensification of existing settlements by reusing brownfields and creating infill development.

5.5 Outlook

It is widely accepted today that extensive urban land use change brings about several unwelcome effects in that it goes along with the loss of open land and natural resources, causes ecological damage, generates automobile dependence, wastes energy, leads to atmospheric pollution, imposes economic costs on local authorities, and implies potential negative social effects such as the exacerbation of spatial social segregation and the exclusion of non-car-owning households from a good deal of work and leisure facilities. The regulation of urban land use change is therefore a key issue of land use policy.

In recent years, however, concerns about the dynamics of urban development and land use change seem to have diminished—at least in the European context. This may be due in part to the wide range of strategies and instruments that exist and can be employed today to bring urban land use change under control. Yet the main reason for reduced worries related to land-demanding developments is probably linked to current reurbanisation trends in many European urban regions. As the demand for inner-city housing has significantly increased in recent years, there is a widespread perception that the problem of extensive urban land use change and urban sprawl has vanished. However, a closer look reveals that in most countries and regions, the dynamic of urban land use change is largely unbroken (e.g. Hierse et al. 2017; Hesse and Siedentop 2018) Therefore, reflecting on how urban land use change can be controlled is still a key element of land use policy and planning. In particular, it seems useful to account for a few challenges in this regard:

1. It appears crucial to embed any attempt to minimise urban land use change in a strategy which at the same time eliminates existing incentives for land-consuming development. It is therefore important to identify and then counteract such incentives in, for instance, tax policies or policies for structural development.
2. Since the rigid control of urban land use change is largely dependent on the political will to achieve this task (which is often lacking), the provision of powerful tools to monitor land use changes and to assess their various impacts is essential. In this vein, ongoing attempts to seek scientifically sound arguments in favour of compact urban development can facilitate efforts to minimise urban land use change with scientific evidence.

3. Given the variety of policy and planning instruments, it is worthwhile to evaluate the effectiveness of these instruments in different contexts and to also scrutinise the feasibility of possible combinations of different instruments. However, such an evaluation of the instruments put in place to curb urban land use change is particularly difficult to credibly carry out due to the countless potential interfering variables. And it is even more difficult to undertake such an evaluation by way of international comparison, because of the differences in administrative and legal structures and cultures that exist in various countries. Against this background, it remains a major research task to keep an eye on practical experiences regarding the applicability of policy and planning strategies and instruments in different contexts, as this will provide a basis for their adaption to the context in which they are to be utilised. Without such efforts to allow for the particularity of different contexts, it hardly seems possible to control, manage and steer the development of fresh land to the most acceptable locations, to minimise urban land use change, and to increase the sustainability of land use patterns.

4. Last but not least, more effective containment of urban sprawl requires social learning and a long-term agenda-setting process. Experience to date with growth management policies has made it clear that a "top-down" strategy operating solely through laws and regulations at the national level cannot be successful. What is needed, instead, is a cooperative political approach that includes a coordinated action programme at the state and local levels based on shared land policy goals. The alliances (*Bündnisse*) that exist in various German *Länder* advocating a more land-saving approach to urban development are an example of this. All the relevant actors and stakeholders (policymakers, administration, chambers of commerce, associations, researchers, NGOs) are represented in them. The aim is to reach a consensus on land-saving targets and to find suitable implementation strategies and instruments. Even though such cooperative approaches have not been consistently successful in the past, there is no doubt that sustainable settlement and land development will not be possible without the mobilisation of actors and their participation in a multi-level decision-making process.

Acknowledgments We would like to thank Stefan Fina and Mine Henki for their intellectual support and stimulating discussions.

References

Abbott, C., & Margheim, J. (2008). Imagining Portland's urban growth boundary: Planning regulation as cultural icon. *Journal of the American Planning Association, 74,* 196–208.

Acioly, C., Jr., & Davidson, F. (1996). Density in urban development.*Building Issues* (8), 3–25.

Alberti, M. (1999). Urban patterns and environmental performance: What do we know? *Journal of Planning Education and Research, 19,* 151–163.

Al-Hathloul, S., & Mughal, M. A. (2004). Urban growth management—The Saudi experience. *Habitat International, 28,* 609–623.

Alonso, W. A. (1964). *Location and land use.* Cambridge, Mass: Harvard University Press.

American Farmland Trust. (1994). *Farming on the edge: A new look at the importance and vulnerability of agriculture near American cities.* Washington DC: American Farmland Trust.

Anas, A., & Rhee, H.-J. (2006). Curbing excess sprawl with congestion tolls and urban boundaries. *Regional Science and Urban Economics, 36,* 510–541.

Angel, S., Sheppard, S. C., & Civco, D. L. (2005). *The dynamics of global urban expansion.* The World Bank, Washington DC: Transport and Urban Development Department.

Anthony, J. (2004). Do state growth management regulations reduce sprawl? *Urban Affairs Review, 39,* 376–397.

Antrop, M. (2004). Landscape change and the urbanization process in Europe. *Landscape and Urban Planning, 67,* 9–26.

Arnold, C. L., & Gibbons, C. J. (1996). Impervious surface coverage. The emergence of a key environmental indicator. *Journal of the American Planning Association, 62,* 243–258.

Bae, C. H. C., & Jun, M. J. (2003). Counterfactual planning—What if there had been no greenbelt in Seoul? *Journal of Planning Education and Research, 22,* 374–383.

Banister, D. (1999). Planning more to travel less: Land use and transport. *The Town Planning Review, 70*(3), 313–338.

Benfield, K. F., Raimi, M., & Chen, D. D. T. (1999). *Once there were greenfields: How urban sprawl is undermining America's environment, economy and social fabric.* New York, NY: National Resource Defense Council.

Bengston, D. N., Fletcher, J. O., & Nelson, K. C. (2004). Public policies for managing urban growth and protecting open space: policy instruments and lessons learned in the United States. *Landscape and Urban Planning, 69,* 271–286.

Bengston, D. N., Potts, R. S., Fan, D. P., & Goetz, E. G. (2005). An analysis of the public discourse about urban sprawl in the United States: Monitoring concern about a major threat to forests. *Forest Policy and Economics, 7,* 745–756.

Bengston, D. N., & Youn Y. C. (2006). Urban containment policies and the protection of natural areas: The case of Seoul's greenbelt. *Ecology and Society, 11.* https://www.ecologyandsociety.org/vol11/iss1/art3

Besecke A., Haensch, R., & Pinetzki, M. (Eds.). (2005). Das Flächensparbuch. Diskussion zu Flächenverbrauch und lokalem Bodenbewusstsein. ISR-Diskussionsbeiträge 56, Institut of Urban and Regional Planning, Technische Universität Berlin.

Bruegmann R. (2005) Sprawl. A compact history. University of Chicago Press, Chicago, IL

Bundesregierung. (2002). *Perspektiven für Deutschland.* Berlin: Unsere Strategie für eine nachhaltige Entwicklung.

Burchell, R. W., Downs, A., McCann, B., & Mukherji, S. (2005). *Sprawl costs. Economic impacts of unchecked development.* Washington DC, Covelo, London: Island Press.

Burchell, R. W., Lowenstein, G., Dolphin, W. R., Galley, C. C., Downs, A., Seskin, S., et al. (2002). *Costs of sprawl—2000.* Washington DC: National Academy Press.

Burchell, R. W., Shad, N. A., Listokin, D., Phillips, H., Downs, A., Seskin, S., et al. (1998). *The costs of sprawl—revisited.* Washington DC: National Academy Press.

Burgess, R. (2000). The compact city debate: a global perspective. In M. Jenks & R. Burgess (Eds.), *Compact cities: sustainable urban forms for developing countries* (pp 9–24). London, New York: Spon Press.

Carlson, T., & Dierwechter, Y. (2007). Effects of urban growth boundaries on residential development in Pierce County, Washington. *Professional Geographer, 59,* 209–220.

Carruthers, J. I., & Ulfarsson, G. F. (2002). Fragmentation and sprawl. Evidence from interregional analysis. *Growth and Change, 33,* 312–340.

Carruthers, J. I., & Ulfarsson, G. F. (2003). Urban sprawl and the cost of public services. *Environment and Planning B—Planning & Design, 30,* 503–522.

Cervero, R. (2003). The built environment and travel. Evidence from the United States. *European Journal of Transport and Infrastructure Research, 3,* 119–135.

Champion, A. G. (2001). A changing demographic regime and evolving polycentric urban regions: consequences for the size, composition and distribution of city populations. *Urban Studies, 38,* 657–677.

Chin, N. (2002). Unearthing the roots of urban sprawl: a critical analysis of form, function and methodology. CASA Working Paper 47, University College London

Churchman, A. (1999). Disentangling the concept of density. *Journal of Planning Literature, 13,* 389–411.

Cieslewicz, D. J. (2002). The environmental impacts of sprawl. In G. D. Squires (Ed.), *Urban sprawl. Causes, consequences and policy responses* (pp. 23–38). Washington DC: The Urban Institute Press.

Clawson, M. (1962). Urban sprawl and speculation in suburban land. *Land Economics, 38,* 99–111.

Clingermayer, J. C. (2004). Heresthetics and happenstance: Intentional and unintentional exclusionary impacts of the zoning decision-making process. *Urban Studies, 41,* 377–388.

Committee on Physical Activity, Health, Transportation, and Land Use, Transportation Research Board, Institute of Medicine of the National Academies. (2005). *Does the built environment influence physical activity? Examining the evidence.* Washington DC: Transportation Research Board.

Cooper, L. M. (2004). *Guidelines for cumulative effects assessment in SEA of plans.* London: Imperial College.

Couch, C., Fraser, C., & Percy, S. (2003). *Urban regeneration in Europe.* Oxford: Blackwell.

Couch, C., & Karecha, J. (2006). Controlling urban sprawl. *Some experiences from Liverpool. Cities, 23,* 353–363.

Creutzig, F., Bren, D. C., Weddige, U., Fuss, S., Beringer, T., Gläser, A., et al. (2019). Assessing human and environmental pressures of global land-use change 2000–2010. *Global Sustainability, 2*(e1), 1–17. https://doi.org/10.1017/sus.2018.15

Criterion Planners/Engineers, Fehr & Peers Associates. (2001). *INDEX 4D Method. A quick-response method of estimating travel impacts from land-use changes.* Portland OR: Technical Memorandum

Davoudi, S. (2003). Polycentricity in European spatial planning: from an analytical tool to a normative agenda. *European Planning Studies, 11,* 979–999.

Deng, X., Huang, J., Rozelle, S., & Uchida, E. (2010). Economic growth and the expansion of urban land in China. *Urban Studies, 47,* 813–843.

Dielemann, F., & Wegener, M. (2004). Compact city and urban sprawl. *Built Environment, 30,* 308–323.

Downs, A. (1998). How America's cities are growing: the big picture. *Brookings Review, 16,* 8–12.

Downs, A. (1999). Some realities about sprawl and urban decline. *Housing Policy Debate, 10,* 955–974.

EEA (European Environment Agency). (2005). *The European environment. State and outlook 2005.* EEA, Copenhagen: State of the Environment Report No 1/2005

EEA (European Environmental Agency). (2006). *Urban sprawl in Europe—The ignored challenge.* Copenhagen: EEA Report 10/2006

Emison, G. A. (2001). The relationship of sprawl and ozone air quality in United States' metropolitan areas. *Regional Environmental Change, 2,* 118–127.

EPA (U.S. Environmental Protection Agency). (2000). Projecting land-use change: A summary of models for assessing the effects of community growth and change on land-use patterns. Cincinnati OH

Ewing, R. (1994). Causes, characteristics, and effects of sprawl: a literature review. *Environmental Planning and Urban Issues, 21,* 1–15.

Ewing, R. (1997). Is Los-Angeles-style sprawl desirable? *Journal of the American Planning Association, 63,* 107–126.

Ewing, R., & Cervero, R. (2010). Travel and the built environment. A meta-analysis. *Journal of the American Planning Association, 76,* 256–294.

Ewing, R., Pendall, R., & Chen, D. T. (2002). *Measuring sprawl and its impacts*. Washington DC: Smart Growth America.

Foley, J. A., DeFries, R., Asner, G. P., Barford, C., Bonan, G., Carpenter, S. R., et al. (2005). Global consequences of land use. *Science, 309,* 570–574.

Freilich, R. H. (1998). The land-use implications of transit-oriented development: Controlling the demand side of transportation congestion and urban sprawl. *Urban Lawyer, 30,* 547–572.

Frenkel, A. (2004). A land-consumption model - Its application to Israel's future spatial development. *Journal of the American Planning Association, 70,* 453–470.

Frenkel, A., & Ashkenazi, M. (2008). Measuring urban sprawl: How can we deal with it? *Environment and Planning B, 35,* 1–24.

Fulton, W., Pendall, R., Nguyen, M., & Harrison, A. (2001). *Who sprawls most?* The Brookings Institution, Washington DC: How growth patterns differ across the U.S.

Gallent, N. (2006). The rural-urban fringe: a new priority for planning policy? *Planning Practice & Research, 21,* 383–393.

Galster, G., Hanson, R., Ratcliffe, M. R., Wolman, H., Coleman, S., & Freihage, J. (2001). Wrestling sprawl to the ground: Defining and measuring an elusive concept. *Housing Policy Debate, 12,* 681–717.

Ganser, R., & Williams, K. (2007). Brownfield development. Are we using the right targets? Evidence from England and Germany. *European Planning Studies, 15,* 603–622.

Gant, R. L., Robinson, G. M., & Fazal, S. (2011). Land-use change in the 'edgelands': Policies and pressures in London's rural–urban fringe. *Land use policy, 28*(1), 266–279.

Garreau, J. (1991). *Edge city: Life on the New Frontier*. New York: Doubleday.

Gihring, T. A. (1999). Incentive property taxation - A potential tool for urban growth management. *Journal of the American Planning Association, 65,* 62–79.

Gillham, O. (2002). *The limitless city: A primer on the urban sprawl debate*. Washington DC, Covelo, London: Island Press.

Glaeser, E. L., & Kahn, M. E. (2003). *Sprawl and urban growth*. Harvard Institute of Economic Research: Harvard University, Cambridge, MA.

Gordon, P., Kumar, A., & Richardson, H. W. (1989). The spatial mismatch hypothesis: some new evidence. *Urban Studies, 26,* 315–326.

Gordon, P., & Richardson, H. W. (2001). The sprawl debate. Let markets plan. *The Journal of Federalism, 31,* 131–149.

Haase, A., Kabisch, S., & Steinführer, A. (2005). Reurbanisation of inner-city areas in European cities. In I. Sagan & D. Smith (Eds.), *Society, economy, environment—Towards the sustainable city* (pp. 75–91). Gdansk/ Poznan: Bogucki Wydawnictwo Naukowe.

Haase, D., & Nuissl, H. (2007). Does urban sprawl drive changes in the water balance and policy? The case of Leipzig (Germany) 1870–2003. *Landscape and Urban Planning, 80,* 1–13.

Handa, V. K. (1996). Construction engineers driving into the 21st century. *Journal of Construction Engineering and Management—ASCE, 122,* 2–6

Handy, S. (2005). Smart Growth and the transportation-land use connection: What does the research tell us? *International Regional Science Review, 28,* 146–167.

Harvey, R. O., & Clark, A. V. (1965). The nature and economics of urban sprawl. *Land Economics, 41,* 1–9.

Hasse, J. E., & Lathrop, R. G. (2003). Land resource impact indicators of urban sprawl. *Applied Geography, 23,* 159–175.

Haughton, G. (1999). Information and participation within environmental management. *Environment and Urbanization, 11,* 51–62.

Healey, P. (1992). A planner's day: Knowledge and action in communicative practice. *Journal of the American Planning Association, 58,* 9–20.

Hersperger, A. M., Oliveiraa, E., Pagliarina, S., Palkaa, G., Verburg, P., Bolligera, J., & Grădinarua, S. (2018). Urban land-use change: The role of strategic spatial planning. *Global Environmental Change, 51,* 32–42. https://doi.org/10.1016/j.gloenvcha.2018.05.001

Hesse, M., & Siedentop, S. (2018). Suburbanisation und suburbanisms—Making sense of continental european developments. *Raumforschung und Raumordnung/Spatial Research and Planning, 76,* 97–108.

Hierse, L., Nuissl, H., Beran, F., & Czarnetzki, F. (2017). Concurring urbanizations? Understanding the simultaneity of sub- and re-urbanization trends with the help of migration figures in Berlin. *Regional Studies, Regional Science, 4,* 189–201.

Hirt, S. (2007). The devil is in the definitions. *Journal of the American Planning Association, 7,* 436–450.

Huang, J., Lu, X. X., & Sellers, J. M. (2007). A global comparative analysis of urban form: Applying spatial metrics and remote sensing. *Landscape and Urban Planning, 82,* 184–197.

Jacobs, J. (1961). *Death and life of great American cities.* New York: Random House.

Jaeger, J. (2000). Landscape division, splitting index, and effective mesh size: New measures of landscape fragmentation. *Landscape Ecology, 15,* 115–130.

Johnson, M. P. (2001). Environmental impacts of urban sprawl: a survey of the literature and proposed research agenda. *Environment and Planning A, 33,* 717–735.

Kain, J. F. (1992). The spatial mismatch hypothesis: three decades later. *Housing Policy Debate, 3,* 371–460.

Kelly-Schwartz, A. C., Stockard, J., Doyle, S., & Schlossberg, M. (2004). Is sprawl unhealthy? A multilevel analysis of the relationship of metropolitan sprawl to the health of individuals. *Journal of Planning Education and Research, 24,* 184–196.

Köck, W., Bovet, J., Gawron, T., & Hofmann, E. (2007). Activating spatial planning law: options for the reduction of land consumption. *Journal for European Environmental and Planning Law, 4,* 2–16.

Kolankiewicz, L., & Beck, R. (2001). Weighing sprawl factors in large U.S. cities. A report on the nearly equal roles played by population growth and land use choices in the loss of farmland and natural habitat to urbanization. Numbers USA, Arlington, VA

Korthals-Altes, W. K. (2009). Taxing land for urban containment. Reflections on a Dutch debate. *Land Use Policy, 26,* 233–241.

Kuemmerle, T., Levers, C., Erb, K., Estel, S., Jepsen, M. R., Müller, D., Plutzar, C., Stürck, J., Verkerk, P. J., Verburg, P. H., Reenberg, A. (2016). Hotspots of land use change in Europe. *Environmental Research Letters, 11.* 10.1088/1748-9326/11/6/064020

Ladd, H. (1991). Population growth, density and the costs of providing public services. *Urban Studies, 29,* 273–295.

Lambin, E. F., Turner, B. L., Geist, H. J., Samuel, B. A., Angelsen, A., Bruce, J. W., et al. (2001). The causes of land-use and land-cover change: moving beyond the myths. *Global Environmental Change, 11,* 261–269.

Lang, R. E., & LeFurgy, J. (2003). Edgeless cities: Examining the noncentered city. *Housing Policy Debate, 14,* 427–460.

Lichtenberg, E., & Ding, C. (2006). Land use efficiency, food security, and farmland preservation in China. Land Lines 18 (online source)

Lopez, R., & Hynes, H. P. (2003). Sprawl in the 1990s. Measurement, distribution, and trends. *Urban Affairs Review, 38,* 325–355.

Matas, A., Ramond, J.-L., & Roig, J.-L. (2010). Job accessibility and female employment probability: The cases of Barcelona and Madrid. *Urban Studies, 47,* 769–787.

McCann, B. A., & Ewing, R. (2003). *Measuring the health effects of sprawl. A national analysis of physical activity, obesity and chronic disease.* Washington DC: Smart Growth America.

McGranahan, G., Marcotullio, P. (2005). Urban systems. In R. Hassan, Scholes, & N. Ash (Eds.), *Ecosystems and human well-being: current state and trends. The Millenium Ecosystem Assessment Project Vol. 1* (pp 797–825). Washington DC, Covelo, London: Island Press.

Meadows, D. H., Meadows, D. L., Randers, J., & Behrens, W. W., III. (1972). *The limits to growth.* New York: Universe Books.

Merenlender, A. M., Brooks, C., Shabazian, D., Gao, S., & Johnston, R. (2005). Forecasting exurban development to evaluate the influence of landuse policies on wildland and farmland conservation. *Journal of Conservation Planning, 1,* 64–88.

Mieszkowski, P., & Mills, E. S. (1993). The causes of metropolitan suburbanization. *Journal of Economic Perspectives, 7,* 135–147.

Mills, E. S. (1972). *Studies in the structure of the Urban economy.* Baltimore, MD: Johns Hopkins University Press.

Moglen, G. E., & Kim, S. (2007). Limiting imperviousness: Are threshold-based policies a good idea? *Journal of the American Planning Association, 73,* 161–171.

Muth, R. (1969). *Cities and housing.* Chicago, IL: Chicago University Press.

Naess, P. (2003). Urban structures and travel behaviour: Experiences from empirical research in Norway and Denmark. *European Journal of Transport and Infrastructure Research, 3,* 155–178.

Naess, P. (2007). The impacts of job and household decentralization on commuting distances and travel modes: Experiences from the Copenhagen region and other Nordic urban areas. *Informationen zur Raumentwicklung, 2*(3), 149–168.

Nechyba, T. J., & Walsh, R. P. (2004). Urban sprawl. *Journal of Economic Perspectives, 18,* 177–200.

Nelson, A. C. (1992). Characterizing exurbia. *Journal of Planning Literature, 6,* 350–368.

Newman, P. (1992). The compact city: An Australian perspective. *Built Environment, 18,* 285–300.

Newton, P. W. (2010). Beyond greenfield and brownfield. The challenge of regenerating Australia's greyfield suburbs. *Built Environment, 36,* 81–104.

Nilsson, K., Pauleit, S., Bell, S., Aalbers, C., & Sick, N. T. (Eds.). (2013). *Peri-urban futures: Scenarios and models for land use change in Europe.* Heidelberg: Springer.

Nuissl, H., Haase, D., Lanzendorf, M., & Wittmer, H. (2009). Environmental impact assessment of urban land use transitions—A context-sensitive approach. *Land Use Policy, 26,* 414–424.

Nuissl, H., & Rink, D. (2005). The 'production' of urban sprawl in eastern Germany as a phenomenon of post-socialist transformation. *Cities, 22,* 123–134.

Nuissl, H., & Schröter-Schlaack, C. (2009). The economic approach towards the containment of land consumption. *Environmental Science and Policy, 12,* 270–280.

Oueslati, W., Alvanides, S., & Garrod, G. (2015). Determinants of urban sprawl in European cities. *Urban Studies, 52*(9), 1594–1614.

Pauleit, S., Ennos, R., & Golding, Y. (2005). Modeling the environmental impacts of urban land use and land cover change—A study in Merseyside, UK. *Landscape and Urban Planning, 71,* 295–310.

Peiser, R. (1989). Density and urban sprawl. *Land Economics, 65,* 193–204.

Peiser, R. (2001). Decomposing urban sprawl. *Town Planning Review, 72,* 275–298.

Pendall, R. (1999). Do land-use controls cause sprawl? *Environment and Planning B: Planning and Design, 26,* 555–571.

Persky, J., & Kurban, H. (2001). *Do federal funds better support cities or suburbs? A spatial analysis of federal spending in the Chicago metropolis.* Washington DC: The Brookings Institution (The Discussion Paper Series)

Poelmans, L., & Van Rompaey, A. (2010). Complexity and performance of urban expansion models. *Computers, Environment and Urban Systems, 34,* 17–27.

Power, A. (2001). Social exclusion and urban sprawl: Is the rescue of cities possible? *Regional Studies, 35,* 731–742.

Pruetz, R. (2003). *Beyond takings and givings: Saving natural areas, farmland and historic landmarks with transfer of development rights and density transfer charges.* Marina Del Ray, CA: Arje Press.

Razin, E., & Rosentraub, M. (2000). Are fragmentation and sprawl interlinked? North American evidence. *Urban Affairs Review, 35,* 821–836.

Real Estate Research Corporation. (1974). The costs of sprawl: Detailed cost analysis. Prepared for the Council on Environmental Quality, the Office of Policy Development and Research, Department of Housing and Urban Development, the Office of Planning and Management, Environmental Protection Agency, Washington DC

Richardson, H. W., Bae, C. H. C., & Baxamusa, M. H. (2000). Compact cities in developing countries: Assessment and implications. In M. Jenks & R. Burgess (Eds.), *Compact cities: Sustainable urban forms for developing countries* (pp. 25–36). London, New York: Spon Press.

Rietdorf, W. (Ed.). (2001). *Auslaufmodell Europäische Stadt? Neue Herausforderungen und Fragestellungen am Beginn des 21.* Jahrhunderts: Verlag für Wissenschaft und Forschung, Berlin.

Salvati, L., & Carlucci, M. (2016). The way towards land consumption: Soil sealing and polycentric development in Barcelona. *Urban Studies, 53*(2), 418–440.

Schneider, A., & Woodcock, C. E. (2008). Compact, dispersed, fragmented, extensive? A comparison of urban growth in twenty-five global cities using remotely sensed data, pattern metrics and census information. *Urban Studies, 45,* 659–692.

Seto, K. C., Güneralp, B. & Hutyra, L. C. (2012). Global forecasts of urban expansion to 2030 and direct impacts on biodiversity and carbon pools. *PNAS, 109,* 16083–16088. https://doi.org/10.1073/pnas.1211658109.

Shoup, D. (2008). Graduate density zoning. *Journal of Planning Education and Research, 28,* 161–179.

Siedentop, S. (2016). Geplante Schrumpfung: vom Paradoxon zum Paradigma? In A. Nordrhein-Westfalen (Ed.), *Megacity* (pp. 51–55). Das Phänomen Schrumpfung und Wachstum. Köln: Ghosttown und Suburbia.

Siedentop, S., & Fina, S. (2010). Monitoring urban sprawl in Germany: Towards a GIS-based measurement and assessment approach. *Journal of Land Use Science, 5,* 73–104.

Siedentop, S., Junesch, R., Straßer, M., Zakrzewski, P., Samaniego, L., & Weinert, J. (2009). *Einflussfaktoren der Neuinanspruchnahme von Flächen.* Bonn: Bundesamt für Bauwesen und Raumordnung.

Song, Y., & Knaap, G. J. (2004). Measuring urban form. Is Portland winning the war on sprawl? *Journal of the American Planning Association, 70,* 210–225.

Song, Y., & Zenou, Y. (2006). Property tax and urban sprawl. Theory and implications for US cities. *Journal of Urban Economics, 60,* 519–534.

Speir, C., & Stephenson, K. (2002). Does sprawl cost us all? Isolating the effects of housing patterns on public water and sewer costs. *Journal of the American Planning Association, 68,* 56–70.

Squires, G. D. (Ed.). (2002). *Urban Sprawl: Causes consequences and policy responses.* Washington DC: The Urban Institute Press .

Talen, E. (2005). *New urbanism and American Planning: The conflict of cultures.* New York and Milton Park, Abingdon: Routledge.

Theobald, D. M. (2001). Land-use dynamics beyond the American urban fringe. *Geographical Review, 91,* 544–564.

Theobald, D. M. (2005). Landscape patterns of exurban growth in the USA from 1980 to 2020. *Ecology and Society, 10,* 32.

Theobald, D. M., Miller, J. R., & Thompson, H. N. (1997). Estimating the cumulative effects of development on wildlife habitat. *Landscape and Urban Planning, 39,* 25–36.

Tiebout, C. (1956). A pure theory of local expenditures. *The Journal of Political Economy, 64,* 416–424.

Torrens, P. M. (2008). A toolkit for measuring sprawl. *Applied Spatial Analysis and Policy, 1,* 5–36.

Torrens, P. M., & Alberti M. (2000). Measuring sprawl. CASA paper 27. University College London

Ulfarsson, G. F., & Carruthers, J. I. (2006). The cycle of fragmentation and sprawl: a conceptual framework and empirical model. *Environment and Planning B: Planning and Design, 33,* 767–788.

Vance, C., & Hedel, R. (2007). The impact of urban form on automobile travel: Disentangling causation from correlation. *Transportation, 34,* 575–588.

Voogt, J. A. (2002). Urban heat Island. In I. Douglas (Ed.), *Encyclopedia of global environmental change.Volume 3: Causes and consequences of global environmental change* (pp. 660–666). Chichester: Wiley.

Wang, Y. M., & Scott, S. (2008). Illegal farmland conversion in China's urban periphery: Local regime and national transitions. *Urban Geography, 29,* 327–347.

Wassmer, R. W. (2002). *Defining excessive decentralization in California and other Western states: An economist's perspective on urban sprawl, part 1*. Sacramento, CA: California Senate Office of Research.

Wates, N. (2000). *The community planning handbook*. London: Earthscan.

Watkins, R., Palmer, J., & Kolokotroni, M. (2007). Increased temperature and intensification of the urban heat island: Implications for human comfort and urban design. *Built Environment, 33*, 85–96.

Weber, J., & Sultana, S. (2005). *The impact of sprawl on commuting in Alabama*. Birmingham, AL: University of Alabama, University Transportation Center for Alabama.

Weitz, J., & Moore, T. (1998). Development inside urban growth boundaries—Oregon's empirical evidence of contiguous urban form. *Journal of the American Planning Association, 64*, 424–440.

Whyte, W. H. (1958). Urban sprawl. In W. H. Whyte (Ed.), *The exploding metropolis*. Garden City, NY: Doubleday

Williams, K., Burton, E., & Jenks, M. (Eds.). (2000). *Achieving sustainable Urban form*. London: E & FN Spon.

Wolff, M., Haase, D., & Haase, A. (2018). Compact or spread? A quantitative spatial model of urban areas in Europe since 1990. *PLoS ONE, 13*(2), e0192326. https://doi.org/10.1371/journal.pone.0192326

Chapter 6
Urban-Rural Interrelations—A Challenge for Sustainable Land Management

Alexandra Doernberg and Thomas Weith

Abstract Although the relevance of urban-rural interrelations is widely acknowledged in science and practice, there is as yet no feasible theoretical or theory-driven approach or model that addresses the multiple interconnections in urban-rural spaces at the regional level and makes them applicable to actors in policy and planning practice. In this chapter, we give a short overview of how the topic has developed; present core concepts for urban-rural interrelations; and discuss their applicability and potential improvements by integrating other concepts and connecting governance debates. In the process, we develop new ideas for the analysis and governance of regional functional interrelations in a bid to improve sustainable land management.

Keywords Urban-rural interrelations · Land management · Telecoupling · Regional governance

6.1 Introduction

For a long time, theory and practice focused on either "urban" or "rural" issues of spatial development and land management (e.g. rural development policy). Planning and policy focused on the urban-rural dichotomy, which is also reflected in dualistic definitions and classifications such as statistical and spatial-functional (Tacoli 1998; Dick 2011; OECD 2013). Although scientific discussions have emphasised the social construction of such spatial categories over the past few years, all practical applications in planning and policy have used this differentiation in the past. In

A. Doernberg (✉)
Working Group "Land Use Decisions in the Spatial and System Context", Leibniz Centre for Agricultural Landscape Research, Eberswalder Str. 84, 15374 Müncheberg, Germany
e-mail: alexandra.doernberg@zalf.de

T. Weith
Institute of Environmental Science and Geography, University of Potsdam, Campus-Golm, Karl-Liebknecht-Str. 24-25, 14476 Potsdam, Germany
e-mail: thomas.weith@zalf.de

Working Group "Co-Design of Change and Innovation", Leibniz Centre for Agricultural Landscape Research, Eberswalder Str. 84, 15374 Müncheberg, Germany

© The Author(s) 2021
T. Weith et al. (eds.), *Sustainable Land Management in a European Context*,
Human-Environment Interactions 8, https://doi.org/10.1007/978-3-030-50841-8_6

addition, a growing number of studies have shown the variety of material and non-material interlinkages between distinct urban and rural areas (e.g. Tacoli 1998; Stead 2002; Buciega et al. 2009; Gebre and Gebremedhin 2019). Many observers consider urban-rural linkages to be an important element for economic development and for achieving greater sustainability, which has led to them identifying policy fields and agendas for promoting urban-rural linkages, cooperation and integrated partnerships (Akkoyunlu 2015; OECD 2013; UN-Habitat 2017; MUFPP 2015; Piorr et al. 2011; Wolff and Mederake 2019; DV 2013).

In the face of growing global challenges (e.g. climate change; intensive resource and land use; food, water and energy insecurity; and social and environmental injustice), greater relevance and urgency are attached to the questions of how cities, city regions and rural areas (and their various interfaces and system intersections) can be sustainable, as well as effectively developed and managed (Bock et al. 2013; UN-Habitat 2017: New Urban Agenda, Seto and Reenberg 2014; WBGU 2016). Urban-rural spaces (especially in peri-urban) exhibit a close interdependence of spatial functions, different demands on land, and a complex constellation of actors and governance schemes, which often results in land use conflicts and requires different modes of research and governance (Repp et al. 2012; PURPLE 2014; Augère-Granier 2016; Zscheischler et al. 2017; Piorr et al. 2018).

No conceptual approaches exist at present that are simultaneously (1) comprehensive in terms of current fields of action, (2) able to capture the complexity of urban-rural interrelationships, and (3) suitable for practice, also in the context of regional governance.

This chapter pursues three major objectives. First, it gives a short overview of societal discourses and the relevance of urban-rural interrelations (Sects. 6.1 and 6.2). The second objective is to examine concepts that describe and analyse urban-rural interrelations. The two analytical concepts presented in Sect. 6.3 can be interpreted as the two ends of a range that displays an increase in complexity from one extreme to the other. The model by Stead (2002), which is easy to understand and is self-evident, is positioned at one end of the range, while the highly sophisticated concept of telecoupling (Liu et al. 2013; Friis et al. 2016) represents the other extreme. Assuming a general science-policy gap (Taylor and Hurley 2016; Diller and Thaler 2017), we then explore how the models can be applied in the practice of regional governance and sustainable land use. The third objective, explored in Sect. 6.4, is to present initial ideas about how to rethink these analytical models to address the practical needs of planners, administrative professionals and other actors involved in land management in urban-rural areas. Finally, we link the discussion about urban-rural interrelations to the question of governance and knowledge for governance and innovation (Sect. 6.5) and provide an outlook (Sect. 6.6).

6.2 Societal Discourses About Urban-Rural Spaces and Interlinkages

Seeking to define urban and rural spaces as distinct places or regions, various attempts have been made to define their characterising factors (Stead 2002; Smith and Courtney 2009; Schnore 1966). In the European Union, for example, each Member State has its own definition of a rural area (Smith and Courtney 2009). Variables such as population size, population density, employment density and land use type are often used to this end, either individually or in combination, depending on the country, organisation and purpose (Stead 2002; Schaeffer et al. 2013, p.81). Nonetheless, there is a "lack of a single standard definition of urban and rural areas", as Stead (2002: 299–300) ascertained.

Looking back to the history of definitions of rural and urban areas reveals an interesting development. While compact cities prevailed up to the nineteenth century, implying a clear dichotomy between the rural and the urban area, the ensuing industrialisation and expansion of urban spaces softened this division towards a rural-urban continuum (Repp et al. 2012; Borsdorf and Bender 2010). By the mid twentieth century at the latest, the progress of urbanisation and of social and economic development had resulted in an approximation of these previously opposing areas (Schaeffer et al. 2013). Seto et al. 2012 confirmed this transformation of the understanding of the term by calling the rural-urban dichotomy a "false idea" (p. 7687). Schaeffer et al. even talked about "one system" with regard to this interconnected area (Schaeffer et al. 2013, p. 81), coming back to Jefferson (1931, p. 446) who was convinced, as early as 1931, that "[u]rban and rural, city and country are one thing, not two things."

Although there are ongoing discussions about "blurred boundaries" (Woods 2009), "middle landscapes" or "hybrid geographies" (Ulied et al. 2010), a recent publication issued by the European Parliament acknowledges that "the traditional division is not completely gone and despite nowadays being urbanised and largely made up of 'hybrid geographies', Europe retains clearly recognisable rural and urban areas" (Augère-Granier 2016, p. 2). This dichotomy is questionable, but the underlying assumption of two relational spaces with reciprocal flows is useful when urban-rural interrelations need to be investigated or governed (Repp et al. 2012).

Only a few efforts have been made to address the interlinkages between the rural and the urban area in their entirety, although many specific areas (e.g. commuting) have been thoroughly investigated as stand-alone phenomena (Stead 2002; Smith and Courtney 2009; Schulze Bäing 2007; Schaeffer et al. 2013; Tacoli 1998). However, the complexity of interactions between both spaces appear to have been underestimated (Smith and Courtney 2009, see Caffyn and Dahlström 2005; Hoggart 2005). In fact, at the urban–rural interface, land use planning and land management in general remain quite challenging (Woods 2009; Geneletti et al. 2017).

Not only scientists and planners have recognised the increasing need to take into account rural-urban interrelations—the political sphere has also become aware of the considerable importance of these ties.

Back in 1999, the European Spatial Development Perspective (ESDP) called for more attention to be paid to urban-rural relationships in order to balance disparities in Europe via measures of spatial planning and regional development (European Commission 1999). An important shift towards an integrative view of this interconnected space is currently discernible.

Several policy documents highlight the relevance of urban–rural interlinkages (UN-Habitat 2017; Wolf and Mederake 2019; Piorr et al. 2018; OECD 2013) to sustainable development in order to promote a balance between human welfare, economic activities, and environmental quality. This new paradigm is expressed in a series of international resolutions and agendas, such as the Sustainable Development Goals (SDG) and Habitat III, both adopted in 2016.

The UN SDGs entered into force, determining the development goals to be achieved globally by 2030 (United Nations 2017). Although they cover a wide range of topics, several objectives of the SDGs point to urban-rural linkages. Both the regional level and the national level of development planning are addressed in the SDGs. In the process, the positive mutual effects generated between the urban, rural and peri-urban space should be promoted with the aim of meeting all three pillars of sustainability—social, economic and environmental sustainability. Furthermore, the SDGs call for city-related issues such as sustainable urbanisation processes and settlement development linked with participatory methods, a reduction of environmental pollution, and the provision of green areas for all city dwellers. Furthermore, the SDGs cover people-centred topics such as combating poverty and hunger, promoting health, education and gender issues, and economic themes such as city development, and climate and environmental protection (United Nations General Assembly 2016). In SDG 11 "Sustainable cities and communities", the UN calls for the support of positive economic, social and environmental links between urban, peri-urban and rural areas by strengthening national and regional planning and development activities.

While the SDGs take a broad approach, Habitat III, the third United Nations Conference on Housing and Sustainable Urban Development, held in Quito/Ecuador in October 2016, focuses on urbanisation processes. The Habitat III resolution gives considerable weighting to rural-urban interlinkages, especially in connection with their role in sustainable urban development. Spatial planning instruments are considered to be ideal for exploiting the promising potential of rural-urban interconnections. Relationship and exchange should be integrative in favour of both areas. One form of exchange is organised such that it reflects supply and demand along the value chain. Other forms of exchange include communication, technology, transport and infrastructure, which may help increase productivity and ensure the better utilisation of spatial, social and economic potential. One aspect of Habitat III is the target of achieving the equal development of regions within the urban-rural-continuum, based on enhanced productivity as a result of synergies between both areas. Partnerships and regional infrastructure projects should support this pathway. Another challenge addressed by the resolution is assuring the consistency of sectorial policies in consideration of the functions of areas, regardless of administrative boundaries. Habitat III also attaches importance to the role of land resources, including their ecological and social features. It calls for the conservation and sustainable management of natural resources, especially ecosystems and biodiversity. In sum, polycentric and balanced territorial development should contribute to achieving sustainability (United Nations General Assembly 2016).

In continuation and advancement of the Habitat process, the UN published "Urban-Rural Linkages: Guiding Principles and Framework for Action to Advance Integrated Territorial Development" (UN Habitat 2019) in 2019. This document broadens and deepens the view on urban-rural interrelations. The ten guiding principles, based on human rights, are integrated governance, balanced partnership, participative, environmentally sensitive, data-driven and evidence-based. Interventions should be locally grounded, functional and spatial system-based, and financially inclusive; they should not harm, and provide social protection. The last point refers to the need to "strengthen urban-rural linkages to overcome conflicts […] and reduce inequalities" (p. 11).

These three resolutions prove the timeliness and societal relevance of addressing rural-urban interlinkages, and show which big challenges linked with this issue must be tackled in future.

A better understanding of the interconnections in urban-rural spaces as socio-ecological systems, and hence **feasible analytical and conceptual approaches** with an adequate complexity, are needed to strengthen urban-rural linkages, and to create options to influence the management of natural resources such as land (management) towards sustainability at different spatial scales and policy levels.

6.3   Simple Models Versus Complex Models: Two Opposite Approaches

Several scientists endeavoured to design models for explaining spatial phenomena and interdependencies, starting with von Thünen, Launhardt and Lösch in Europe in the late nineteenth century (Schöler 2005). Scientists from various disciplines contributed models to this development process. Since the 1970s at the latest, scientists have been searching for more sophisticated models to further their understanding of the manifold interactions and interdependencies between rural and urban areas. There are now several models with different starting points and varying levels of complexity (Repp et al. 2012; Akkoyunlu 2015; Kasper and Giseke 2017). In the following, we compare the model by Stead (2002) with the very complex model of telecoupling (Liu et al. 2013) in order to present two concise examples from a multitude of models and to illustrate the extreme bandwidth of complexity.

6.3.1 Stead's Model of Urban-Rural Flows

According to Stead (2002), urban and rural areas are regarded as being distinct (however they are defined), but interconnected and interdependent. In the model, functional relations (flows) are used to describe urban-rural relationships. Based on the theoretical framework of Preston (1975), Stead (2002) places the functional relations between urban and rural spaces in the centre of his model, and differentiates between flows of people and flows of materials/goods (Fig. 6.1).

For the "flows of people" category, the model presents six sub-categories: Work; Education and training; Migration; Recreation and tourism; Cultural activities; and Commerce (see Fig. 6.1). The second category ("flows of materials") has two sub-categories: Waste and pollution, and Food, Water, Resources, Environmental Benefits. In addition, the flows are attributed with **directions** that specify the direction in which the flows move within the interrelationships between both spatial units. As Fig. 6.1 shows, flows may be unidirectional (e.g. waste) or bidirectional (e.g. recreation and tourism).

This framework is able to cover "observable and quantifiable" exchange between rural spaces and urban spaces, and vice versa (Preston 1975: 173; Stead 2002). Stead (2002) used this framework to examine typical urban-rural interrelationships in the West of England. However, this examination was limited to flows for which

Fig. 6.1 Original model of flows of people and materials by Stead (own illustration based on Stead 2002)

quantitative data was available, and ignored difficult-to-measure interdependencies such as information and financial flows, due to lack of data (Repp et al. 2012).

Stead's model is useful for theoretically analysing the interlinkages between rural and urban areas. However, revisions must be made to reflect current spatial current dynamics and to grasp the even more complex interactions (Repp et al. 2012). One option is to update the model (see Fig. 6.2). First, flow directions are revised based on new scientific knowledge (e.g. about multi-local working places or re-urbanisation processes). In some cases, a bidirectional dimension needs to be added to the existing flows or directions even need to be changed, given that current situations do not reflect the situation in West England over 15 years ago. The second option is to introduce new categories of interlinking "immaterial flows" that are better at reflecting current developmental trends, which go beyond the exchange of goods and people. Adding the categories "Knowledge and Innovation", "Habitation" and "Lifestyle, Consumption Patterns and Social Values" emphasises the greater importance of linkages relating to information, knowledge and network integration (Repp et al. 2012). The grey arrows in Fig. 6.2 show how the original model has been modified, and highlight the greater complexity of urban-rural interdependencies (e.g. reciprocal relations in housing and work) as well as the existence of non-material flows.

In the revised version Repp 2012 recognizes that urban-rural interlinkages are also shaped by power constellations and distant effects (see also Pütz 2004).

From the authors' point of view, the flow-based model provides a useful systematisation scheme that is adaptable and can therefore be applied more generally to show the different development stages (e.g. urbanisation cycle of cities) or, more specifically, the interrelations of a particular urban-rural area under investigation. Many policy and planning documents refer implicitly to the concept (e.g. regional development concepts for retail, transport or housing), without explicitly naming

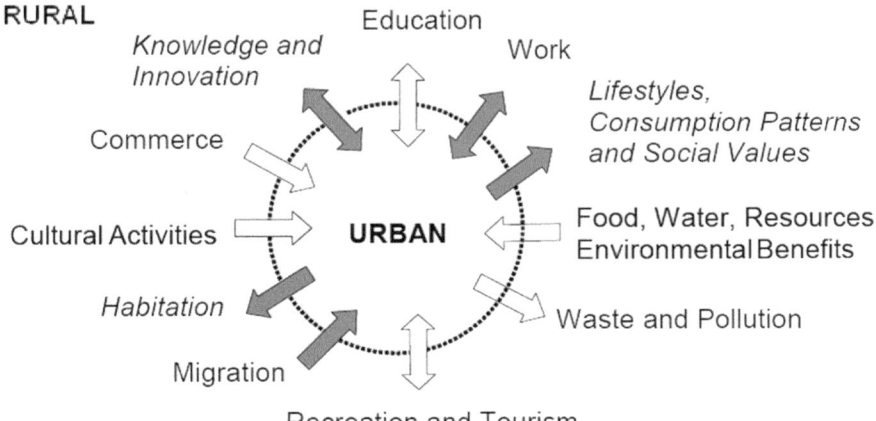

Fig. 6.2 Adapted model of rural-urban interlinkages by Repp et al. 2012 (own illustration, modified and extended based on Stead 2002)

it. The Stead model and its adaptation by Repp et al. 2012 are not well known in the scientific community (examples include Eppler et al. 2015; Kasper and Giseke 2017).

6.3.2 Complex Models of Teleconnection and Telecoupling

Focusing on social-ecological and human-environmental systems, the framework and concept of telecoupling offers a different perspective on urban-rural relations as well as land use systems. Telecoupling is an umbrella concept that merges relevant frameworks from land use science such as teleconnection and globalisation (Liu et al. 2019). "It enables researchers to explore interrelationships among various distant interactions and feedbacks across multiple scales. It also captures the complexity of increasingly prevalent distant environmental and socioeconomic interactions, as well as their diverse drivers and effect" (Liu et al. 2019: 21). In a nutshell, the concept means that human–environmental systems are increasingly interlinked and therefore interact over large distances. Human-induced processes in one part of the world affect another distant part. The feedback between social processes and land outcomes in the interacting systems make the concept highly interesting for researchers in land use science and beyond (Eakin et al. 2014; Friis et al. 2016).

The concept was first applied on thematic areas related to vulnerabilities in connection with environmental changes (Adger et al. 2009) and later in broader contexts such as urban studies (Seto et al. 2012; Haase 2019). In the urban context, the conceptual frameworks of urban teleconnections and telecouplings (UT) were used to study urban-rural relations (Seto et al. 2012; Friis et al. 2016). They represent the shift from place-based to process-oriented conceptualisation along a continuum of land systems (Seto et al. 2012).

These concepts address the observation that land use changes induced by urbanisation and the use of environmental, economic and cultural resources by the urban population affect not only the cities and their surroundings, but also distant places (e.g. rural areas in other regions of the world) and are interlinked (Seto et al. 2012; Liu et al. 2019). These interrelations can be illustrated using the example of meat-based diets in European cities, which cause deforestation in the Amazonas region (Haase 2019).

Two major strands of telecoupling concepts have co-evolved in recent years (Friis et al. 2016): first, the original structured approach that was developed by Liu et al. (2013, 2014) and has been represented by scholars such as Friis et al. 2016 and Garrett and Rueda 2019; second, the heuristic and actor-oriented approach following Eakin et al. (2014), which include social and functional distances in the concept alongside geographical distance.

The idea of telecoupling is based on a set of human and natural **systems** as the core of a telecoupled system (Liu et al. 2013; Friis et al. 2016). In the initial concept by Liu et al. (2013), the systems address **agents, causes and effects** that are interconnected. These causes and effects are inseparably interlinked with each other by "feedback loops", meaning that a cause induces an effect, and vice versa (Liu et al. 2013, p. 3). The systems are linked by **flows** that can transport material (e.g. goods, natural resources, organisms), energy and information (e.g. land titles, agricultural techniques). These flows may be unidirectional or bidirectional (Liu et al. 2014). At the system level, a distinction can be made between sending, receiving and spillover systems (Fig. 6.3). While the functions of sending and receiving systems are obvious—one system acts as the sender of a flow and the other as the recipient—the spillover systems require further explanation. Spillover systems are involved in the interaction between sending and receiving systems influencing or being influenced by the connecting flows in mainly three different ways: they can either act as an intermediate stop to the flow medium (e.g. transportation hub), as a point on the route being affected by the passing flow (e.g. local pollution by passing traffic), or as an active part of the hitherto bilateral interaction, turning it into a trilateral issue (e.g. as an additional trade partner) (Liu et al. 2013). The role played by systems (sending, receiving, spillover) depends on the type of flow, and is therefore not determined a priori (Friis et al. 2016).

The initial concept and its further developments (Liu et al. 2014) have been applied in various studies dealing with aspects such as threats to biodiversity and conservation (Kuemmerle et al. 2019), transnational land deals (Liu et al. 2014) or flows and processes in global agri-food systems (Garrett and Rueda 2019).

The following situation is conceivable when applying the telecoupling concept to the context of close urban-rural interrelations. A power plant located at the rural site

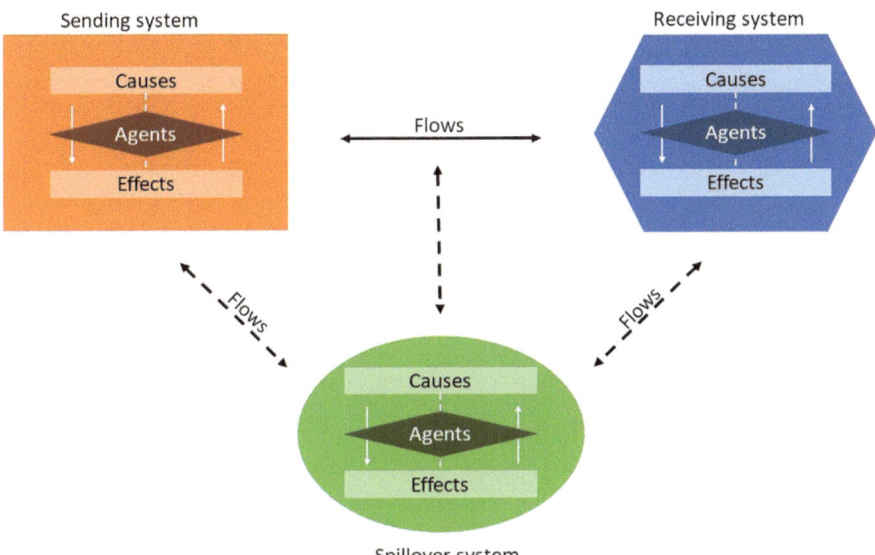

Fig. 6.3 Initial telecoupling framework (own illustration according to Liu et al. 2013; Friis et al. 2016)

where the raw material is converted into energy represents the spillover system. The agents involved are the producers of the energy carrier (e.g. farmers who produce short rotation coppice), the power plant operator, the network operator, the energy supplier, and the receiving households in the city. Besides flows of energy, raw material and money as payments for the traded commodities, information is also exchanged between the agents about energy prices, demand and trade details. The causes of this urban-rural relationship include energy demand by the city population, the limited possibility of fully producing one's own urban energy, the rural capacity of producing renewable energy carriers, and the political orientation towards renewable energy within the overarching framework of the energy transition. In this telecoupled system, the effects are the expansion of the production of renewable raw materials, entailing the avoidance of nuclear risks, climate-friendly energy production, the diversification of income for rural farmers, a land-use change in the production area, and energy supplies for city dwellers.

In contrast to Liu et al. 2013, the heuristic and more actor-centred approach by Eakin et al. (2014) (see Fig. 6.4) assumes that there are spatial and social distances between the systems, and that they are not connected a priori. These distances are considered to be geographically separate, and are viewed in terms of social networks, institutions and governance. Place-based socio-ecological systems are governed independently and within a given governance frame (1), which may cause unexpected and indirect effects (2) if they interact (e.g. via market transactions). The outcome of a telecoupling process can be positive or negative (e.g. biodiversity loss, population placement, reforestation). Land change-related problems occur when the actors (institutions) and governance mechanisms fail to recognise the effects of telecoupling or are unable to account for the consequences (Eakin et al. 2014; Friis et al. 2016).

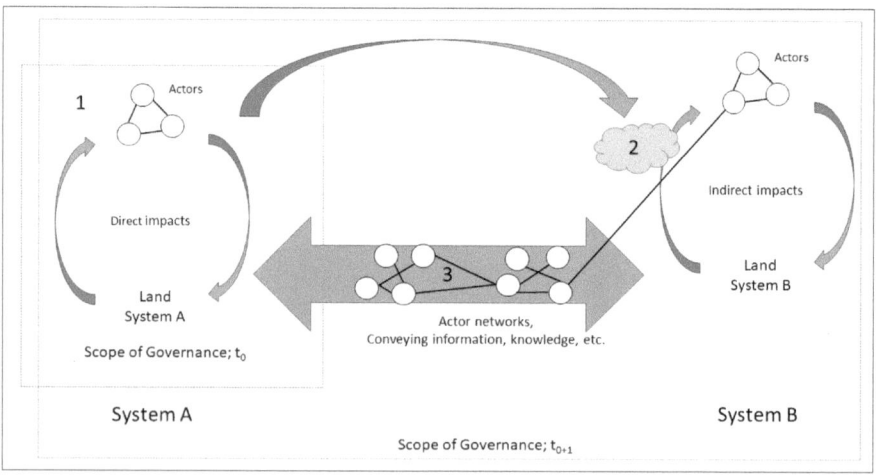

Fig. 6.4 Heuristic approach of the telecoupling framework by Eakin (own illustration based on Eakin et al. 2014: 147)

Feedback processes in (distantly) coupled systems can influence the existing governance structure or create new institutional arrangements (3), and have therefore been assessed as having the potential for institutional change (Friis et al. 2016). Another relevant distinction from the initial concept by Liu et al. (2013) is the recognition of differences of power, influence, and possible asymmetrical relationships in terms of material, capital, information, etc., which "create[s] asymmetries in the responsibility and nature of response" (Eakin et al. 2014: 149). Due to globalisation, processes of economic intensification and connection can be observed. The material flows induced (goods, people, capital) often go hand in hand with non-material flows such as information, knowledge, ideologies or discourses facilitated by information technologies and social networks (Eakin et al. 2014; Persson and Mertz 2019). These flows enable actors in the systems to interact or even to change scales (Friis et al. 2016).

A prominent example of indirect land use changes is biofuel production, such as induced by the European Union's Renewable Energy Directive or by US energy policy. These measures have caused negative environmental and social impacts in many countries around the globe, including increased land competition and international trade, a decrease in food crops, a loss of biodiversity. In the case of biofuel production, land use changes occur not only in the countries where the policy decisions were made, but also in other countries such as Mexico or Brazil (Eakin et al. 2014,2017), which also raises questions about the governance of food systems.

6.3.3 Preliminary Summary and Discussion

The two core concepts and their adaptations (flows of people and materials, and telecoupling) provide useful analytical perspectives on urban-rural interrelations. As yet, however, both concepts have mainly been applied in science and tested in a number of empirical studies with different thematic fields (Stead 2002; Friis 2019).

Table 6.1 summarises the strengths and weaknesses of the different concepts in terms of urban-rural relations, land management and governance.

Both concepts display the variety of urban-rural interrelations (represented by flows). Whereas the Stead model and its adaption by Repp et al. 2012 focus on the type and direction of functional interlinkages, the models' explanatory power regarding land use changes remains rather limited. Although the models make no direct reference to land use management and governing urban-rural relations, they enable subject areas, actors and programmes for urban-rural relations and regional planning to be identified and operationalised by differentiating functions and flows (Repp et al. 2012; Kasper and Giseke 2017). They are less complex and easier to understand than the teleconnection and telecoupling approaches, making them more feasible for practitioners.

The telecoupling model by Liu et al. (2013) is similar to Stead's concept in that different material and non-material flows are also an important element for analysis. However, the telecoupling concept provides an additional analytical layer

Table 6.1 Overview of the key concepts concerning urban-rural interrelations

Concept/model	Background/focus of the (case) study	Explicit reference to urban-rural relations and spaces	Space and scale	Reference to land management	Reference to (regional/functional) governance	Key references
Urban-rural relationships (flows of goods and people)	Relationships between urban and rural areas	Yes	Urban/regional scale Nearby spaces	No, focus is on functions	Yes, key issues for managing relationships are identified	Stead (2002)
Urban-rural interrelations	Land use and land management	Yes	Urban/regional scale Nearby spaces	Yes, feasibility of the concept for sustainable land management was examined	Yes, implicit discussion of governance instruments	Repp et al. (2012)
The urban land teleconnections (ULT) and urban telecoupling (UT) framework	Land use change (urban-rural) resulting from urbanisation links processes and places, allows the consequences of urbanisation and land use change to be identified, and connects urban functions with rural land uses	Short distance (peri-urban, suburban, regional) and long distance (interregional and international) But concepts aim to overcome the local and global aspects of land use change and typical urban or rural classifications (based on density, form of building space or administrative boundaries)	Short distance (peri-urban, suburban, regional) and long distance (interregional and international)	No, focus is on urbanisation processes and land change and related impacts	No (ULT) UT concepts can be linked to government and agency Haase (2019)	Seto et al. (2012); Güneralp et al. (2013); Haase (2019)

(continued)

Table 6.1 (continued)

Concept/model	Background/focus of the (case) study	Explicit reference to urban-rural relations and spaces	Space and scale	Reference to land management	Reference to (regional/functional) governance	Key references
Structured telecoupling approach	Land use changes driven by globalisation	No, but possibility to combine local couplings and telecoupling	Usually global, distant telecoupling In Liu et al. (2013), no explicit reference to urban and rural spaces or to local couplings	Yes, but not in detail	Partly, a link to governance is possible, but is not an element of analysis Recognition that local coupled and telecoupled systems require different governance approaches due to multiple flows, agents, causes and effects that go beyond different scales and administrative and political borders in telecoupled systems Liu et al. (2013)	Liu et al. (2013, 2014); Friis et al. (2016)
Actor-centred telecoupling approach	Changes in land use and land ownership, as well as the actors and institutions involved		Usually global, distant telecoupling	Land resources as part of the food system, problem of land grabbing Eakin et al. (2017)	Yes, explicit reference to governance and institutional change: national governance (Eakin et al. 2014), international governance (Eakin et al. 2017) Link to multi-scalar agency	Eakin et al. (2014); Eakin et al. (2017); Friis et al. (2016)

such as causes, effects and feedback loops. The analysis of causes in particular can provide useful information for land use policy and management (Liu et al. 2019). One major feature of the telecoupling concept is that it offers a structured and processual perspective on urban-rural relations and widens the perspective "from rural-urban interactions to wider human–environment interactions" (Friis 2019: 58).These two analytical perspectives can be also found in approaches that are used in ecosystem research (Liu et al. 2019). The concept by Liu et al. (2013) provide a kind of checklist for analysing land change (Friis et al. 2016). This makes it possible to describe the components and entities of telecoupling, enabling different entry points for analysis. Finally, the model can be used for multiple scales and to study temporal dynamics and system changes.

However, the model's explanatory features focus primarily on distant interactions in the original concept of telecoupling. Potential expansions of the concept arise from the possibility to combine local couplings and telecoupling (Liu et al. 2019). This could be very useful when studying urban-rural relations or land change processes in peri-urban spaces.

The urban telecoupling (UT) concept offers another possibility "to link decision making, actions, government, agency and land (use) changes at both urban and rural ends of the pathway" (Haase 2019, p. 263). We therefore state that the telecoupling approach can be applied to show the complex interactions and interdependencies between urban and nearby rural spaces. In our opinion, the concept is ideal for application to study urban-rural linkages at the local and global scale where not only the social and spatial distance, but also and the complex interplay of social processes and effects at land level play a significant role.

We assume that the combination of local couplings and telecouplings may bring additional valuable insight when studying urban-rural interrelations in Germany, Europe and elsewhere. The effects of distant actions or of high-level policy decisions and discourses (e.g. EU and national bioenergy strategies) can also be observed in cities and their hinterlands (e.g. renewable energy production, which competes with food production and other land uses that address local needs, land grabbing). On the other hand, this increases the complexity of approaches that are already quite complex.

Finally, we want to highlight the importance of distinguishing between analytical and conceptual approaches. Following Kasper and Giseke (2017), we recommend using the term "analytical approaches" for theories and models that aim to describe, analyse and explain specific phenomena, including relations, functions and under-lying rules, and to create a better understanding and common knowledge base. The term "conceptual approaches" should be reserved for principles, plans or strategies that operationalise the procedure for achieving policy or planning objectives, for instance, and that are more actor and action-oriented.

6.4 Discussion About Potential Improvements

After presenting the strengths and shortcomings of the two concepts of flows (Stead 2002) and telecoupling (Liu et al. 2013; Eakin et al. 2014), we now introduce two additional analytical layers that are of relevance to urban-rural relations. The first layer relates to the functional flow of ecosystem services, and ties in with questions of governance and environmental justice. The second layer refers to land uses, and draws attention to the relation between multiple land uses (multi-functionality) and interconnections.

6.4.1 The Ecosystem Service Concept and Urban-Rural Relations

Both the Stead (2002) model and the telecoupling model by Liu et al. 2013 refer more or less explicitly to ecosystem services (ESS) as an essential flow or interaction between urban and rural areas. However, ESS may also provide a good reference frame when discussing the questions of sustainability, the quality of life and environmental justice in regional and global contexts.

A common definition of ESS is that introduced by Constanza et al. (1997), who defined them as "the benefits human populations derive, directly or indirectly, from ecosystem functions" (Constanza et al. 1997: 253). Examples of ESS include food, raw material production, flood protection and cultural services such as aesthetic and recreational services (MEA 2005).

While the concept of ESS has become very prominent in science over the last two decades, practitioners and policy-makers have difficulty putting the concept into practice, due, among other things, to a lack of a framework that links the valuation of ESS with "effective policy instruments and governance arrangements" (Bouma and van Beukering 2015: 4).

Since the delivery of ESS and the distribution of social benefits and costs occur in different spatial units and at different scales (local, regional and global) (Bouma and van Beukering 2015), the concept is quite useful when discussing questions concerning urban–rural relations, land use competition around land-based ecosystem services (Müller et al. 2016) or environmental justice (Agyeman, et al. 2016). Its potential for supporting decision-making in planning and policy is widely acknowledged (Fürst et al. 2017). On the other hand, the multi-level nature of ESS and the spatial mismatch between many ESS and administrative boundaries impede the governance of ESS (Bouma and van Beukering 2015) and the implementation into practice. Moreover, criteria are still required to decide in which planning contexts are conducive to applying the ESS concept (Fürst et al. 2017). This could also weaken the applicability of the concept in the management of urban-rural relations.

Many authors and policy documents refer to the role that rural areas play in providing ESS to cities (Augère-Granier 2016; Schröter-Schlaack et al. 2016). For example, the German TEEB report (Schröter-Schlaack et al. 2016) highlights the role played by rural areas in providing ESS, and the use of ESS by urban dwellers. The report also provides examples of cooperation between rural and urban actors involved in agriculture and water management. In contrast, Gebre and Gebremedhin (2019) stress the mutual benefits from ecosystem-based interlinkages between urban and rural areas, but call for good management (including the protection of rural services) in the face of an increased urban demand for rural resources.

6.4.2   Multi-Functionality

The concept of multi-functionality reflects the ability to use a site for multiple different purposes. Adopting an economic perspective, the production process of a commodity, e.g. the cultivation of wheat, always creates side effects in the form of non-commodities. These are outputs with an economic, social and/or ecological benefit, such as food security, recreation or education (Wüstemann 2005). The connection between both types of products is described as synergetic and joint. At the farm level, multifunctional agriculture is seen as a means of diversification (Zasada 2011). Multi-functionality is often associated with agricultural activities, but "it is not specific to agriculture; it is a property of many economic activities" (OECD 2001, p. 9). According to this statement and following Wüstemann et al. (2008), the concept of multi-functionality can also be applied to reflecting on urban-rural interlinkages. Changes in society and lifestyle in recent decades have increased the importance of non-productive outcomes from rural areas compared to traditional agricultural commodities (Zasada 2011; see also Marsden 1999; Brandt and Vejre 2004; Luttik and van der Ploeg 2004). Urban dweller therefore have a greater demand for rural non-commodities such as enjoying the countryside, experiencing farm tourism and buying locally produced food from the farm (Zasada 2011). Consequently, this concept is ideal for depicting this new trend, enabling functional interrelations to be analysed in a much more complex setting. This includes the necessity to reflect the fact that one type of land use causes a variety of interconnections.

   We assume that integrating these two analytical layers may make a valuable contribution to the research and governance of urban-rural relations.

6.5   Governance of Interrelations: Knowledge for Governance

Until now, interrelational models have mainly focused on an analytical understanding of functional connections and spatial relations. However, causes, effects and flows also simultaneously affect the options for influencing and changing land use. There has been little debate about modes of regional governance, particularly in land use issues (Nölting and Mann 2018), although knowledge about new types of "governance" has been developed and used more frequently since the 1990s.

Whereas in the past the (national) state was regarded as an (assumed) central actor that influences and controls land use and spatial development, various different actors and their interactions are now coming to the fore. The key aspects are the varying forms of interaction and coordination of different social actors from the state, economy, civil society and science (Benz et al. 2007, p. 13). Consideration is therefore given to the entire organisational and regulatory system, which coordinates interaction between actors of all kinds. "It is … about how we establish goals, how we define rules for reaching the defined goals, and finally how we control outcomes following from the use of these rules" (Vatn 2010, p. 1246).

The introduction of collectively effective regulations will lead to the minimisation of land use conflicts and to the achievement of common goals. This means ensuring a target-oriented perspective for action that includes a process-oriented view of the various steps from policy formulation to implementation and the analysis of effectiveness (Ostrom 2011). The complexity of governance results especially (1) from vertical interconnectivity, the "multi-level system" (Benz 2009) and (2) consideration of cross-sectoral horizontal interconnections. This is in part represented by spatial planning at the national, regional and regional level, municipal land use planning, environmental planning and various forms of sectoral planning (e.g. transport, waste) in Europe (Reimer et al. 2015). In addition, regional development approaches (e.g. development concepts, networks), financial subsidies (e.g. tax incentives, Common Agricultural Policy) and other project-driven activities need to be included.

In the context of land use, Gentry et al. (2014) additionally request that the perspective is opened for international governance instruments since distant relationships often cross international borders. In the light of the original notion of the telecoupling concept—the distance between the interrelated systems—the need to develop governance instruments becomes even more urgent so that their increased tasks can be matched. There have also been calls for greater consideration of flows especially, between urban and rural areas, as a proper component of this analytical concept (Gentry et al. 2014). Land use changes in one place can be induced by social, economic or political processes and changes elsewhere.

At the same time, this perspective refers to the dimension of functional governance, which goes hand in hand with functional relations, complementing traditional forms of place-based and territorial governance. For this reason, the "construction of space by governance" has also been coined (Kilper 2010: 16). Functional governance takes up the above-mentioned forms of spatial interrelations in a space of flows (Massey 2005), and has so far been regarded primarily as a challenge without a comprehensive solution (cf. also Friies and Nielsen 2014). The interrelations themselves may influence land differently and may change over time, e.g. due to feedback loops. Hence, dynamics are an additional challenge.

The discussion of dynamics is quite often linked to debates about innovation and transition. Since spatial activities in one system can cause pressure in other systems, reactions such as change or rejection must be expected. In addition to unintended flows, explicit and intended flows of knowledge (e.g. about effects, impacts, etc.) may also influence actors. This can cause direct or indirect pressures on an institutional system and, as a result, on land use and land management. Although research about change and transition processes is still in its infancy (cf. Oberlack et al. 2019), knowledge about social innovation processes (Sovacool and Hess 2017; Petterson and Huitema 2019) and change models (e.g. Kristof 2010; Oberlack et al. 2019) exists in various forms. For example, the concept of the regional innovation system (RIS) highlights the generation, transfer and application of knowledge. The RIS seeks to draw attention to regional conditions for establishing innovation (Tödtling and Trippl 2005). Organisations, institutions and actors involved in the generation, diffusion and use of knowledge are important (Arnold et al. 2014; Doloreux and Parto 2005; Fritsch 2013). Regions have different potential for change, innovation and adaptation. In consequence, account must be taken of specific regional constellations and the permanent change of interrelations (innovation).

Conversely, forms of governance require comprehensive knowledge about the objects and processes to be influenced. New forms of knowledge are therefore directly linked (Rydin 2007) and are required in the scientific debate on telecoupling (Zaehringer et al. 2019). The interaction between knowledge for governance (e.g. functional interrelations) and knowledge about governance (e.g. multilevel governance experiences) is of particular importance. This means, on the one hand, integrating and reflecting on mutual learning processes, depending on places and people. On the other hand, "distance learning" is necessary, reflecting spatial interrelations and feedback loops. To this end, additional resources are needed to realise internal and external exchange. In consequence, co-design processes must reflect not only inner-regional knowledge generation and dissemination, but also the appropriate inclusion of knowledge flows from other regions.

At the same time, this requires a differentiated understanding of knowledge transfer. Going beyond a simple loading dock approach, answers must be found about what kind of knowledge can be transferred from or to other spaces from different contexts with different and heterogeneous networks of actors and a range of institutional settings, and then applied and implemented there (Rogga et al. 2014).

6.6 Outlook

From the authors' perspective, the main tasks for the future are (1) to further develop a model to integrate regional interrelations based on functional interrelations, and (2) to improve regional governance and transition processes, based on the analytical model.

The challenge here is to adequately consider the aforementioned complexities of the various approaches.

To reflect current discussions about the role of science in solving real-world problems by using co-design approaches, developers of models must also consider the need to find applicable models that allow a broad practical application for the solution of real-world problems. This is also a consequence of a decade-long ongoing debate about gaps between theory and practice in the field of spatial development and land use planning, addressed by a number of scientists (Alexander 2010, 2015; Taylor 2016; Sanyal 2000; Thomson 2000; Vogelij 2014). On the one hand, scientists state that theories are needed to broaden planners' minds (Diller and Thaler 2017); on the other hand, theoretical approaches are thought to be too far from practical planners' working reality, and unhelpful in their daily tasks (Sanyal 2000; Hellmich et al. 2017).

A further development and linkage with governance approaches (such as urban-rural partnerships, DV 2013) should not only include governance modes (Rydin 2007) and their application. It also requires a first reference to the objectives to be pursued and the underlying values and norms. This also comprises the presentation of value conflicts and the distribution of benefits and costs. Thus, interregional cooperation and exchange are generally thought to increase prosperity, especially economically, due to achieving comparative cost advantages. At the same time, this may be accompanied by negative environmental impacts that cause spatially one-sided pressures. Such one-sided pressures are now frequently criticised in the discussion about environmental justice. Although there are some parallels between the development of approaches in environmental justice and land use science (especially telecoupling research), adoption of the normative dimension of environmental justice is a major methodological challenge, and could politicise telecoupling research (Corbera et al. 2019).

Nonetheless, the authors wish to go one step further, and see the need for a societal and scientific debate about spatial interrelations (and, in particular, urban-rural interrelations) in the broader context of spatial justice, which goes beyond the environmental dimension. In Germany, for example, initial approaches exist in the form of political and scientific discourses on equal living conditions or in the context of research projects on a just urban-rural equilibrium (see www.regerecht.de). A new cross-regional debate on globally accepted values such as justice (Höffe 1989) could change the framework conditions for global and regional governance.

Acknowledgments Thanks to Jana Zscheischler, Annegret Repp and Sarah Keutmann for developing the topic and for contributing to previous versions of this chapter.

References

Adger, W. N., Eakin, H., & Winkels, A. (2009). Nested and teleconnected vulnerabilities to environmental change. *Frontiers in Ecology and the Environment, 7*(3), 150–157.

Agyeman, J., Schlosberg, D., Craven, L., & Matthews, C. (2016). Trends and directions in environmental justice: From inequity to everyday life, community, and just sustainabilities. *Annual Review of Environment and Resources, 41*(1), 321–340. https://doi.org/10.1146/annurev-environ-110615-090052.

Akkoyunlu, S. (2015). The potential of rural-urban linkages for sustainable development and trade. *International Journal of Sustainable Development & World Policy, 4*(2), 20–40. https://doi.org/10.18488/journal.26/2015.4.2/26.2.20.40.

Alexander, E. R. (2010). Introduction: Does planning theory affect practice, and if so, how? *Planning Theory, 9*(2), 99–107.

Alexander, P., Rounsevell, M., Dislich, C., Dodson, J., Engström, K., & Moran, D. (2015). Drivers for global agricultural land use change: The nexus of diet, population, yield and bioenergy. *Global Environmental Change, 35,* 138–147.

Arnold, M., Mattes, A., & Sander, P. (2014). Regionale Innovationssysteme im Vergleich. DIW Wochenbericht Nr. 5.2014 vom 29. Januar 2014, DIW Berlin — Deutsches Institut für Wirtschaftsforschung e. V., Berlin.

Augère-Granier, M. -L. (2016). Bridging the rural-urban divide: Rural-urban partnerships in the EU. *European Parliamentary Research Service.* https://www.europarl.europa.eu/RegData/etu des/BRIE/2016/573898/EPRS_BRI(2016)573898_EN.pdf.

Benz, A. (2009). *Politik in Mehrebenensystemen.* Governance 5. Wiesbaden: Springer.

Benz, A., Lütz, S., Schimank, U., & Simonis, G. (2007). *Handbuch Governance. Theoretische Grundlagen und empirische Anwendungsfelder.* Wiesbaden: Springer.

Bock, S., Hinzen, A., Libbe, J., Preuß, T., Simon, A., & Zwicker-Schwarm, D. (2013). *Urbanes Landmanagement in Stadt und Region. Urbane Landwirtschaft, urbanes Gärtnern und Agrobusiness.* Berlin: Deutsches Institut für Urbanistik (Difu-Impulse, 2013, 2).

Borsdorf, A., & Bender, O. (2010). *Allgemeine Siedungsgeographie.* Wien: Böhlau.

Bouma, J. A., & van Beukering, P. J. H. (2015). Ecosystem services: from concept to practise. In Jetske A. Bouma, & Pieter J. H. van Beukering (Eds.), *Ecosystem services. From concept to practice* (pp. 3–21). Cambridge: Cambridge University Press.

Brandt, J., & Vejre, H. (2004). Multifunctional landscapes—motives, concepts and perspectives. In J. Brandt & H. Vejre (Eds.), *Multifunctional Landscapes—Theory* (Vol. I, pp. 3–31). Southampton, WIT Press: Values and History.

Buciega, A., Pitarch, M.-D., & Esparcia, J. (2009). The context of rural-urban relationships in Finland, France, Hungary, The Netherlands and Spain. *Journal of Environmental Policy & Planning, 11*(1), 9–27. https://doi.org/10.1080/15239080902774929.

Caffyn, A., & Dahlström, M. (2005). Urban-rural interdependencies: Joining up policy in practice. *Regional Studies, 39*(3), 283–296.

Corbera, E., Busck-Lumholt, L. M., Mempel, F., & Rodriguez-Labajos, B. (2019). Environmental justice in telecoupling research. In C. Friis & J. Ø. Nielsen (Eds.), *Telecoupling* (pp. 213–232). Cham: Springer International Publishing.

Costanza, R., d'Arge, R., Groot, R. de, Farber, S., Grasso, M., Hannon, B. et al. (1997). The value of the world's ecosystem services and natural capital. *Nature, 387* (6630), 253-260. 10.1038/387253a0.

Dick, E. (2011). Rural-urban linkages and their implications for new forms of governance. In *N-AERUS Conference*, 20–22 October. Spain: Madrid. https://n-aerus.net/web/sat/workshops/2011/PDF/N-AERUS_XII_Dick_Eva_RV.pdf.

Diller, C., & Thaler, Th. (2017). Zum Gap zwischen theoriebasierter Planungsforschung und Planungspraxis. *Raumforschung und Raumordnung, 75*(1), 57–69. https://doi.org/10.1007/s13 147-016-0431-6.

Doloreux, D., & Parto, S. (2005). (2005): Regional innovation systems: Current discourse and unresolved issues. *Technology in Society, 27*(2), 133–153.

DV—Deutscher Verband für Wohnungswesen, Städtebau und Raumordnung e. V. (2013): Partnership for sustainable rural-urban development. Existing evidences. *Final report.* https://op.europa. eu/en/publication-detail/-/publication/21ba8cd7-7436-4347-bf62-2f179a0e8747.

Eakin, H., DeFries, R., Kerr, S., Lambin, E. F., Liu, J., Marcotullio, P. J., & Zimmerer, K. (2014). Significance of telecoupling for exploration of land-use change. In K. C. Seto & A. Reenberg (Eds.), *Rethinking global land use in an urban era* (pp. 141–161). Cambridge, MA: MIT Press.

Eakin, H., Rueda, X., & Mahanti, A. (2017). Transforming governance in telecoupled food systems. *E&S, 22*(4). 10.5751/ES-09831-220432 .

Eppler, U., Fritsche, U. R., & Laaks, S. (2015). Urban-rural linkages and global sustainable land use. GLOBALANDS Issue Paper. https://www.researchgate.net/publication/277556667_Urban-Rural_Linkages_and_Global_Sustainable_Land_Use.

European Commission (1999). European Spatial Development Perspective—Towards Balanced and Sustainable Development of the Territory of the European Union. Agreed at the Informal Council of Ministers responsible for Spatial Planning. European Communities: Luxembourg.

Friis, C., Nielsen, ØJ., Otero, I., Haberl, H., Niewöhner, J., & Hostert, P. (2016). From teleconnection to telecoupling: taking stock of an emerging framework in land system science. *Journal of Land Use Science, 11*(2), 131–153.

Friis, C. (2019). Telecoupling: A new framework for researching land-use change in a globalised world. In: C. Friis & J. Ø. Nielsen (Eds.) *Telecoupling* (pp. 49–67). Cham: Springer International Publishing.

Fritsch, M. (2013). Das regionale Innovationssystem. In *Set HoF-Handreichungen 2. Beiheft „die hochschule" 2013*, pp. 15–18.

Fürst, C., Luque, S., & Geneletti, D. (2017). Nexus thinking—how ecosystem services can contribute to enhancing the cross-scale and cross-sectoral coherence between land use, spatial planning and policy-making. *International Journal of Biodiversity Science, Ecosystem Services & Management, 13*(1), 412–421. https://doi.org/10.1080/21513732.2017.1396257.

Garrett, R. & Rueda, X. (2019). Telecoupling and consumption in agri-food systems. In: C. Friis, & J. Ø. Nielsen (Eds.), *Telecoupling* (pp. 115–137). Cham: Springer International Publishing.

Gebre, T., & Gebremedhin, B. (2019). The mutual benefits of promoting rural-urban interdependence through linked ecosystem services. In *Global ecology and conservation, 20*, e00707. 10.1016/j.gecco.2019.e00707 .

Geneletti, D., La Rosa, D., Spyra, M., & Cortinovis, C. (2017). A review of approaches and challenges for sustainable planning in urban peripheries. *Landscape and Urban Planning, 165,* 231–243. https://doi.org/10.1016/j.landurbplan.2017.01.013.

Gentry, B. S., Sikor, T., Auld, G., Bebbinton, A. J., Benjamininsen, T. A., Hunsberger, C. A., et al. (2014). Changes in land-use governance in an Urban Era. In K. C. Seto & A. Reenberg (Eds.), *Rethinking global land use in an urban era* (pp. 239–271). Cambridge, MA: MIT Press.

Güneralp, B., Seto, K. C., & Ramachandran, M. (2013). Evidence of urban land teleconnections and impacts on hinterlands. *Current Opinion in Environmental Sustainability, 5*(5), 445–451. https://doi.org/10.1016/j.cosust.2013.08.003.

Haase, D. (2019). Urban telecouplings. In C. Friis & J. Ø. Nielsen (Eds.), *Telecoupling* (pp. 261–280). Cham: Springer International Publishing.

Hellmich, M., Lamker, C. W., & Lange, L. (2017). Planungstheorie und Planungswissenschaft im Praxistest: Arbeitsalltag und Perspektiven von Regionalplanern in Deutschland. *Raumforschung und Raumordnung - Spatial Research and Planning, 75*(1), 7–17.

Höffe, O. (1989). *Politische Gerechtigkeit*. Suhrkamp: Frankfurt/M.

Hoggart, K. (2005). *The city's hinterland: dynamism and divergence in Europe's peri-urban territories*. Aldershot: Ashgate.

Jefferson, M. (1931). Distribution of the world's city folks: A study in comparative civilization. *Geographical Review, 21,* 446–465.

Kasper, C., & Giseke, U. (2017). Arbeitspapier: Analytische und konzeptionelle Ansätze für die Entwicklung von Stadt und Land. Technische Universität Berlin. https://rural-urban-nexus.org/sites/default/files/RUN_AP1%202_v4_190612.pdf.

Kilper, H. (2010). Governance und die soziale Konstruktion von Räumen. In H. Kilper (Ed.), *(2010) Governance und Raum* (pp. 9–24). Baden-Baden: NOMOS-Verlag.

Kristof, K. (2010). *Wege zum Wandel*. München: Oekom-Verlag.

Kuemmerle, T., Kastner, T., Meyfroidt, P. & Qin, S. (2019). Conservation telecoupling. In: C. Friis & J. Ø. Nielsen (Eds.), *Telecoupling* (pp. 281–302). Cham: Springer International Publishing.

Liu, J., Herzberger, A., Kaspar, K., Carlson A., & Connor, T. (2019). What is telecoupling? In C. Friis & J. Ø. Nielsen (Eds.), *Telecoupling* (pp. 19–48). Cham: Springer International Publishing.

Liu, J., Hull, V., Batistella, M., DeFries, R., Dietz, T., Fu, F., et al. (2013). Framing sustainability in a telecoupled world. *Ecology and Society, 18*(2), 26. https://doi.org/10.5751/ES-05873-180226.

Liu, J., Hull, V. & Moran, E. F. (2014). Applications of the telecoupling framework to land-change science. In K. C. Seto, A. Reenberg, E. F. Lambin, J. Lupp, M. Turner & U. Dettmar (Eds.), *Rethinking global land use in an urban era* (pp. 119–139). Cambridge, MA: MIT Press.

Luttik, J., & van der Ploeg, B. (2004). Functions of agriculture in urban society in the Netherlands. In F. Brouwer (Ed.), *Sustaining agriculture and the rural economy: Governance, policy and multifunctionality* (pp. 204–222). Cheltenham: Edward Elgar.

Marsden, T. (1999). Rural futures: The consumption countryside and its regulation. *Sociologia Ruralis, 39,* 501–526.

Massey, D. (2005). *For Space.* London: Sage.

MEA—Millenium Ecosystem Assessment. (2005). *Millenium ecosystem assessment: Ecosystems and human well-being: Synthesis.* Washington, DC: Island Press

MUFFP—Milan Urban Food Policy Pact. (2015). Milan urban food policy pact and framework for action. https://www.foodpolicymilano.org/en/the-text-of-the-milan-urban-food-policy-pact/.

Müller, D., Haberl, H., & Bartels, L. E. et al. (2016): Competition for land-based ecosystem services. trade-offs and synergies. In J. Niewöhner (Ed.). *Land use competition. Ecological, economic and social perspectives* (pp. 127–147). Cham: Springer.

Nölting, B., & Mann, C. (2018). Governance strategy for sustainable land management and water reuse: Challenges for transdisciplinary research. *Sustainable Development., 26*(6), 691–700.

Oberlack, C., Breu, T., Giger, M., Harani, N., Herweg, K., Mathez-Stiefel, S.-L., et al. (2019). Theories of change in sustainability science. *GAIA, 28*(2), 106–111.

OECD—Organization for Economic Cooperation and Development (2013). *Rural-urban partnerships: An integrated approach to economic development.* https://www.oecd-ilibrary.org/urban-rural-and-regional-development/rural-urban-partnerships_9789264204812-en.

OECD—Organization for Economic Cooperation and Development. (2001). *Multifunctionality: Towards an analytical framework.* OECD: Paris.

Ostrom, E. (2011). Background on the institutional analysis and development framework. *Policy Studies Journal, 39/1,* 7–27. 10.1111/j.1541-0072.2010.00394.x

Patterson, J. J., & Huitema, D. (2019). Institutional innovation in urban governance: The case of climate change adaptation. *Journal of Environmental Planning and Management, 62*(3), 374–398.

Persson, J., & Mertz, O. (2019). Discursive telecouplings. In: C. Friis & J. Ø. Nielsen (Eds.), *Telecoupling* (pp. 313–336). Cham: Springer International Publishing.

Piorr, A., Zasada, I., Doernberg, A., Zoll, F. & Ramme, W. (2018). *Research for AGRI committee– urban and peri-urban agriculture in the EU, European parliament, policy department for structural and cohesion policies, Brussels.* https://www.europarl.europa.eu/RegData/etudes/STUD/2018/617468/IPOL_STU(2018)617468_EN.pdf.

Piorr, A., Ravetz, J., & Tosics, I. (2011). *Peri-urbanisation in Europe: Towards a european policy to sustain urban-rural futures.* Academic Books Life Sciences: University of Copenhagen.

Preston, D. A. (1975). Rural-urban and inter-settlement interaction: Theory and analytical structure. *Area, 7*(3), 171–174.

PURPLE—Peri-urban regions platform Europe (2014). *The Peri-urban charter.* https://www.purple-eu.org/uploads/General%20Assembly%2029%20Jan%202014/Peri-Urban%20Charter%20-%20PURPLE.pdf.

Pütz, M. (2004). Regional Governance. Theoretisch-konzeptionelle Grundlagen und eine Analyse nachhaltiger Siedlungsentwicklung in der Metropolregion München. Oekom-Verlag: München.

Reimer, M., Getimis, P., & Blotevogel, H. (2015). *Spatial planning systems and practices in Europe.* London, Routledge: Taylor

Repp, A., Zscheischler, J., Weith, T., Strauß, C., Gaasch, N., & Müller, K. (2012). *Urban-rurale Verflechtungen: Analytische Zugänge und Governance-Diskurs.* Müncheberg: Diskussionspapier Nr. 4, Leibniz-Zentrum für Agrarlandschaftsforschung (ZALF) e.V.

Rogga, S., Weith, T., Aenis, T., Müller, K., Köhler, T., Härtel, L. & Kaiser, D. B. (2014). *Wissenschaft-Praxis-Transfer jenseits der "Verladerampe": zum Verständnis von Implementation und Transfer im Nachhaltigen Landmanagement.* Müncheberg : Diskussionspapier Nr. 8. Leibniz-Zentrum für Agrarlandschaftsforschung Müncheberg.

Rydin, Y. (2007). Re-examining the role of knowledge without planning theory. *Planning Theory, 6*(1), 52–68.

Sanyal, B. (2000). Planning's three challenges. In Rodwin, L. & Sanyal, B. (Eds.), The profession of city planning: Changes, images and challenges: 1950–2000 (pp. 312–333). CUPR/Transaction.

Schaeffer, P. V., Kahsai, M. S., & Randall W. J. (2013). Beyond the rural–urban dichotomy: Essay in honor of professor A. M. Isserman. *International Regional Science Review*, 36(1), pp. 81–96.

Schnore, L. F. (1966). The rural-urban variable: an urbanite's perspective. *Rural Sociology, 31*(2), 131–143.

Schöler, K. (2005). *Raumwirtschaftstheorie*. München: Vahlen.

Schröter-Schlaack, C. et al. (2016). Naturkapital Deutschland – TEEB DE, 2016. Ökosystemleistungen in ländlichen Räumen – Grundlage für menschliches Wohlergehen und nachhaltige wirtschaftliche Entwicklung. Schlussfolgerungen für Entscheidungsträger. Leibniz Universität Hannover, Hannover, Helmholtz-Zentrum für Umweltforschung – UFZ, Leipzig. Available online at https://www.ufz.de/export/data/global/190505_TEEB_DE_Landbericht_Langfassung.pdf.

Schulze Bäing, A. (2007). Rural-urban relationships: The search for the evidence base. Paper contribution to the Association of European Schools of Planning (AESOP) 21st Congress Planning in the Risk Society; 11–14 July 2007, Naples, Italy.

Seto, K. C., Reenberg, A., Boone, C. G., Fragkias, M., Haase, D., Langanke, T., et al. (2012). Urban land teleconnections and sustainability. *Proceedings of the National Academy of Sciences of the United States of America (PNAS), 109*(20), 7687–7692. https://doi.org/10.1073/pnas.1117622109.

Seto, K. C., & Reenberg, A. (2014). Rethinking global land use in an Urban Era. In K. C. Seto, A. Reenberg, E. F. Lambin, J. Lupp, M. Turner & U. Dettmar (Eds.), *Rethinking global land use in an urban era* (pp. 1–7). Cambridge, Massachusetts: The MIT Press.

Smith, I., & Courtney, P. (2009). *Preparatory study for a seminar on rural-urban linkages fostering social cohesion.* DG Regional Policy: Final paper.

Sovacool, B. K., & Hess, D. J. (2017). Ordering theories: Typologies and conceptual frameworks for sociotechnical change. *Social Studies of Science, 47*(5), 703–750.

Stead, D. (2002). Urban-rural relationships in the West of England. *Built Environment, 28*(4), 299–310.

Tacoli, C. (1998). Rural-urban interactions: A guide to the literature. *Environment and Urbanization, 10*(1), 147–166.

Taylor, J., & Hurley, J. (2016). "Not a lot of people read the stuff": Australian urban research in planning practice. *Urban Policy and Research, 34*(2), 116–131. https://doi.org/10.1080/08111146.2014.994741.

Thompson, R. (2000). Re-defining planning: The roles of theory and practice. *Planning Theory & Practice, 1*(1), 126–133. https://doi.org/10.1080/14649350050135248.

Tödtling, F., & Trippl, M. (2005). One size fits all? Towards a differentiated regional innovation policy approach. *Research Policy, 34*(2005), 1203–1219.

Ulied, A., Biosca, O., & Rodrigo, R. (2010). Urban-rural narratives and spatial trends in Europe: The state of the question. Commissioned by: Ministry of Environment, and Rural and Marine Affairs https://81.47.175.201/urban_rural/220710_URBANO_RURAL_ING.pdf.

UN-Habitat (2017). United nations conference on housing and sustainable urban development habitat III: Issue paper on urban-rural linkages. https://www.researchgate.net/publication/321168294_Habitat_III_ISSUE_PAPER_10_ON_URBAN-RURAL_LINKAGES/link/5a12ef53a6fdcc717b522655/download.

UN-Habitat. (2019). Urban-rural linkages: Guiding principles to advance integrated territorial development. Nairobi. https://urbanrurallinkages.files.wordpress.com/2019/04/url-gp.pdf

United Nations. (2017). The sustainable development agenda. https://www.un.org/sustainabledeve lopment/development-agenda (Download: 30.01.2017).

United Nations General Assembly. (2016). United nations conference on housing and sustainable urban development (Habitat III): Draft outcome document of the united nations conference on housing and sustainable urban development (Habitat III), September 29, 2016. https://nua.unh abitat.org/uploads/DraftOutcomeDocumentofHabitatIII_en.pdf.

Vatn, A. (2010). An institutional analysis of payments for environmental services. *Ecological Economics, 69* (2010), 1245–1252.

Vogelij, J. (2014). Does ESPON support planning practice? *Planning Theory & Practice, 15* (1), 139–143. https://doi.org/10.1080/14649357.2013.873232.

WBGU—German Advisory Council on Global Change. (2016). Humanity on the move: Unlocking the transformative power of cities. Berlin: WBGU. https://www.wbgu.de/fileadmin/user_upload/wbgu/publikationen/hauptgutachten/hg2016/pdf/hg2016_en.pdf.

Wolff, F., & Mederake, L. (2019). Rahmenbedingungen und Instrumente für die Gestaltung nachhaltiger Stadt-Land-Verknüpfungen. Im Auftrag des Umweltbundesamtes. (UBA Texte 86/2019). https://www.umweltbundesamt.de/sites/default/files/medien/1410/publikationen/2019-08-15_texte_86-2019_run-bericht_ap3-1_3-2.pdf.

Woods, M. (2009). Rural geography: Blurring boundaries and making connections. *Progress in Human Geography, 33*(6), 849–858. https://doi.org/10.1177/0309132508105001.

Wüstemann, H. (2005). Multifunktionalität von landwirtschaftlichen Betrieben in Elbe-Elster. In: Dannenberg, P., Schleyer, C., & Wüstemann, H. (Eds.), *Regionale Vernetzungen in der Landwirtschaft: Beiträge eines teilprojektübergreifenden regionalen Workshops am 13.01.2005 in Bad Liebenwerda (Landkreis Elbe-Elster)* SUTRA-Working Paper Nr. 6. pp. 14–22.

Wüstemann, H., Mann, S., & Müller, K. (2008). Multifunktionalität: Von der Wohlfahrtsökonomie zu neuen Ufern. München: Oekom-Verlag.

Zasada, I. (2011). Multifunctional peri-urban agriculture—A review of societal demands and the provision of goods and services by farming. *Land Use Policy, 28,* 639–648.

Zaehringer, J., Schneider, F., Heinimann, A., & Messerli, P. (2019). Co-producing knowledge for sustainable development in telecoupled land systems. In C. Friis & J. Ø. Nielsen (Eds.), *Telecoupling* (pp. 357–381). Cham: Springer International Publishing

Zscheischler, J., Rogga, S., & Busse, M. (2017). The adoption and implementation of transdisciplinary research in the field of land-use science—a comparative case study. *Sustainability, 9,* 11, Article: 1926.

Part II
Co-Production of Knowledge

Chapter 7
Transdisciplinary Research in Land Use Science—Developments, Criticism and Empirical Findings from Research Practice

Jana Zscheischler

Abstract The particular importance of transdisciplinarity (TD) is emphasised against the backdrop of urgent complex real-world challenges and a changed societal demand for knowledge. It is no longer just a matter of producing new scientific insights, but also of achieving the solution-oriented goals and producing the action knowledge that support sustainable development and land management. Transdisciplinary research (TDR) projects have been supported in Germany over the past two decades. However, critical questions are increasingly being raised about the extent to which such projects have been successful. This chapter introduces the development of the TDR concept; describes the current criticism of TDR; and presents empirical findings from research practice. The results reveal a number of implementation deficits that can be traced back to a misfit with academic structures and a lack of knowledge.

Keywords Co-production of knowledge · Sustainable land management · Mode 2

7.1 Land Use Science—From Land Cover to Global Change Research

For a long time, the natural environment was predominantly an object of natural sciences and the belief that progress is a matter of technology development prevailed. The environmental crisis, arising ethical discourses on responsibility (Jonas 1984), the postulation of a risk society (Beck 1986) and the ensuing awareness of the increasing vulnerability of modern societies have modified society's opinion on progress and science (Gibbons et al. 1994; Nowotny et al. 2001). As a result, more integrative concepts such as "human–environment interactions" started evolving in the early 1970s (Crumley 2007 cit. in Palsson et al. 2013). This development was

J. Zscheischler (✉)
Working Group "Co-Design of Change and Innovation", Leibniz Centre for Agricultural Landscape Research—Land Use & Governance, Eberswalder Str. 84, 15374 Müncheberg, Germany
e-mail: jana.zscheischler@zalf.de

reflected in the emergence of new disciplines, such as ecological economics, environmental sociology and sustainability science (Costanza 1989; Catton & Dunlap 1978; Komiyama and Takeuchi 2006).

In this context, land use science also gradually developed an integrated socio-ecological systems perspective over the past two decades. While researchers initially focused on monitoring and modelling biophysical characteristics and land cover changes (Verburg et al. 2013), they now seek for a more integrative understanding. In addition, land use science has become an integral part of global change research and sustainability science (Braimoh and Osaki 2010; Meyfroidt et al. 2013), in which the urgency to change actions is often emphasised (Palsson et al. 2013).

This growing awareness of the decisive role played by human activities is especially reflected in the concept of the "Anthropocene", which considers that human activities have become a major geological factor (Crutzen 2002). Jahn et al. (2015, p. 92) regarded the diagnosis of the "Anthropocene" as one of the most fundamental changes of perspective over the last one hundred years: "Society and nature are so closely interwoven that they can no longer be independently investigated'.[1]

Since land use dynamics are simultaneously affected by biophysical, ecological, economic and socio-cultural drivers, data and knowledge generated from land cover analyses are insufficient for our understanding and for providing answers to many of the urgent questions posed by society. What is also required is knowledge about the actors involved, and their values, beliefs and motivations for decision-making. A "radical change in perspective and action" is required, as new research questions arise and necessitate new ways of thinking and action (Palsson et al. 2013).

This view has resulted in a new relation between knowledge and action (knowledge for action), and consequently a new role for science. Science is expected to provide not only more "systems knowledge", but also knowledge about societal targets and opportunities for transformation. The evolution of the transdisciplinary research (TDR) approach in land use science can be directly related to these developments.

7.2   The Development of the Concept of TDR

The concept of transdisciplinarity (TD) can still be regarded as a relatively young one. Although the term "transdisciplinarity" was used by Jantsch (1970, 1972) and Piaget (1972) in reference books of philosophy of science back in the early 1970s, it rarely appeared until the 1990s (cf. Völker 2004). TD evolved from the concept of "interdisciplinarity", which was further clarified after the organisation and quality of interdisciplinary research was unable to keep pace with the success generated by the dissemination and use of the term (ibid.). For this reason, Mittelstraß introduced the concept of "transdisciplinarity" at the "Bielefelder Symposium" in 1986, although its definition was almost identical to the term "interdisciplinarity" (e.g. Mittelstrass

---

[1]Translated from the German "Gesellschaft und Natur sind so eng verwoben, dass sie nicht mehr unabhängig voneinander untersucht werden können," p. 92.

2000). The term "transdisciplinarity" was used to elaborate the concept of "interdisciplinarity", because the latter was previously inadequately explained and superficially used. Völker (2004) called it a "terminological rescue attempt".[2]

The resulting notion of TDR as "perfected interdisciplinarity" persists to this day, which is especially apparent in regional differences between Europe and the US (Klein 2008). In the North American debate, the notion of TDR refers back to the "taxonomy of cross disciplinary research" after Rosenfield (1992) used the lexical morpheme "trans" to describe a collaborative research approach differing from interdisciplinarity where researchers "work jointly but still from disciplinary-specific basis", transcending disciplinary boundaries by "using shared conceptual framework drawing together disciplinary-specific theories, concepts, and approaches to address common problems." In contrast, the meaning of "trans" in the "European" concept is related to the North American concept of transdisciplinarity,[3] but it is extended by the science-to-society transgression. Hence, the main difference between the two definitions lies in the involvement of non-academics, which is a distinguishing aspect of the "European" definition. Finally, these different meanings of TD and the relating confusion partly resulted in neologisms such as the "co-design" and the "co-production of knowledge" (e.g. Mauser et al. 2013). In addition, it can be claimed that these terms more clearly illustrate the core idea of TD as it is mainly understood today, namely an equal collaboration between science and practice in the development and design of a research project, and in the production and dissemination of knowledge.

This extended meaning of "transdisciplinarity" can be traced back to the diagnosis of Gibbons et al. (1994), and later Nowotny et al. (2001), who described a new type of knowledge production resulting from a changed relationship between science and society. According to the authors, this new mode of knowledge production (Mode-2) differs from the traditional Mode-1 by the context of application and the involved relation to societal problems. The heterogeneity and organisational diversity of societal responsibility was highlighted via the TDR approach. While Mode-1 science was characterised by academia having a monopoly on knowledge production, Mode-2 science also allowed the integration of further knowledge types from extra-scientific actors. Thus, the claim for the existence of a different type of knowledge production stressed not only the integration of different disciplinary perspectives but also perspectives from outside academia. In addition, the authors regarded the two modes of knowledge production as not mutually exclusive, but as complementary.

At around the same time that Gibbons et al. (1994) observed and described a "new production of knowledge", the concept of "post-normal science" outlined by Funtowicz and Ravetz (1993) gained attention. Funtowicz and Ravetz consider science as in an ongoing process of change that is primarily shaped by the focused problem constellations and definitions. The demand for a new type of knowledge production is made plausible against the backdrop of new political challenges by a perceived global ecological crisis and new societal risks[4]: *"To characterize an issue*

---

[2]Translated from German "terminologischer Rettungsversuch".

[3]Often referred to as "transdisciplinary team science".

[4]In 1986, Ulrich Beck published his highly acclaimed book "Risk Society".

*involving risk and environment, in what we call 'post normal science', we can think
of it as one where facts are uncertain, values in dispute, stakes high and decisions
urgent."* (Funtowicz and Ravetz, 1993).

The authors argue that these challenges and problems are virtually impossible to explain by the dominating reductionist research approaches in science. Instead, systemic and synthesising approaches are required to tackle problems with a high degree of "unpredictability, incomplete control and plurality of legitimate perspectives."

In this context, the authors postulate that science has a strong responsibility for societal development. They refer to the history of progress that has been successfully pushed by scientific knowledge. However, they also voice criticism: *"...After centuries of triumph and optimism, science is now called on to remedy the pathologies of the global industrial system of which it forms the basis."* (ibid.)

Many aspects of "post-normal science", such as "grasping complexity", "dealing with uncertainty" or "accounting a diversity of perceptions" (e.g. Mobjork 2010; Pohl and Hirsch Hadorn 2008), have been adopted and incorporated in the discourse of TD.

To date, practical applications of TDR can be found in the field of integrated environmental or sustainability science, as well as in health science (e.g. Klein 2008; Bammer 2005). In fact, sustainability science appears to be the ideal designated field for TDR (Hirsch Hadorn et al. 2006; Scholz and Steiner 2015). In this field, TDR is based on the derivation of a changed perception of great challenges and political objectives such as the Sustainable Development Goals (SDGs); it is backed by politically motivated funding programmes[5] for sustainability research.

Today, science is not only expected to understand and explain phenomena, but also to provide guidance for action. Hence, knowledge production is called on to handle normative orientation and interrelate "descriptive, normative and practice-oriented forms of knowledge" (Pohl and Hirsch Hadorn 2008). This differentiation into the above three types of knowledge was discussed by several authors, who divided topics into (i) systems knowledge, (ii) target knowledge, and (iii) transformation knowledge (Jantsch 1972; Wiek 2007; Zierhofer and Burger 2007; Schäfer et al. 2010). Systems knowledge refers to questions about characteristics and dynamics of a problem, considering complex human–environment interactions and diverse interpretations (Know what?). Target knowledge represents normative knowledge, and captures desired goals and the needs and direction for change (Know where?). Transformation knowledge incorporates support for the development of strategies for societal transformation processes and concrete action (Know how?).

A similar differentiation into knowledge types can also be found in the concept put forward by Max-Neef (2005), who outlined his idea of a "transdiscipline" by interrelating the specialised disciplines taught by modern-day universities (see Fig. 7.1). Max-Neef distinguished between four different levels: At the basic "empirical" level

---

[5]In Germany, this is especially supported and funded by the Federal Ministry of Education and Research.

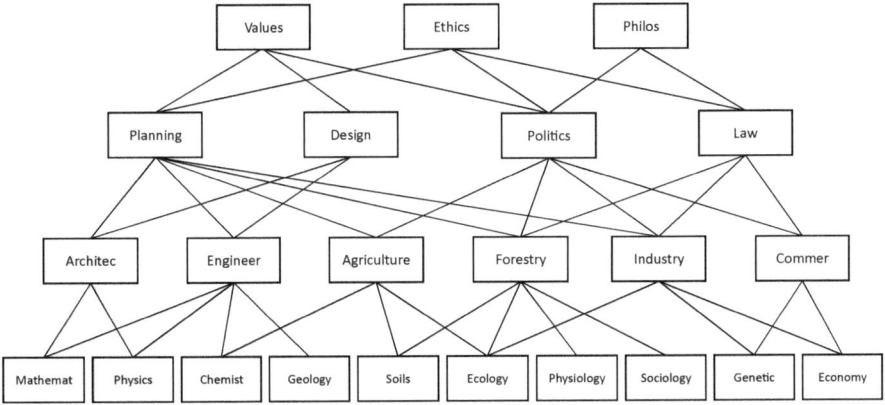

Fig. 7.1   The concept of a "transdiscipline" (based on Max-Neef 2005)

of his pyramid, he placed the basic disciplines that are capable of answering questions on "What exists?" The next level, called the "pragmatic level", covers the "technological disciplines" that contribute to the question "What are we capable of doing?" This level was named pragmatic level, which is headed by the normative level, with disciplines based on the question "What is it we want to do?" He called the highest level the "value level", with disciplines that ask and answer the question "What should we do, and how?" Transdisciplinarity based on Max-Neef (2005) results from coordination between all four hierarchical levels of this pyramid.

In the recent discourse, however, TD is less understood as a discipline in its own right,[6] but rather as a problem-oriented research principle that integrates different disciplinary and sectoral viewpoints, as well as knowledge types, also from outside academia. In this regard, TD is also strongly linked with (participatory) action research approaches (see also Cornwall and Jewkes 1995).[7]

Nonetheless, the concept of a "transdiscipline" based on Max-Neef illustrates that the different knowledge types are reflected by highly specialised science (university) but must be united to tackle "real-world" problems. In addition, the concept of Max-Neef clarifies that Mode-1 science represents the basis for Mode-2 science, which can be regarded as complementary by the additional integrative capacity.

A milestone in the conceptual development of TD was the "td-net" conference in Zurich in 2000. Here, different epistemological and research-practical discussions were brought together for the first time. A key result of this conference was a common definition of TD, which was subsequently broadly disseminated.

---

[6]According to Mittelstraß (2005), the identity of disciplines is determined by certain objects of research, theories, methods and aims of research. He argued that transdisciplinarity is not a theory principle, but rather a research guiding principle and a form of organisation.

[7]Scholz (2011) provided a good overview of the differences and similarities between TD, action research, participatory research and community-based research.

The Zurich definition describes TD "as a new form of learning and problem solving involving cooperation among different parts of society and academia in order to meet complex challenges of society" (Häberli et al. 2001).

7.3   The Impact of TDR, Criticism and Open Questions

The increasing launch and funding of TDR projects over the last two decades (see Bergmann et al. 2005; Defila et al. 2008) sparked an ongoing debate on adequate evaluation approaches. A large part of the literature on TDR is now dedicated to questions about quality criteria, impact measurement and evaluation frameworks (e.g. Carew and Wickson 2010; Jahn et al. 2012; Klein 2008; Roux et al. 2010; Walter et al. 2007; Wickson et al. 2006; Zscheischler et al. 2018). Nonetheless, since Klein (2008) claimed that the evaluation of TDR is "one of the least-understood aspects", this statement continues to be highly relevant.

Difficulties in identifying consistent evaluation criteria result from the high degree of context specificity of TDR, the non-projectable changes, the high degree of uncertainty and the comparability of various projects (Wickson et al. 2006). Non-linear interdependencies, multiple interacting drivers of change and long-time lags (Roux et al. 2010) require varying approaches and criteria compared to conventional research efforts. Consequently, several articles have discussed potential evaluation principles for TDR (e.g. Klein 2008; Loibl 2005; Spaeth 2008). Nonetheless, no generally accepted quality standards have been implemented to date.

This lack of quality standards is a major criticism against TDR (Goebel, Hill, Fincham and Lawhon 2010). Although the development of quality criteria, as a kind of guiding principle, makes a valuable contribution to supporting researchers who apply TD, greater effort needs to be made to prove the effects of TDR. The highly complex question of verifying societal effects becomes especially important considering the backdrop of an increasingly output-oriented, competitive science. Greater focus on the societal effects of TDR and a more outcome-oriented approach are regarded as important (Russell et al. 2008; Walter et al. 2007) to substantiate the added value of TDR. Until recently, there were few considerations of the effects and outcomes beyond the TDR process.

Beyond numerous plausible explanations justifying TDR, there is little knowledge about its (true) performance. Doubts and a critical attitude are reflected by the question of whether transdisciplinarity is simply a "word à la mode" (Lawrence 2004; Mittelstraß 2005). In this context, the broad interest in adequate evaluation approaches can be interpreted as a legitimacy crisis.

TDR is time-consuming, and requires a large amount of resources. Effect analyses that justify these higher levels of effort are lacking. The additional expense incurred by TDR has not yet been justified by an established improvement in results. This deficit can be linked to a lack of empirical findings (e.g. Lieven & Maasen 2007; Tress et al. 2007). The few empirical studies on TDR projects are dispersed over several disciplines, application fields and case studies.

7.4   The Role of Transdisciplinary Research in the Field of Land Use Science—Results from a Comparative Case Study in Germany

To narrow the aforementioned knowledge gap, I conducted a number of studies among researchers and actors from practice who address questions concerning sustainable land management (SLM) and apply TDR in Germany (see Zscheischler et al. 2014, 2017, 2018; Zscheischler and Rogga 2015). In the following, I synthesise the results and present some key findings.

7.4.1   Sustainable Land Management Can Be Seen as a Designated Field for TDR

The results show that SLM is regarded as a designated field for the TDR approach. This is evidenced by the discussions in the field of land use science (Zscheischler et al. 2014), by a review of the literature (Zscheischler and Rogga 2015), and from ontological analyses of "complex real-world" problems such as the question of "increasing land use competition" (Zscheischler et al. 2016).

Sustainable land management can be seen as an upcoming conceptual framework that includes different perspectives on land as an increasingly valuable resource with regard to global challenges, such as climate change, demographic change, value change, economic change, and loss of biodiversity. It includes a strong orientation for action under the normative goal of sustainability, integrating ecological with economic and societal demands, and in this regard, it integrates knowledge, sectoral viewpoints and values. As one of the major challenges linked to SLM, a deeper understanding of human-nature interrelations as well as spatial interdependencies of land use decisions is mentioned (Zscheischler et al. 2014). Land use and land use change are driven by actors with diverse interests and demands on different scales. Thus, SLM faces many challenges with respect to integrating these competing interests. In addition, the core concept of sustainability is normatively driven, and values play a guiding role when taking action.

Against this backdrop, TDR poses a rationalisation potential, and it can provide "socially robust" knowledge to tackle these very complex real-world problems. This relation between SLM and TDR is also supported by the results of a meta-synthesis. It can be shown that many case studies that have applied TDR are related to the "management" of natural resources or land use (Zscheischler and Rogga 2015).

7.4.2   TDR Plays an Increasing Role, and the Concept Is Being Consolidated

There has been a boost in publication output over the past decade, reflecting the increased importance of TDR for land use science. Over a longer period, the concept of TDR has been highly debated in terms of its epistemological, theoretical and ontological foundations; its methodological approach; and its function in science and practice.

A review of this literature shows that, while the concept of TDR can be considered as being in a "consolidating phase", it has recently been used mainly to describe a collaborative process of knowledge production that involves multiple disciplines and stakeholders aiming at enhancing the capabilities to tackle highly complex real-world problems (Zscheischler et al. 2014; Zscheischler and Rogga 2015). In line with other scholars, three core features of TDR can be identified: (i) complex real-world problems, (ii) collaborations and (iii) evolving methodologies that can be differentiated into further key concepts.

In particular, the concepts of "mutual learning" and "knowledge integration" are broadly discussed, specifying the quality of TDR. However, despite the emphasis placed on the idea of "knowledge integration" in the context of TDR, it is depicted only inconsistently and vaguely.

Nonetheless, it should also be noted that the interpretation of TDR in its current form, as a participative, problem-focused and action-oriented research approach, is strongly driven by a Central European perspective. In fact, a large proportion of publications come from countries such as Germany, Switzerland, Austria and the Netherlands.

The results also show that the debate on TDR is dominated by theoretical and conceptual contributions.

7.4.3   Attitudes Toward the TDR Approach Are Appreciative

My empirical studies revealed that researchers who apply TD are generally apprecia-tive of the TDR approach and have a positive basic attitude towards it (Zscheischler et al. 2017, 2018).

In contrast to scientists and practitioners with idealistic motives, other scholars considered TD primarily as an alternative way to attract external funding. Researchers in Germany are under increasing pressure to raise external funding. Thus, empirical findings have revealed that the primary reason for launching a TDR project is to attract external funding. This entails modifying the in design and wording of the project proposal to satisfy the call for proposals without any deeper methodological proficiency. As a consequence, some researchers also reported that they considered TDR to be something of a "necessary evil" to be fulfilled to secure funding and that they sought to reach a TD threshold that had to be passed to obtain funding; others changed their minds during the course of projects.

Nevertheless, TDR appears to be welcomed as an opportunity to pursue and test the transformation measures intended to achieve sustainability goals. Scientists are highly motivated to contribute to more sustainable land use.

Practitioners exhibited motivations and interests that differed greatly to those of scientists. The involvement of practitioners in the research process was considered by scientists to be one of the main difficulties in TDR. Interview partners occasionally criticised practice partners. They complained about practice partner saturation, about their disinterest in integrated and abstract approaches, and their sole focus on solution-based results. The science partners therefore made a greater effort to involve stakeholders with an affinity for risk and an interest in experimentation.

7.4.4 Understanding of TDR Remained Vague

The studies (ibid.) showed that understanding of TDR among researchers who apply TD differs from the conceptual advancing discussion led by scientists who advocate TD in the literature. Although an increase in consistency is apparent in the TD literature, there is only a vague understanding of the concept itself among the relevant researchers.

In one of the very few empirical studies on TDR, Tress et al. (2005) demonstrated that 81 per cent of researchers who performed TD had only a vague understanding of the concept. Almost one decade later, this vague understanding still prevails, as shown by the results. Nonetheless, the analysis of 13 TDR projects (Zscheischler et al. 2017) indicates that a shared notion of TD as a form of "science-practice collaboration" starting with a "real-world problem" is common to all projects. Other central features of TDR, as discussed in the theoretical literature, such as "mutual learning" and "knowledge integration", exhibit very little consideration, or none at all.

Moreover, TD appears to be conceived as an instrument of transfer, meaning the application of real-world problem solutions from academia to practice. In this context, empirical studies revealed a common understanding of TDR as an approach for harmonising research results with the requirements of practice, which is a rather "shortened" notion of TDR.

The general uncertainty among researchers with regard to the concept of TDR is also documented by the key terms used in project proposals – they are hesitant when asked about their understanding of TDR and the success of their project, or when this was openly discussed.

7.4.5   The Application of TDR Is Often Shortened

Furthermore, echoing this rather vague understanding of TDR, a shortened application of TDR resulting in multiple differences from the proposed ideal–typical concept
was found (Zscheischler et al. 2017). The majority of the investigated projects showed
no element of an ideal–typical co-design process for the initial phase. Instead, project
issues and composition were strongly shaped by pragmatism, following the logic of
temporary projects. "Feasibility" and "efficiency" preclude ideal–typical proceeding.
Thus, individuals or core groups of a few people (mainly representing science) determined the project objectives and desirable partners from science and practice alone.
Scientific and practice partners and the project objectives were mainly selected based
on previous projects, pre-existing contacts and networks. This pattern can be observed
in all projects, which thus stresses the importance of mutual trust and network reliance
in the selection of project partners.

7.4.6   Multidisciplinarity Prevailed

Interdisciplinary collaboration that integrates conceptual frameworks and theory
from different disciplines remained an exception; there was often no strategic planning or management (Zscheischler et al. 2017). Some scientists met, while other
sub-projects remained separate from each other in the projects under investigation.
Although dependencies led to greater exchange between sub-projects, there were
attempts to avoid such dependencies owing to fears of a delay in project organisation.
    The projects were shaped by a strong natural science orientation from the very
beginning. When asked about interdisciplinary collaboration in their projects, interviewees often cited adjacent disciplines in the field of land-related ecosystem
research, such as agricultural science, forest science and hydrology.
    As such, alternating perspectives on sustainable land issues from the humanities
and social sciences were marginalised. Social science contributions were generally
recognised in principle, but were rarely placed at the heart of the project from an
overall perspective. Thus, collaboration between methodologically and ontologi-
cally "distanced" disciplines remained rather additive. In most projects, social scientists were not only outnumbered, but also classified as "interface specialists" who
routinely worked as transfer agents or science communicators.

7.4.7   Involvement of Practitioners Aims at Acceptance
         and Implementation

There were different roles played by actors from practice, and their incorporated
knowledge bases. One emphasis was placed on the involvement of such actors for

the production of spatial development models and scenarios ("Leitbild" processes), for acceptance analyses, and for the testing of technical innovations at practitioners' facilities or sites (Zscheischler et al. 2017).

The quality of the involvement processes under investigation differed widely, ranging from projects with highly intensive co-operation and co-creation of knowledge to projects that conducted information activities as participative measures only. The overall picture reveals that information and consultation events clearly outweighed more integrative approaches and methods. Stakeholder and public acceptance of science activities and implementation appeared to be the prevailing goal of stakeholder involvement. Smaller projects (in terms of the number of institutions and partners involved) reported a noticeably more intensive exchange of information and perspectives among project members. However, one of the largest consortiums (approximately 35 partners) experienced a very successful stakeholder dialogue process involving more than 60 actors from practice. Unfortunately, these activities remained unattached to the core item of the project, which led to frustration for both the dialogue moderators (in this case scientists) and their non-academic project partners.

As a general observation, many scientists consider their research entirely separately from the stakeholder process. Others displayed a rather limited conception of stakeholder involvement. As an example, specific work packages were outsourced to providers that had been termed "non-scientific project partners". Another frequent expression of that scientific "services" mentality was the provision of testing areas by landowners.

Non-academic actors were frequently classified as "partners" (bound to the project through contracts) and "actors" (involved through interviews and surveys, focus groups or workshops). Practice partners from municipalities, public authorities, NGOs and so forth were often bound to the project via (co-funded) employment at their respective institution.

One observed strategy was that many projects implemented "regional coordinators" who were assigned to establish or strengthen multiple communication processes involving (a) science and practice, (b) horizontal cross-sectoral communication (e.g. land management-related local authorities such as planning, environment, economy), and (c) vertical actor-based communication (e.g. micro-level to macro-level actors). This demanding position was frequently occupied by novices from universities who started from scratch, often without the expert knowledge of regional networks and peculiarities. Thus, project coordinators emphasised that the success of participative action depended strongly on the personality of "regional coordinators".

Instead of a strategic concept of knowledge integration, the composition of objectives and results in the form of summaries and in the manner of multidisciplinary research was observed.

7.4.8 Challenges and Barriers to Applying TDR Are Often Underestimated, but Need Professionalisation

Most of the few empirical contributions found in the literature are dedicated to barriers and facilitators of applying TDR. An examination of the material and the results from participatory observation during many events leads to the assumption that the implementation of TDR remains a difficult challenge; therefore, the practice of TDR cannot keep pace with the progress of the theoretical discourse. Most mentioned barriers and challenges generally result from a lack of resources (time, labour) and problems that arise from habitual, mental and cognitive differences among people who are part of heterogeneous research teams, i.e. differences in ontological models, expectations, levels of commitment to and engagement in the process, and levels of skill and experience. In a small number of cases, the lack of the capacity to adapt to a changing context (personnel issues) is noted. Communication is widely considered to facilitate the transdisciplinary process and is therefore accepted as a strategy to overcome interpersonal differences. In this regard, the implementation gap appears to be twofold, characterised by both operational and cognitive inconsistencies.

The observed gap between theory and practice can be attributed to the short history of the TDR concept. As evident from the statistical distribution of the literature sample (Zscheischler and Rogga 2015), the publication output merely started to increase during the past decade. Another reason can be identified in the heterogeneity of the concept. In fact, over the stretch of the last two decades, the (mainly normative) debate on TDR revealed a highly fragmented discourse that led to multiple understandings.

Finally, the demands for the application of TDR in a real-world setting may be underestimated on a regular basis. According to our investigations and others (cf. Tress et al. 2007), researchers' struggles to implement TDR do not depend on their professional experience with the concept, but might rather be justified by a lack of opportunities for reflection within respective projects.

7.4.9 Scholarliness Runs the Risk of Falling Behind

The results also indicated that scholarliness runs the risk of falling behind in TDR projects (Zscheischler et al. 2017, 2018). There were difficulties in balancing scientific claims against practical orientation in TDR projects. Coordinating scientists appear to have been particularly affected, and they complained about a lack of time to conduct their own research activities and write publications. Moreover, the pressure to publish hampers scholars' openness to a time-intensive transdisciplinary process. In this regard, extra-academic organisations (often spin-off companies) have appeared to be better positioned to assume the coordination of transdisciplinary projects. However, such projects have tended to deprioritise scientific aspects, and risked becoming pure consulting projects.

Additional proof of this finding is that junior scientists put their academic career opportunities at risk by assisting in TDR projects. Many doctoral students tasked with co-ordinating parts of a project had failed to complete their dissertation by the time the project had come to an end. This finding corresponds with results from a comparative study published by Lange and Fuest (2015), who investigated similar projects. Moreover, the scientific quality of doctoral theses evolving from TDR projects is thought to be relatively low. Consequently, some professors have little motivation to supervise such theses.

In addition, the results from a survey (Zscheischler et al. 2018) show a strong "practice tendency" of the perception of TDR success among scientists. Scientists seem to consider TD research mainly as a research form that prioritises practical outcomes.

7.4.10 The Science-Practice Benefits Equilibrium Is Off-Balance

One objective of our study was to assess the extent to which TDR contributes to the goals of SLM and keeps its promises. Since many conceptual papers on the evaluation of TDR have shown that it is virtually impossible to prove a direct impact, we asked researchers and practitioners about their perspectives on TDR project success.

The results (see Zscheischler et al. 2018) showed that there is a clear conceptual deficit regarding success dimensions and criteria when asking about the success and benefits of recent TDR projects in the form of an open question. Nonetheless, a quantitative survey, which asked respondents to rate prescribed criteria, indicated that there is a kind of basic shared "success profile" among all project participants. This "success profile" highlights criteria with a high relevance to practice, whereas typically scientific success criteria were rated as less important. This assessment indicates a significant imbalance within the science-practice benefits equilibrium, as advocated in the literature, which leans toward the practice-oriented side of the TD ideal.

As shown (Zscheischler et al. 2018), many criteria are simultaneously important and that must be considered for the successful execution of TDR projects. It can be assumed that a deficit in fulfilling one criterion cannot be compensated by overperforming in another.

Our results show that the "output performance" and "process quality" of projects are important for the overall success perception of a project, while personal benefits for "career opportunities" seem to have no influence. This finding underlines the assumption of a high degree of idealism among participating scientists.

In general, the results revealed that the overall success assessment of TDR is rather moderate. On the one hand, this finding can be traced back to the often sub-optimally realised TDR process (see Zscheischler et al. 2017). On the other hand, the study revealed both high and vague expectations.

7.5 Conclusion and Outlook

The results highlight that TDR is an approach with a high potential for complex land use issues, where normative discourses, conflicting interests, sectoral and disciplinary viewpoints, and different knowledge types are increasingly integrated in the search for sustainable solutions. Scientists who apply TD mainly have an appreciative attitude towards the concept, when asked about their opinion on TD. However, in spite of this acknowledgement, a crucial gap between the theoretically described "ideal type" of TDR and its "real-world" application was identified. This gap is accompanied by a vague understanding of what TD precisely constitutes among researchers who apply TD. The simplified conceptualisation of TD, meaning an instrument of transfer that harmonises research results with requirements from practice, may help explain the observation of a prevailing multidisciplinarity and an often-found low level of involvement of practitioners. In this context, it emerged that the benefits, especially regarding the scientific knowledge gain, are rarely reflected on, and therefore go unnoticed. As a social innovation in the academic system, it can be argued that TDR is currently undergoing an upscaling process, risking "rhetorical mainstreaming". It will therefore be important in future to assure the quality of TDR processes, and to narrow the misfit between prevailing academic structures and increasing professionalisation and profound knowledge among researchers who apply TD. And yet TDR methods and the underlying theoretical foundations are rarely taught at university. Greater consideration of these methods and foundations in higher education would make a significant contribution to the better adoption and quality of TDR processes and outcomes.

References

Bammer, G. (2005). Integration and implementation sciences: Building a new specialization. In *Ecology and Society* (Vol *10* No 2). Baumgartner 2010.

Beck, U. (1986). *Risikogesellschaft: auf dem Weg in eine andere Moderne*. Frankfurt a M: Suhrkamp.

Bergmann, M., Brohmann, B., Hoffmann, E., Loibl, M. C., Rehaag, R., Schramm, E., & Voß, J. P. (2005). Quality criteria of transdiciplinary research. *A guide for the formative evaluation of research projects. ISOE-Studientexte* (13).

Braimoh, A. K., & Osaki, M. (2010). Land-use change and environmental sustainability. *Sustainability Science, 5*(1), 5.

Carew, A.L. & Wickson, F., 2010. The TD Wheel: A heuristic to shape, support and evaluate transdisciplinary research. *Futures* 42 (10, SI), 1146–1155.

Catton Jr, W. R. & Dunlap, R. E. (1978). Environmental sociology: A new paradigm. *The American Sociologist*, 41–49.

Goebel, A., Hill, T., Fincham, R. A., & Lawhon, M. (2010). Transdisciplinarity in urban South Africa. *Futures, 42*(5), 475–483.

Cornwall, A., & Jewkes, R. (1995). What is participatory research? *Social Science & Medicine, 41*(12), 1667–1676.

Costanza, R. (1989). What is ecological economics? *Ecological Economics, 1*(1), 1–7.

Crumley, C. (2007). Historical ecology: integrated thinking at multiple temporal and spatial scales. In A. Hornborg & C. Crumley (Eds.), *The World System and the Earth System: Global Socio-Environmental Change and Sustainability Since the Neolithic* (pp. 15–28). CA: Left Coast Press, Walnut Creek.

Crutzen, P.J. (2002). Geology of mankind nature 415. 6867, 23.

Defila, R., Di Giulio, A., & Scheuermann, M. (2008). *Management von Forschungsverbünden.* Weinheim: Möglichkeiten der Professionalisierung und Unterstützung. Wiley-VCH.

Funtowicz, S. O., & Ravetz, J. R. (1993). Science for the post-normal age. *Futures, 25*(7), 739–755.

Gibbons, M., Limoges, C., Nowotny, H., Schwartzman, S., Scott, P. & Trow, M. (1994). *The new production of knowledge: The dynamics of science and research in contemporary societies.* Sage.

Häberli, R., Grossenbacher-Mansuy, W. & Klein, J. T. (2001). Synthesis. In J.T. Klein, W. Grossenbacher-Mansuy, R. Häberli, A. Bill, R. W. Scholz & M. Welti (Eds.), Transdisciplinarity: Joint problem solving among science, technology, and society. An effective way for managing complexity. Basel: Birkhäuser (Synthesebücher Schwerpunktprogramm Umwelt), pp. 6–22.

Hadorn, G.H., Bradley, D., Pohl, C., Rist, S. & Wiesmann, U. (2006). Implications of transdisciplinarity for sustainability research. *Ecological Economics, 60* (1), 119–128.

Jahn, T., Bergmann, M., & Keil, F. (2012). Transdisciplinarity: Between mainstreaming and marginalization. *Ecol. Economics, 79,* 1–10.

Jahn, T., Hummel, D., & Schramm, E. (2015). Nachhaltige Wissenschaft im Anthropozän. *GAIA-Ecological Perspectives for Science and Society, 24*(2), 92–95.

Jantsch, E. (1970). Inter-Disciplinary and Transdisciplinary University. Systems Approach to Education and Innovation. *Pol. Sci., 1,* 403–428.

Jantsch, E. (1972). *Towards interdisciplinarity and transdisciplinarity in education and innovation* (pp. 97–121). Interdisciplinarity: Problems of teaching and research in universities.

Jonas, H. (1984). *Das Prinzip Verantwortung* (p. 426). Suhrkamp: Versuch einer Ethik für die technologische Zivilisation.

Klein, J. T. (2008). Evaluation of Interdisciplinary and Transdisciplinary Research. *American Journal of Preventive Medicine, 35*(2), 116–123.

Komiyama, H., & Takeuchi, K. (2006). Sustainability science: building a new discipline. *Sustainability Science, 1*(1), 1–6.

Lange, H., & Fuest, V. (2015). Optionen zur Stärkung inter-und transdisziplinärer Verbundforschung: Abschlussbericht. *Bremen: artec.*

Lawrence, R. J. (2004). Housing and health: from interdisciplinary principles to transdisciplinary research and practice. *Futures, 36*(4), 487–502.

Lieven, O., & Maasen, S. (2007). Transdisciplinary research: Heralding a 'new deal' between science and society? *GAIA – Ecol. Perspectives for Sci. & Soc., 16*(1), 35–40.

Loibl, M. C. (2005). Recommendations for the evaluation of transdisciplinary research. *GAIA—Ecology Perspectives for Science & Society, 14* (4), 351–353.

Max-Neef, M. A. (2005). Foundations of transdisciplinarity. *Ecology Economics 53* (1), 5–16.

Mauser, W., Klepper, G., Rice, M., Schmalzbauer, B. S., Hackmann, H., Leemans, R., & Moore, H. (2013). Transdisciplinary global change research: the co-creation of knowledge for sustainability. *Current Opinion in Environmental Sustainability, 5*(3–4), 420–431.

Meyfroidt, P., Lambin, E. F., Erb, K.-H., & Hertel, T. W. (2013). Globalization of land use: Distant drivers of land change and geographic displacement of land use. *Current Opinion in Environmental Sustainability, 5*(5), 438–444.

Mittelstrass, J. (2000). Transdisciplinarity–new structures in science. In *Innovative Structures in Basic Research. Ringberg-Symposium* (Vol. 4, No. 7).

Mittelstraß, J. (2005). *Methodische Transdisziplinarität. Technikfolgenabschätzung. Theorie und Praxis, 14*(2), 18–23.

Mobjork, M. (2010). Consulting versus participatory transdisciplinarity: A refined classification of transdisciplinary research. *Futures 42* (8, SI), 866–873.

Nowotny, H., Scott, P. & Gibbons, M. (2001). Re-Thinking Science: Knowledge and the Public in an Age of Uncertainty. Polity: Cambridge.

Palsson, G., Szerszynski, B., Sörlin, S., Marks, J., Avril, B., Crumley, C., & Buendía, M. P. (2013). Reconceptualizing the 'Anthropos' in the Anthropocene: Integrating the social sciences and humanities in global environmental change research. *Environmental Science & Policy, 28,* 3–13.

Piaget, J. (1972). L'épistémologie des relations interdisciplinaires. In G. Berger, A. Briggs, & G. Michaud (Eds.), *L'interdisciplinarité—Problèmes d'enseignement et de recherche* (pp. 127–139). Paris: Organization for economic cooperation and development.

Pohl, C., & Hirsch, G. (2008). Methodological challenges of transdisciplinary research. *National Science and Societies, 16*(2), 111–121.

Rosenfield, P. L. (1992). The potential of transdisciplinary research for sustaining and extending linkages between health and social sciences. *Social Science and Medicine, 35,* 1343–1357.

Roux, D. J., Stirzaker, R. J., Breen, C. M., Lefroy, E. C., & Cresswell, H. P. (2010). Framework for participative reflection on the accomplishment of transdisciplinary research programs. *Environmental Science & Polity, 13*(8), 733–741.

Russell, A. W., Wickson, F., & Carew, A. L. (2008). Transdisciplinarity: Context, contradictions and capacity. *Futures, 40*(5), 460–472.

Schäfer, M., Ohlhorst, D., Schon, S., & Kruse, S. (2010). Science for the future: challenges and methods for transdisciplinary sustainability research. *African Journal of Science, Technology, Innovation and Development, 2*(1), 114–137.

Scholz, R. W. (2011). *Environmental literacy in science and society: From knowledge to decisions.* Cambridge University Press.

Scholz, R. W., & Steiner, G. (2015). The real type and ideal type of transdisciplinary processes: part II—what constraints and obstacles do we meet in practice? *Sustainability Science, 10*(4), 653–671.

Spaeth, P. (2008). Learning ex-post: Towards a simple method and set of questions for the self-evaluation of transdisciplinary research. *GAIA—Ecology Perspectives for Science & Soc*iety *17* (2), 224–232.

Tress, G., Tress, B., & Fry, G. (2005). Clarifying integrative research concepts in landscape ecology. *Landscape Ecology, 20*(4), 479–493.

Tress, G., Tress, B., & Fry, G. (2007). Analysis of the barriers to integration in landscape research projects. *Land Use Pol., 24*(2), 374–385.

Verburg, P. H., Erb, K.-H., Merz, O., & Espindola, G. (2013). Land System Science: between global challenges and local realities. *Current Opinion Environment Sustainability, 5*(5), 433–437.

Völker, H. (2004). *Von der Interdisziplinarität zur Transdisziplinarität* (pp. 9–28). Transdisziplinarität. Bestandsaufnahme und Perspektiven. Göttingen: Universitäts-Verlag.

Walter, A. I., Helgenberger, S., Wiek, A., & Scholz, R. W. (2007). Measuring societal effects of transdisciplinary research projects: Design and application of an evaluation method. *Evaluation & Programme Planning, 30*(4), 325–338.

Wickson, F., Carew, A. L., & Russell, A. W. (2006). Transdisciplinary research: characteristics, quandaries and quality. *Futures, 38*(9), 1046–1059.

Wiek, A. (20070. Challenges of transdisciplinary research as interactive knowledge generation experiences from transdisciplinary case study research. *GAIA—Ecology Perspectives for Science & Society 16* (1), 52–57.

Zierhofer, W. & Burger, P. (2007). Transdisciplinary research—A distinct mode of knowledge production? Problem-orientation, knowledge integration and participation in transdisciplinary research projects. *GAIA—Ecology Perspectives for Science & Soc*iety *16* (1), 29–34.

Zscheischler, J., Gaasch, N., Manning, D. B., & Weith, T. (2016). Land use competition related to woody biomass production on arable land in Germany. In *Land Use Competition* (pp. 193–213) Cham: Springer.

Zscheischler, J., & Rogga, S. (2015). Transdisciplinarity in land use science–A review of concepts, empirical findings and current practices. *Futures, 65,* 28–44.

Zscheischler, J., Rogga, S., & Weith, T. (2014). Experiences with transdisciplinary research: sustainable land management third year status conference. *Systems Research and Behavioral Science, 31*(6), 751–756.

Zscheischler, J., Rogga, S., & Busse, M. (2017). The adoption and implementation of transdisciplinary research in the field of land-use science—a comparative case study. *Sustainability, 9*(11), 1926.
Zscheischler, J., Rogga, S., & Lange, A. (2018). The success of transdisciplinary research for sustainable land use: individual perceptions and assessments. *Sustainability Science, 13*(4), 1061–1074.

Chapter 8
Innovations for Sustainable Land Management—A Comparative Case Study

Jana Zscheischler and Sebastian Rogga

Abstract There continues to be a poor understanding of how transformation and socio-technological change in the specific field of sustainable land use and management can be effectively governed and supported. The aim of this article is to contribute to this knowledge gap by presenting the findings from a comparative case study of nine local innovation projects that sought solutions for sustainable land management (SLM). For each of the nine projects, we examined the (i) problem definitions and framings, (ii) the type and degree of innovation, (iii) the different approaches taken to manage innovation processes, and (iv) the leverage points of these solutions in the governance system of SLM. The results show that SLM innovations start from diverse problem framings and emerge from distinct action fields. We found a broad variety of innovation types following distinct solution strategies that can be clustered into (i) multiple land use, (ii) knowledge-based decision support tools, (iii) co-management approaches, and (iv) new organisations and institutions. All nine projects applied multi-actor approaches to facilitate reflexive processes of social learning and cognitive reframing by embedding experimental innovation management approaches such as real-world laboratories (thus optimising the solution) into larger transdisciplinary and participatory processes (to adjust to societal discourses and normative orientations).

Keywords Sustainability innovation · Transdisciplinary research · Governance of land

J. Zscheischler (✉) · S. Rogga
Working Group "Co-Design of Change and Innovation"; Research Area "Land Use and Governance", Leibniz Centre for Agricultural Landscape Research (ZALF) e.V., Eberswalder Str. 84, 15374 Müncheberg, Germany
e-mail: jana.zscheischler@zalf.de

S. Rogga
e-mail: sebastian.rogga@zalf.de

© The Author(s) 2021
T. Weith et al. (eds.), *Sustainable Land Management in a European Context*,
Human-Environment Interactions 8, https://doi.org/10.1007/978-3-030-50841-8_8

8.1 Sustainable Land Management—A Normative Orientation for Transformation

Land is an essential but limited "resource" to humans. Demand for land and land-based goods is increasing, and will continue to increase in the future due to a growing world population, economic growth, the energy transition, changes in consumption patterns, and, not least, climate change. As a result, it is assumed that there will be greater land use competition and more environmental degradation in the future (Haberl et al. 2014; Niewöhner et al. 2016).

Land has therefore become a key issue of sustainable development, since land use causes many sustainability problems. In the search for sustainable and socially responsible solutions that take into account complex interactions between different demands, associated actors and their (often conflicting) interests, perceptions and values, many scientists increasingly promote and discuss the concept of sustainable land management (SLM).[1]

Although SLM is not a clearly defined term, there are some common features that can be summarised as follows: SLM provides a normative orientation for policymaking and management; it takes a holistic systems perspective and addresses complex socio-ecological interactions and dynamics. SLM also takes into account multi-level and cross-sectoral approaches that stress social learning, experimentation, negotiation and the harmonisation of different goals; SLM involves multiple actor groups (e.g. Hurni 2000; Schwilch et al. 2012; Weith et al. 2013; Fritz-Vietta et al. 2017; Nölting and Mann 2018).

In general, SLM can be regarded as a concept of change and transformation building on the idea of adaptive resource co-management (see Armitage et al. 2009). It aims to achieve "*a change in understanding that goes beyond the individual to become situated within wider social units or communities of practice*" through social interactions between actors within social networks (Reed et al. 2010). In this context, the role of transdisciplinarity is highlighted by several authors (e.g. Hurni 2000; Nölting and Mann 2018).

However, beyond underlining the importance of transdisciplinarity, there is an "apparent lack of a practical, structured (yet flexible) methodology for fostering SLM in diverse contexts" (Schwilch et al. 2012). Thus, it is an open question how transformation and socio-technical change towards SLM can be effectively designed and supported.

The aim of transdisciplinary (research) processes (TDR) is to provide socially robust orientation for sustainable solutions. However, the initiation and management of innovation processes plays a central role in developing, testing and implementing such solutions. And yet very few studies explicitly address SLM innovations. A lot of scholarly work has been conducted on innovation processes in general, but it can be

---

[1]A literature review shows that there has been an increase in the number of publications on SLM over the past decade. A "scopus" search for the term "sustainable land management" generated 168 articles in April 2019.

assumed that SLM innovations differ considerably from the "usual" types of innovation. While market-based (business) innovations are generated against the backdrop of the interests of individual economic actors who often accept the externalisation of costs, SLM innovations pursue a general interest and the idea of common goods in a bid to avoid the externalisation of costs.

Thus, SLM innovations are based on a different set of push factors and are often dependent on policy-driven innovation systems and the funding of collaborative actors from science and practice (from different administrative levels and sectors).

The aim of this article is to identify specifics of SLM innovations and associated innovation processes by undertaking a comparative case study of nine research projects[2] conducted within the German funding programme entitled "Innovation Groups for Sustainable Land Management". We argue that understanding these processes is highly relevant in the bid to improve the steering and design of effective innovation and transformation processes for SLM.

8.2 The "Sustainable Land Management" Innovation System

Knowledge is considered a central resource in innovation processes (e.g. Howells 2002; Thornhill 2006); therefore, attention was increasingly focused on exchange and cooperation between actors and organisations for developing and disseminating innovations. This idea can be found in the discourse on "innovation systems", where innovations arise as a result of the interplay between different levels of an institutional structure (Edquist 1997). In this context, the meaning of "regional innovation systems" (RIS) was increasingly discussed. RIS are "typically understood to be a set of interacting private and public interests, formal institutions, and other organizations that function according to organizational and institutional arrangements and relationships conducive to the generation, use, and dissemination of knowledge" (Doloreux and Parto 2005, p. 134). So far, the concept of regional innovation systems has been discussed as a concept of spatial positioning and clustering (national/regional; increasing competitiveness of regions; regional development).

By offering various support measures to promote research and innovations in SLM, national innovation policy seeks to initiate regional innovation systems via innovation policy programmes. In Germany, for example, the Federal Ministry of Education and Research (BMBF) launched a range of consecutive funding programmes[3] dealing thematically with the integrated consideration of different uses

---

[2]All nine research projects were accompanied by a scientific coordinating research project over a period of five years. As part of this scientific coordinating project, the authors had very good access to the documents and team members of all nine projects. The results of the comparative case study are based on the analysis of qualitative interviews with coordinating scientists, and of documents, participatory observations during numerous events (workshops, conferences), and informal talks.

[3]https://www.fona.de/en/topics/land-management.php.

of land and natural resources, their interactions, and the corresponding development of solutions: REFINA (2006–2012),[4] Sustainable Land Management (2010–2016), Innovation Groups for Sustainable Land Management (2014–2020)[5] and Stadt-Land-Plus (2017–2026).[6] This funding agency regularly issues thematic calls—similar to EU funding structures—and sets important framework conditions for research (e.g. action field of research, disciplinary focus). Within the FONA framework programme (Research for Sustainable Development), the BMBF also encourages and often requires research projects to pursue research modes based on societally relevant research goals (Newig et al. 2019). The term "Sustainable Land Management" is thus largely supported (top-down) by research policy and predefined (science plays an advisory role but with a control component), but at the same time influenced (bottom-up) by the concrete thematic design of each research project.

Looking at these calls for tender and the available literature on RIS reveals that the "Sustainable Land Management innovation system" differs from the RIS concept in that it is heavily dependent on external funding and is only an innovation system for a certain period of time (demolition after the end of the eligibility period). Thus one of the core challenges is to achieve the continuity and transfer of processes to practice beyond the project duration. At the same time, there are a number of similarities with the RIS concept, such as.

- the importance of knowledge and the concept of the learning region, individual and collective learning, exchange of knowledge (e.g. Blättel-Mink 2006)
- cooperation between different actors: companies, politics, research, administration and a combination of public and private interests (Doloreux and Parto 2005)
- the regional dimension of innovation processes and development; the region as the locus of innovation
- the policy focus: systematic support of regional development (capacity building in regions, local comparative advantages, etc.).

All SLM funding programmes[7] are framed by political strategies of the German Federal Government.[8] They promote the development and testing of innovative concepts and strategies as well as "knowledge bases, technologies, instruments and system solutions" to (1) "reduce land consumption", (2) "protect livelihoods", and (3) mitigate "increasing competition for land and natural resources". The declared aim is to protect nature and the climate; secure energy and food supplies; promote health, social justice and the balancing of interests; and assure a high level of life

---

[4]https://www.fona.de/en/measures/funding-measures/archive/research-for-the-reduction-of-land-consumption-refina.php.

[5]https://www.fona.de/en/measures/funding-measures/innovation-groups-for-sustainable-land-man agement-fuer-ein-nachhaltiges-landmanagement_copy.php.

[6]https://www.fona.de/en/measures/funding-measures/city-countryside-plus.php.

[7]The results are based on an analysis of the call for proposals of REFINA, SLM, IG SLM and Stadt-Land-Plus.

[8]e.g. The National Sustainable Development Strategy, the German Hightech Strategy, and specified by the FONA framework (research for sustainable development).

quality. As a level of consideration, regions are regarded as a particularly promising spatial intervention level (e.g. "sustainable development of regions", regional circular economy", "regional value chains"). Emphasis is also placed on the central role of holistic systems approaches, reflected by the demand for "system solutions", which consider the "system context", complex interactions and urban–rural linkages.

In sum, SLM is regarded as a "key issue" and a highly "complex area of activities" that integrates different sectors, demands and interests such as water, soil, biodiversity, regional value creation, urban and rural areas, and others. Consequently, an interdisciplinary, cross-sectoral and transdisciplinary collaboration between actors from practice (politics, administration, economy and civil society) and science was mandatory.

8.3 Analysis of Innovations in Sustainable Land Management

We sought adequate frameworks and theoretical models to identify the specifics of SLM innovations, to better understand the underlying innovation processes, and to structure our study. Despite a thorough review, we found no conceptual framework with a consistent explanation or guidance for the very diverse processes and projects that we accompanied.

Not only does the literature distinguish between different types of innovation (technical, social and process innovations) that require different management approaches, innovation processes are also embedded in very complex contexts. Due to this complexity, there is often doubt as to the extent to which such complex processes can be controlled and purposefully managed (e.g. Sauer 1999; Kristof 2010).

However, it can also be stated that change and innovation processes are more than random events that do not happen out of nothing, but that can and need to be stimulated. In this context, many authors emphasised the role of conditions and innovation contexts that can be managed to support innovations. Particularly concerning sustainability innovations, the "multi-impulse hypothesis" has increasingly come to the fore (e.g. Kramer 2010). Here, the foci can be very diverse, as demonstrated by a collection of theories about socio-technical change with "the most explanatory power" or applicability by Sovacool and Hess (2017). The authors interviewed 35 experts from different disciplinary backgrounds and found 96 distinct theories. The most frequently mentioned theories were the socio-technological transition approach (after Geels 2002, Geels and Schot 2007), social practice theory (Shove et al. 2012), discourse theory, social construction of technology, and sociotechnical imaginations. Sovacool and Hess (ibid.) were able to show that most theories focus on the categories (i) agency, (ii) structure, (iii) meaning, (iv) relations, and (v) norms.

For our analysis,[9] we followed these main categories and chose a rather research pragmatic approach, adopting the suggestion by Heideloff and Radel (1998) to focus on a higher degree of abstraction. Thus, we focused on reasonings and meanings by considering **problem framings and definitions** and normative orientations of the desired **solutions** as well as the underlying **strategies**. We additionally applied established categories from the innovation literature, such as the **type of innovation** and **innovation degree** (e.g. Hauschildt et al. 2016; Kasmire et al. 2012; Baregheh et al. 2009) to determine whether there are any specific properties relevant to SLM innovations. We also sought to determine the **main barriers** to the successful implementation and distribution of these innovations, and how attempts were made to manage and control **innovation processes**.

8.4   Case Study: The German Funding Programme "Innovation Groups for Sustainable Land Management"

As described above, the "Innovation Groups for Sustainable Land Management" (IG) funding programme (2014–2020) is part of a land-focused funding line initiated by the BMBF. In contrast to the preceding programme "Sustainable Land Management" (2010–2016), the IG programme focuses on developing systemic solutions to complex real-world issues, analysing innovation conditions, and building up capacities and competences among the practice actors and scientists involved in order to drive innovative solutions that have been outlined and defined during the application process.

In total, nine joint research projects (the "Innovation Groups", or IGs) were funded over a period of five years (2014–2019) with Germany as the geographical focus of application. The IGs differed in size, actor composition (i.e. academic disciplinary background, practitioners' background) and research questions. In terms of topics, the IGs can be clustered in at least four different application fields: (1) research on regional energy transition in the context of land consumption and land competition; (2) research on multiple or diversified land use options on the same plot of land; (3) research on cultural landscape development; and (4) research on interlinkages between urban and rural spaces. Numerous IGs conducted their research activities in more than one cluster, e.g. by combining Clusters 3 and 4. Table 8.1 provides more detailed information on the projects under investigation, along with various characteristics.

---

[9]Our findings result from accompanying the projects over a period of five years. They are based on the analysis of project proposals and websites, participatory observations during events, informal talks and semi-structured interviews with coordinating scientists.

Table 8.1 Overview of the "Innovation Groups" under investigation

Project	Problem framing and definition	Solution approach/innovation	Innovation process design/methodology	Degree of innovation	Type of innovation	Main barriers
APV-Resola	Growing demand for renewable energy; growing competition for (agrarian) land resources	Development of agrophotovoltaic test sites (APV) that integrate solar power generation into agrarian (crop) production	Living Lab (according to project proposal)	First conceptual ideas in the 1980s; new in-situ realisation of APV	Technological, procedural	Legal framework conditions; entrepreneurial risks; lack of social acceptance at the local level; loss of area payment grants by CAP
AUFWERTEN	Negative environmental effects on agrarian land resources (degradation, loss of biodiversity) under conditions of growing productivity demands on land	Agro-forestry systems—cultivation of agricultural woods and field crops on the same parcel of land	"Trial and error" (lab character, open innovation techniques, multidimensional)	Realisation on test sites since the 1990s	Technological, procedural	Legal framework conditions, disadvantages in CAP area payments, lack of infrastructure along the value chain
EnAHRgie	Land use competition and conflicts in the development of "energy transition landscapes"	Integrated regional energy concept, tool development, policy development	Participatory process with regional (organised) stakeholders backed up by scientific evidence	Novelty in the regional context	Social innovation (action-guiding concept)	General inertia; lack of acceptance of certain technologies

(continued)

Table 8.1 (continued)

Project	Problem framing and definition	Solution approach/innovation	Innovation process design/methodology	Degree of innovation	Type of innovation	Main barriers
Ginkoo (two regional case studies)	(A) Loss of a traditional cultural landscape (B) Ethically questionable poultry production	(A) Collaborative landscape management (B) dual purpose chicken	Real-World Lab	Pre-existing "niche" solutions (A) 30 yrs. of regional activities (B) "Re-innovated" technique	(A) Technological (hay), institutional innovation (B) Social innovations	Legal and economic framework conditions, stakeholder conflicts
INOLA	Land use competition and conflicts in the development of "energy transition landscapes"	Tools for participatory process design; organisation development ("regional citizens foundation" as the lead organisation of the change process)	Participatory process with stakeholders and citizens, backed up by scientific evidence	Novelty in the regional context	Social innovation (institutional, procedural)	Financing, inertia of institutions, acceptance issues among citizens
REGIO-BRANDING	Marginalised rural areas in metropolitan regions	Development of a regional brand by means of a collaborative, bottom-up research process	Participatory research design, transdisciplinarity, CBPAR	New approach (to regional marketing)	Social innovation, institutional innovation	Continuation of activities requires commitment by regional decision-makers, cross-sectoral collaboration

(continued)

Table 8.1 (continued)

Project	Problem framing and definition	Solution approach/innovation	Innovation process design/methodology	Degree of innovation	Type of innovation	Main barriers
Render	Land use competition and conflicts in the development of "energy transition landscapes"	Application of a regional, inter-municipal energy transition concept	Transdisciplinary learning and application process	Novelty in the regional context	Social innovation, procedural innovation	Political will, inertia of institutions
Stadt PARTHE land	Growing land use competition in peri-urban regions with changing cultural landscapes	Concept for cultural landscape management	Incremental management approach ("muddling through")	New approach (to cultural landscape management)	Integration of technological and social innovations	Hosting institution, financing, acceptance
Urban–Rural Solutions	Demographic change challenges the existing provision of public services in rural regions	New instruments/options as a result of regional urban–rural co-operation	Design thinking approach	New for the specific application field	Social and technological innovations	Continuation of activities requires commitment by decision-makers, cross-sectoral collaboration, multi-level governance, local/regional

8.4.1 Problem Definition of Projects and Societal Pressure for Action

One central and frequently described success factor for innovation processes is the consideration of concrete application needs raised by prospective appliers. As a consequence, the question of how the problem is defined at the beginning plays a decisive role for the design and development of solutions (e.g. Enkel 2009; de Jong et al. 2016).

The majority of projects used the narrative of "increasing land use competition" in order to prove the relevance of their research project and the desired solutions. Here, the projects align their narrative with that of the funding agency, and address the problem definition described in the announcement.

Applicants additionally formulate five other problem areas and fields of action: (i) the loss of cultural landscapes, (ii) environmental degradation, (iii) increasing land use conflicts, (iv) disparities between urban and rural areas and the challenges of demographic change, and (v) the challenges of spatial justice. Thus, the projects reflect a rather wide range of fields of action and current problems, and illustrate the diversity of challenges in SLM. At the same time, it must be acknowledged that the projects represent only some of the current land use challenges. Thus, other urgent issues such as landscape fragmentation, climate change impacts, urban sprawl and contamination remained unconsidered (see EEA 2019).

One striking aspect is that many of the nine research projects focus on the challenge of changing the energy system (APV-Resola, AUFWERTEN, INOLA, render, EnAHRgie). This is not surprising, since the spatial dimension of renewable energy production and distribution in Germany is decisive (see Meyer and Priefe 2015). There is considerable pressure to act and transform the energy supply system, which is discussed extensively by many societal actors.

8.4.2 Solution Strategies and Types of Innovation

It is striking that the concept of innovation, which is traditionally often equated with the understanding of technological innovation, is significantly expanded in the context of SLM. While there are still projects that focus on new technologies and processes (such as agroforestry systems and agrophotovoltaic plants), most projects place a strong emphasis on "social innovations".

In the literature, there are different interpretative patterns of the term "social innovation" and corresponding heterogeneous theoretical approaches. In accordance with others (e.g. Taylor 1970; Brooks 1982; Schubert 2016), we consider the term as an extension of the notion of technological innovation and define "social innovation" as "new" practices that provide alternative solutions to persisting problems (Zapf 1989), driven by specific actors in specific operating contexts (Rogers 2003; Howaldt and Schwarz 2010).

Such "new" practices were the aim of many of the nine projects under investigation. These practices include new participation concepts, decision support tools, the initiation and strengthening of new collaborations and networks, and new organisations.

It emerges that the solution approaches adopted by the nine projects can be grouped into four clusters that follow different strategies:

1. ***Multiple land use*** (Example: agroforestry systems, agrophotovoltaic plants)
   These projects primarily focused on the development of technological innovations that follow the strategy of "multiple land use" to mitigate land use competition. In contrast to concept of "multifunctional land use", which advocates heterogeneity at the landscape level (see Mander et al. 2007), "multiple land use" means that one land plot is used simultaneously for two purposes such as agrophotovoltaic plants that are installed above crops, generating two harvests: the harvest of solar power and of crops (see Photo 8.1). The innovation revolves around a new technology that was developed in strong interaction with farmers on a specific plot, applying an experimental living-lab approach.
   A similar research setting was applied by the AUFWERTEN innovation group, which investigated the applicability of agroforestry systems to "regular" farming practices. As in the first example, a farmer provided test sites where the applicability of the innovation was tested in a real-world laboratory setting. In contrast to the agrophotovoltaic test site, the agroforestry project allowed for a rather open innovation research approach because the installation of woody structures on

Photo 8.1 Multiple land use by an agrophotovoltaic plant that enables solar power and crops to be harvested simultaneously from one plot (*source* ISE Fraunhofer)

arable land requires little monetary investment and, thus allows experimentation and many variations. In contrast, the APV plant was rather cost-intensive. Thus, the margin for processual adaption and innovation was narrow.

2. **Improvement of decision-makers' knowledge base** (example: integrated energy concepts)

   Several projects developed instruments to help decision-makers make well-informed and justified decisions when dealing with complex situations where stakeholders often have opposing views and where trade-offs between different societal goals are prevalent. This included integrated and action-oriented concepts for municipal administrations (usually in the context of energy transformation), decision support systems, and associated tools and instruments (EnAHRgie, render and ginkoo).

3. **Co-management approaches** (example: collaborative landscape management)

   Another strategy was to establish new alliances, networks and collaborations for pooling scarce resources, facilitating mutual knowledge exchange and social learning, and to coordinate collective actions. As an example, in one project, where the loss of the traditional cultural landscape was defined as a problem shared by all of the actors involved, the idea of collaborative landscape management was advocated as a promising solution (Zscheischler et al. 2019). In this case, several partial solutions (also technological innovations) were co-designed and co-developed by actors from different sectors such as farmers, tourism providers and nature conservation. These solutions were complemented by the idea of an integrated management concept that includes new forms of coordination, new co-operation and new networks on a community level. The process combined a real-word lab approach (for co-designing and testing several partial solutions) with a transdisciplinary process. This combination facilitated a joint problem framing, the harmonisation of actors' different visions and conflicts by means of a consensual discourse, the initiation of social learning processes, and the strengthening of actors' relationships (Fig. 8.1).

4. **New Organisations and Institutions**

   Another type of outcome and innovation was the development and construction of new organisations and institutions that took responsibilities and (in some cases) mandates to manage or co-ordinate land resources. Here, a general approach was to install an administrative or co-ordination position, which we call the "land use manager".

   These managers worked at the interface between governance and government, as well as between different land use sectors (vertical and horizontal structures). From an organisational point of view, the implementation of "land use managers" can be seen as an adaptation of the current German practice of administrative units that work at the municipal or regional level based on cross-sectoral tasks, defined by policy demands. As an example, "climate protection managers" have been installed in recent decades to accompany the implementation of climate targets set by the government (at different administrative levels). Climate protection managers work horizontally (cross-sectorally) as well as vertically (between hierarchical levels) within the administration. They also perform communication

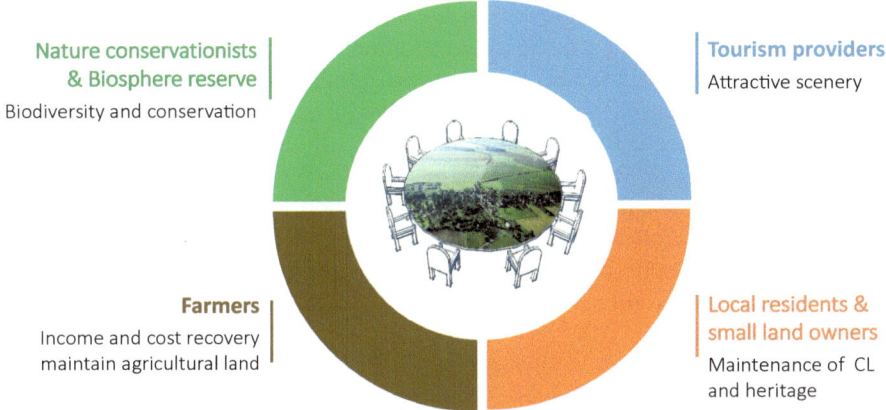

Nature conservationists
& Biosphere reserve
Biodiversity and conservation

Tourism providers
Attractive scenery

Farmers
Income and cost recovery
maintain agricultural land

Local residents &
small land owners
Maintenance of CL
and heritage

Fig. 8.1 Collaborative landscape management integrates different interests and pools resources from actors of different sectors to manage and develop a traditional cultural landscape including the typical attractive scenery, cultural heritage, biodiversity and cost recovery (own source)

tasks involving social learning, information and acceptance by the populace. Although there is also a concrete political goal in Germany to reduce excessive land consumption, there is not yet a corresponding administrative position that addresses "land management" as a cross-sectional issue. Therefore, the projects filled that "blind spot", albeit in different ways. The INOLA project, for example, tasks a citizens' foundation with pursuing the goal of achieving the regional energy transition. The foundation was initiated by citizenship interests and political will, and was eventually given a political mandate by its founding rural districts (Landkreise).

Other projects established citizens' initiatives to coordinate actions of different actor groups towards the common goal/interest of cultural landscape management.

8.4.3   Steering Innovation Processes

All projects were invited by the application call to implement a transdisciplinary research approach that brings together different actors from science and practice, integrating their perspectives and interests. This requirement is based on the assumption that the process design and development of sustainability innovations plays an important role in the transformation towards the socially responsible use of land and natural resources.

Innovations in integrated sustainable land management (or sustainability innovations) also differ from the more classic (product, process, marketing or organisational) innovations studied in economics. Besides addressing economic exploitation structures, they also seek, among other things, to achieve the multifunctionality of agrarian landscapes. Sustainability innovations are thus often associated with competitive disadvantages (Nidumolu et al. 2015). In addition, in order to be able to develop and establish themselves, sustainability innovations usually require "second–order" innovations; they "challenge" conventional paradigms, routines and institutions (Knickel et al. 2009), and eventually reconfigure regimes (Geels and Schot 2007) where technical and economic drivers have proven inadequate (Hegger et al. 2007). Moreover, the development and enforcement (including adoption and impact) of innovations is seen as a social process (Currie et al. 2005) and as a result of co-evolutionary development and learning processes that involve different stakeholder groups and bring together knowledge resources (e.g. Ingram et al. 2015; Kemp et al. 2009; Schot and Geels 2008).

Most of the projects then addressed these conceptual considerations by designing innovation processes with the participation of various stakeholder groups. Different approaches were used in the nine research projects (see Table 8.1). Examples include the "real-world" or "living lab" approach, the "open innovation" concept, the method of "design thinking". Transdisciplinary processes were also conducted, as well as very incremental management approaches that can be described as "muddling through". Combinations of several approaches, such as a transdisciplinary learning process with a real-world lab approach (see also Rogga et al. 2018) were often observed.

Such combinations appear to be pertinent: while transdisciplinary processes enable the joint definition of problems in the consensual discourse, the integration of different stakeholder perspectives, the elaboration of objectives and the identification of suitable innovations, the other methods and approaches enable these solutions and the corresponding transformation knowledge to be tested and experimented with.

8.4.4 Leverage Points in the Governance System of Land (Use)

As stated above, we consider SLM to be a concept of change and transformation. In this context, the term "management" refers to activities and interventions that seek to bring about a shift towards more sustainable land use. Land (use), however, is a highly regulated subject/field with several intervention levels, as shown in Fig. 8.2 (adapted after Hurni 1997).

With regard to the nine projects under investigation, we found that most projects comprised interventions and developed innovations at levels between the land plot (e.g. technological innovations such as agrophotovoltaic plants, agroforestry systems) and the community (e.g. collaborative landscape management, integrated spatial energy concepts). However, we also found that the projects coordinators mentioned legal conditions as main barriers to innovation processes. Thus, it may be worthwhile considering TDR projects that also involve actors from higher intervention levels in the future.

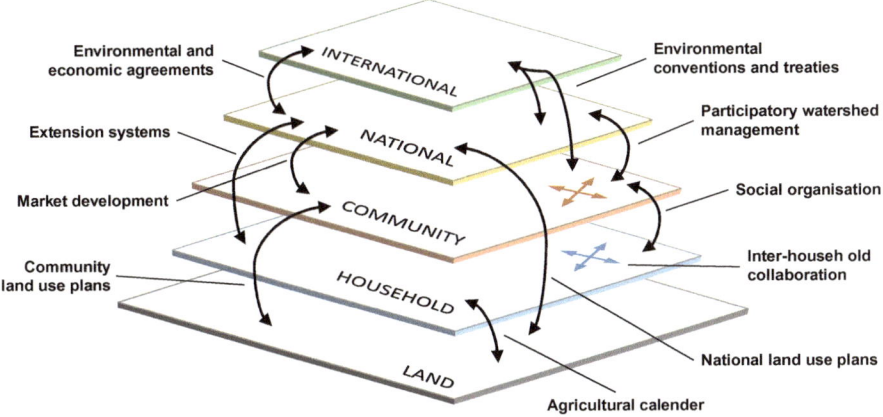

Fig. 8.2   Intervention levels for sustainable land management (adapted after Hurni 1997, 2000)

8.5   Discussion and Conclusion: Three Theses on the Specifics of SLM Innovations

The aim of this article was to identify the specifics of SLM innovations and the underlying innovation processes (see Table 8.1). The comparative analysis of nine transdisciplinary research projects led us to formulate three theses on the specifics of SLM innovations, which are presented and discussed below.

Thesis #1
SLM is a New and Complementary Form of Governance and Normative Orientation

A closer look at the solutions developed in the nine projects under examination shows that SLM and the underlying innovations can be regarded as a strategy to counterbalance the disadvantages and shortcomings of land use that is managed sectorally and organised by the government. Thus, we consider SLM to be a "new" and complementary form of governance that focuses on innovation processes and follows an adaptive co-management approach (Armitage et al. 2009) involving science and practice to iteratively design and test solutions for the more sustainable use of natural resources and land.

This thesis is supported by the types of innovation evolving from the nine projects under examination. Although two of the nine projects focused on technological innovations (such as agroforestry systems and agrophotovoltaic plants), most of the projects centred around social innovations. These innovations included new collaborations and networks between actors from different sectors; the development of new institutions such as the establishment of a coordinating authority (e.g.,"land use managers") or of tools for supporting well-informed and justified decisions in situations of great uncertainty and complexity.

Although other scholars have highlighted the integrative nature of SLM (e.g. Nölting and Mann 2018; Wang and Aenis 2019), comprising the components of technology, policy and land use planning (Hurni 2000), the interaction of different types of innovation and the specific role played by social innovations in SLM has not yet been considered. We argue that taking a deeper look at these conditions would enhance our understanding of sustainability transformations.

Thesis #2
SLM Innovation Processes Follow a Multiple-Objective Strategy and Combine Transdisciplinary Processes with Innovation Management Approaches

Most projects followed a two-fold approach and combined a number of partial solutions (local niche innovations) with a "regional" integration concept. As a consequence, transdisciplinary processes were frequently combined with other innovation management approaches such as real-world labs or the open innovation approach. Thus, transdisciplinary processes were applied to integrate different types of knowledge and perspectives from a wider sphere of actor groups, especially when framing and jointly defining the problem or initial situation (systems knowledge), but also when supporting the development of visions and targeted sustainable futures (target knowledge). This corresponds with the ideas of many scholars who regard transdisciplinarity as a valuable approach to provide "socially robust orientation" for sustainable development (Nowotny et al. 2001). However, it can be argued that complementary approaches that allow the co-design and testing of innovations (transformative knowledge) within this jointly defined framework are required for the development of concrete solutions and interventions.

The combination of transdisciplinary processes with real-world lab approaches has already been observed in other studies (Rogga et al. 2018). It has been shown that embedding the innovation process in a broader transdisciplinary discourse is conductive to estimating unintended effects more effectively, and reducing ethical concerns, but also to considering critical questions of legitimacy.

Thesis #3
Innovations for Sustainable Land Management Are Neither Disruptive nor Radical

The literature on innovation studies often differentiates between incremental and disruptive or "radical" innovations. This differentiation refers to how an innovation influences and challenges the existing system and stimulates follow-up innovations.

Accordingly, radical innovations are innovations that have a "high degree of novelty, being totally or substantially new" and a "profound effect on future development, establishing whole new fields of study" as well as new technological systems, "making dominant rival technologies or processes obsolete" (Kasmire et al. 2012). These innovations are frequently associated with eureka moments, ingenious ideas and flashes of inspiration (e.g. van de Poel 2003).

In contrast, there are "incremental innovations" that are far less ground-breaking, new and original, and more likely to result in changes and modifications in the form of incremental improvements to existing innovations and technological systems.

This differentiation according to the novelty and impact of innovations appears to be rather questionable, because assessing the novelty of an innovation is a very subjective and vague matter. Kasmire et al. (2012, p. 348) argue: "when examined critically, the birth process of many "radical" innovations reveals only logical, even obvious, small steps with no "eureka" moments".

With regard to the projects examined here, it emerged that SLM innovations were neither "sensationally new" nor radical or disruptive. Instead, they can be considered as incremental adaptations towards new normative objectives or paradigms such as the Sustainable Development Goals (SDGs) or growing awareness of the importance of a systems perspective. SLM innovations in the nine projects under investigation were often locally tailored solutions that were mainly based on existing ideas, and were now being implemented.

The extent to which the concept of "radical" innovations applies to SLM innovations is questionable because they often revolve around social innovations such as new forms of coordination and collaboration. However, the concept of "radical" innovations was derived from and is related to the retrospective study of technological innovations (ibid.). Nonetheless, the concept of "radical" innovations seems hardly applicable to SLM innovations, also with regard to the impact of such innovations.

Although some of the projects also focused on technical/product innovations (e.g. hay ovens, biomillers, agrophotovoltaic systems, agroforestry systems), these innovations were closely tied to the well-regulated resource of "land".

For example, agroforestry systems with short-rotation wood strips for energy wood production have been researched and developed in Germany for over 25 years. However, the implementation and dissemination of such systems in practice has so far failed due to the legal framework. In this case, minimal changes (also in terms of subsidy backdrops) may cause considerable land use changes and, in retrospect, appear to be "radical". However, it is unlikely that agroforestry systems will completely replace other agricultural systems. It is more likely that such systems will complement other systems, where appropriate.

The strict regulation of land is not only a strong barrier to SLM innovation. It is very likely that it will influence the innovation process from the very beginning, excluding legally unworkable solutions from the outset. From a critical point of view, the innovations developed within the nine projects under examination are mainly based on the existing system. The extent to which these innovations challenge current regime structures is an open question.

References

Armitage, D. R., Plummer, R., Berkes, F., Arthur, R. I., Charles, A. T., Davidson-Hunt, I. J., & McConney, P. (2009). Adaptive co-management for social–ecological complexity. *Frontiers in Ecology and the Environment, 7*(2), 95–102.

Baregheh, A., Rowley, J., & Sambrook, S. (2009). Towards a multidisciplinary definition of innovation. *Management Decision, 47*(8), 1323–1339.

Blättel-Mink, B. (2006). *Kompendium der Innovationsforschung*. Wiesbaden: VS Verlag für Sozialwissenschaften.

Brooks, H. (1982). Social and technological innovation. In *Managing Innovation* (pp. 1–30). Pergamon Press: New York, NY, USA.

Currie, M., King, G., Rosenbaum, P., Law, M., Kertoy, M., & Specht, J. (2005). A model of impacts of research partnerships in health and social services. *Evaluation and Program Planning, 28*(4), 400–412.

Doloreux, D., & Parto, S. (2005). Regional innovation systems: Current discourse and unresolved issues. *Technology in Society, 27*(2), 133–153.

Edquist, C. (Ed). (1997). Systems of innovation. Technologies, institutions and Organizations, Pinter, London.

EEA. (2019). *Land and Soil in Europe*. Signals, European Union: Why we need to use these vital and finite resources sustainably.

Enkel E. (2009). Chancen und Risiken von Open Innovation. In Zerfaß, A. & Möslein, K.M. (Eds.) Kommunikation als Erfolgsfaktor im Innovationsmanagement. Gabler.

Fritz-Vietta, N. V., Tahirindraza, H. S., & Stoll-Kleemann, S. (2017). Local people's knowledge with regard to land use activities in southwest Madagascar-Conceptual insights for sustainable land management. *Journal of Environmental Management, 199,* 126–138.

Geels, F. W. (2002). Technological transitions as evolutionary reconfiguration processes: A multi-level perspective and a case-study. *Research policy, 31*(8–9), 1257–1274.

Geels, F. W., & Schot, J. (2007). Typology of sociotechnical transition pathways. *Research Policy, 36*(3), 399–417.

Haberl, H., Mbow, C., Deng, X., Irwin, E. G., Kerr, S., Kuemmerle, T., et al. (2014). Finite land resources and competition. *Rethinking Global Land Use in an Urban Era, 14,* 35–69.

Hauschildt, J., Salomo, S., Schultz, C. & Kock, A. (2016). *Innovationsmanagement*. Vahlen.

Hegger, D. L., van Vliet, J., & van Vliet, B. J. (2007). Niche Management and its Contribution to Regime Change. The Case of Innovation in Sanitation. In: *Technology Analysis & Strategic Management, 19* (6), (pp. 729–746). https://doi.org/10.1080/09537320701711215.

Heideloff, F., & Radel, T. (1998). Innovation in Organisationen–ein Eindruck vom Stand der Forschung. *Organisation von Innovation: Strukturen, Prozesse* (pp. 7–39). Interventionen. München: Hampp.

Howaldt, J., & Schwarz, M. (2010). Soziale Innovation–Konzepte, Forschungsfelder und-perspektiven. In *Soziale Innovation* (pp. 87–108). VS Verlag für Sozialwissenschaften.

Howells, J. R. (2002). Tacit knowledge, innovation and economic geography. *Urban Studies, 39*(5–6), 871–884.

Hurni, H. (1997). Concepts of sustainable land management. *ITC Journal*, 210–215.

Hurni, H. (2000). Assessing sustainable land management (SLM). *Agriculture, Ecosystems & environment, 81*(2), 83–92.

Ingram, J., Maye, D., Kirwan, J., Curry, N., & Kubinakova, K. (2015). Interactions between niche and regime. An analysis of learning and innovation networks for sustainable agriculture across Europe. In *The Journal of Agricultural Education and Extension 21*(1), (pp. 55–71).

de Jong, S. P. L., Wardenaar, T., & Horlings, E. (2016). Exploring the promises of transdisciplinary research: A quantitative study of two climate research programmes. *Research Policy, 45*(7), 1397–1409.

Kasmire, J., Korhonen, J. M., & Nikolic, I. (2012). How radical is a radical innovation? An outline for a computational approach. *Energy Procedia, 20,* 346–353.

Kemp, R., Loorbach, D., & Rotmans, J. (2009). Transition management as a model for managing processes of co-evolution towards sustainable development. *International Journal of Sustainable Development & World Ecology, 14*(1), 78–91. https://doi.org/10.1080/13504500709469709.

Knickel, K., Brunori, G., Rand, S. & Proost, J. (2009). Towards a Better Conceptual Framework for Innovation Processes in Agriculture and Rural Development. From Linear Models to Systemic Approaches. In *The Journal of Agricultural Education and Extension 15*(2), (pp. 131–146).

Kramer, M. (Ed.). (2010). *Integratives Umweltmanagement: Systemorientierte Zusammenhänge zwischen Politik, Recht.* Management und Technik: Springer-Verlag.

Kristof, K. (2010). *Models of Change: Einführung und Verbreitung sozialer Innovationen und gesellschaftlicher Veränderungen in transdisziplinärer Perspektive.* VDF Hochschulverlag AG.

Mander, Ü, Helming, K., & Wiggering, H. (2007). Multifunctional land use: meeting future demands for landscape goods and services. *Multifunctional Land Use* (pp. 1–13). Berlin, Heidelberg: Springer.

Meyer, R. & Priefe, C. (2015). Energiepflanzen und Flächenkonkurrenz: Indizien und Unsicherheiten. In: GAIA24/2 (2015), 108–118.

Newig, J., Jahn, S., Lang, D. J., Kehle, J. & Bergmann, M. (2019). Linking modes of research to their scientific and societal outcomes. Evidence from 81 sustainability-oriented research projects. In *Environmental Science and Policy 101* (2019), (pp 147–155). https://doi.org/10.1016/j.envsci.2019.08.008

Nidumolu, R., Prahalad, C. K., & Rangaswami, M. R. (2015). Why sustainability is now the key driver of innovation. *IEEE Engineering Management Review, 43*(2), 85–91.

Niewöhner, J., Nielsen, J. Ø, Gasparri, I., Gou, Y., Hauge, M., Joshi, N., & Shughrue, C. (2016). Conceptualizing distal drivers in land use competition. *Land Use Competition* (pp. 21–40). Cham: Springer.

Nölting, B., & Mann, C. (2018). Governance strategy for sustainable land management and water reuse: Challenges for transdisciplinary research. *Sustainable Development, 26*(6), 691–700.

Nowotny, H., Scott, P., & Gibbons, M. (2001). *Rethinking science. Knowledge and the public in the age of uncertainty.* Polity Press: Cambridge.

Reed, M., Evely, A. C., Cundill, G., Fazey, I. R. A., Glass, J., Laing, A., & Stringer, L. (2010). What is social learning? *Ecology and Society.*

Rogers, E. M. (2003). The Diffusion of Innovations, Free Press Trade Paperback Ed, 5th edit.; Free Press: New York, NY, USA.

Rogga, S., Zscheischler, J., & Gaasch, N. (2018). How Much of the Real-World Laboratory Is Hidden in Current Transdisciplinary Research? *GAIA-Ecological Perspectives for Science and Society, 27*(1), 18–22.

Sauer. (1999). Perspektiven sozialwissenschaftlicher Innovationsforschung—Eine Einleitung. In: Sauer & Lang, 9–22.

Schot, J., & Geels, F. W. (2008). Strategic niche management and sustainable innovation journeys: theory, findings, research agenda, and policy. *Technology Analysis & Strategic Management, 20*(5), 537–554.

Schubert, C. (2016). Soziale innovationen. In: Innovationsgesellschaft Heute; Springer: Wiesbaden, Germany, 403–426.

Schwilch, G., Bachmann, F., Valente, S., Coelho, C., Moreira, J., Laouina, A., & Reed, M. S. (2012). A structured multi-stakeholder learning process for Sustainable Land Management. *Journal of Environmental Management, 107,* 52–63.

Shove, E., Pantzar, M. & Watson, M. (2012). *The dynamics of social practice: Everyday life and how it changes.* Sage.

Sovacool, B. K., & Hess, D. J. (2017). Ordering theories: Typologies and conceptual frameworks for sociotechnical change. *Social Studies of Science, 47*(5), 703–750.

Taylor, J. B. (1970). Introducing social innovation. *J. Appl. Behav. Sci., 6,* 69–77.

Thornhill, S. (2006). Knowledge, innovation and firm performance in high- and low-technology regimes. *Journal of Business Venturing, 21*(5), 687–703.

Van de Poel, I. (2003). The transformation of technological regimes. *Research Policy, 32*(1), 49–68.

Wang J. & Aenis, T. (2019). Stakeholder analysis in support of sustainable land management: Experiences from southwest China. *Journal of Environmental Management 243*(1), 1–11. https://doi.org/10.1016/j.jenvman.2019.05.007.

Weith, T., Besendörfer, C., Gaasch, N., Kaiser, D.B., Müller, K., Repp, A., et al. (2013). Nachhaltiges Landmanagement: Was ist das? Diskussionspapier Nr. 7 des Wissenschaftlichen Begleitvorhabens (Modul B).

Zapf, W. (1989). Über Soziale Innovationen. *Soziale Welt, 40,* 170–183.

Zscheischler, J., Busse, M. & Heitepriem, N. (2019). Challenges to Build up a Collaborative Landscape Management (CLM)—Lessons from a Stakeholder Analysis in Germany. *Environmental Management*, 1–13.

Chapter 9
Knowledge Exchange at Science-Policy Interfaces in the Fields of Spatial Planning, Land Use and Soil Management: A Swiss Case Study

Marco Pütz and Regula Brassel

Abstract In this article, we investigate knowledge exchange at the intersection of science and Swiss public policy in the fields of spatial planning, land use and soil management. Based on a literature review and expert interviews, we identify six types of knowledge exchange, and examine the barriers to and opportunities for knowledge exchange. These six underlying concepts suggest knowledge exchange is a challenging task because different expectations exist on how knowledge should be exchanged.

Keywords Knowledge exchange · Science-policy interface · Spatial planning · Land use · Switzerland

9.1 Introduction

Over the last few decades, the way in which science has been considered to be an instrument to inform policymakers has changed. Science addressing environmental policy issues often has to deal with endemic uncertainties, conflicting values and different goals among actors. This has challenged traditional science, which has been widely understood as a provider of value-free and definite factual knowledge (Funtowicz and Ravetz 1993: 739, 744, 749 et seq.; Ravetz 1999: 647–650). Observing this development, Funtowicz and Ravetz (1993) introduced the concept of "post-normal" science—a "problem-solving strategy [for policy issues], where systems uncertainties or decision stakes are high" (Funtowicz and Ravetz 1993: 749). Systems uncertainties refer to problems that cannot be solved simply by detecting a specific fact, but where a complex reality has to be understood and managed. Decision stakes encompass the diverse interests of different actors concerning an issue (Funtowicz and Ravetz 1993: 744). To deal with these system uncertainties and decision stakes, post-normal science implies that it is not only researchers or

M. Pütz (✉) · R. Brassel
Wirtschafts- und Sozialwissenschaften, Regionalökonomie und –entwicklung, Eidgenössische Forschungsanstalt für Wald, Schnee und Landschaft (WSL), Zürcherstrasse 111, CH-8903 Birmensdorf, Switzerland
e-mail: marco.puetz@wsl.ch

© The Author(s) 2021
T. Weith et al. (eds.), *Sustainable Land Management in a European Context*,
Human-Environment Interactions 8, https://doi.org/10.1007/978-3-030-50841-8_9

165

official experts who should debate the quality of scientific policy inputs, but indeed all the actors affected by the issue and interested in contributing to a solution should be included in the discussion (Funtowicz and Ravetz 1993: 752 et seq.; Ravetz 1999: 651). Gibbons et al. (1994: 1) also observed that the modes of knowledge production had been changing. From their point of view, knowledge is no longer solely produced disciplinarily and in a context of mainly scientific interests (Mode 1 knowledge production), but also transdisciplinarily and in "a context of application" (Mode 2 knowledge production) (Gibbons et al. 1994: 3–5; Zscheischler et al. 2018). However, there has been a shift not only in the way knowledge is produced and who is involved in this process, but also in the way knowledge is exchanged (Bielak, Campbell, Pope, Schaefer, and Shaxson, 2008): referring to the findings of Funtowicz and Ravetz (1993), Gibbons et al. (1994) and Pretty and Chambers (1993), Bielak et al. (2008: 202 et seq.) asserted that

> [i]t is no longer tenable to rely on the notion of a linear progression through an orderly research process driven by scientists, to a dissemination phase driven by communication specialists, to an adoption phase in which end users (whether in policy or management) presumably apply research findings directly in their everyday activities. Rather, science must be socially distributed, application-oriented, transdisciplinary, and subject to multiple accountabilities. From a one-way linear process, science is evolving to a multi-party, recursive dialogue.

The positivist perspective of "knowledge transfer", where knowledge is understood as something that can simply be handed over to other individuals in a one-way exchange process, has been complemented by other (more subjectivist) perspectives (Rogga et al. 2014). Subjectivist perspectives take into account the idea that different kinds of knowledge exist, which are individually and socially constructed (Fazey et al. 2014: 206). Knowledge exchange arising from such a perspective "tend[s] to result in knowledge exchange activities that encourage mutual learning through multi-stakeholder interactions" (Fazey et al. 2014: 206), which is exactly what Bielak et al. (2008: 202 et seq.) postulated. Therefore, today various definitions of knowledge exchange and a broad variety of different terms with diverging underlying assumptions exist including "knowledge sharing, generation, coproduction, comanagement; transfer, brokerage, storage, exchange, transformation, mobilization, and translation" (Fazey et al. 2013: 20; see also Mauser et al. 2013). In this article, we understand knowledge exchange according to Fazey et al. (2013: 20) "as a process of generating, sharing, and/or using knowledge through various methods appropriate to the context, purpose, and participants involved." However, knowledge exchange does not always operate to the satisfaction of all the actors involved. Recently, various scholars have begun to discuss the challenges of knowledge exchange at the intersection of science and public policy (which we refer throughout to as "science/policy interfaces") (Böcher and Krott 2014; Saarela and Söderman 2015; van Enst et al. 2014) and how they can be improved (Böcher and Krott 2014; Saarela and Söderman 2015). Others have focused on how knowledge exchange at science/policy interfaces may be implemented most effectively (Reed et al. 2014).

In the following article, we investigate the different types of knowledge exchange that actors in Switzerland have adopted at the intersection of science and public policy (which we refer to as "science/policy interfaces") in the areas of spatial planning, land use and soil management, and use our investigations to develop a typology of knowledge exchange. Furthermore, we examine the barriers to and opportunities for knowledge exchange in Swiss spatial planning, land use and soil management. The goal of this article is to better understand how knowledge is exchanged, and to find out what impedes and enhances knowledge exchange, drawing on Switzerland as a case study. We assess in what respects the findings from Switzerland have been represented in previous literature, i.e. which concepts of knowledge exchange our empirically developed typology refers to. By looking at the concepts that underlie the types of knowledge exchange actors adopt, it becomes clear why knowledge exchange is a challenging task: differing concepts of knowledge exchange result in different expectations on how knowledge is to be exchanged.

We discuss the following research questions

1. Which types of knowledge exchange do actors adopt at science/policy interfaces in spatial planning, land use and soil management in Switzerland?
2. Which concepts of knowledge exchange do these different types of knowledge exchange refer to?
3. How do actors at the science/policy interfaces in the fields of spatial planning, land use and soil management in Switzerland assess the barriers to and opportunities for knowledge exchange?

In Switzerland, the land and the soil are resources that have come under increasing pressure. Construction and urban sprawl threaten agricultural land, biodiversity and the landscape. A growing population, high mobility (and the resulting consumption of land by transport infrastructure) and growing demand for more per-capita living space have all helped drive this process (Schweizerischer Bundesrat, KdK, BPUK, SSV and SGV, 2012: 1, 4, 6). We understand pressure on the land and the soil to be a typical problem for post-normal science in the sense of Funtowicz and Ravetz (1993): it is an environmental policy issue with high systems uncertainties and high decision stakes. While population growth, mobility and the demand for more personal space have been expressed as reasons for the consumption of land and soil, the role of other drivers (such as zoning policy, tax policy and capital markets) is not fully understood, to say nothing of the interdependencies among the various drivers (Brils et al. 2016: 792; Plieninger et al. 2016; Schweizerischer Bundesrat et al. 2012: 1, 4, 6). Moreover, spatial planning, land use and soil management all epitomise the presence of diverse interests, which are introduced into each issue by a variety of actors. These include the construction industry, real estate companies, private landowners, farmers and conservationists (high-decision stakes). We therefore presume that, in their attempts to find solutions in a complex context like this, actors adopt various types of knowledge exchange, which encompass not only the positivist perspective of "knowledge transfer", but also other types of knowledge exchange that go beyond that.

9.2 Conceptualising "Knowledge Transfer" and "Knowledge Exchange"

The scholarly debate on knowledge exchange in science/policy interaction has been unfolding for quite some time. According to Fazey et al. (2013: 20), knowledge exchange is "a process of generating, sharing, and/or using knowledge through various methods appropriate to the context, purpose, and participants involved." Accordingly, this definition implies two basic understandings of knowledge exchange: (a) knowledge exchange in a wider sense, as an overarching concept encompassing a variety of subjacent concepts (e.g. knowledge transfer, the coproduction of knowledge); (b) knowledge exchange as a specific concept, incorporating "a two- or multiple-path process with reciprocity and mutual benefits, maybe with multiple learning, but not necessarily recognition of the equitable value of the different forms of knowledge being exchanged" (Fazey et al. 2013: 20).

In line with these basic understandings by Fazey et al. (2013), other scholars provide similar overviews of knowledge exchange theories and concepts. Nutley et al. (2014) distinguished rational-linear models, context-focused models, interactive models and post-modern models of practice-policy interaction. Stewart et al. (2014) differentiated among "knowledge transfer", "knowledge exchange" and "knowledge interaction". They regarded "knowledge transfer" to be linear models of knowledge uptake, dominating between the 1960 and 1990s, which required the effective packaging and communication of knowledge. The concept of "knowledge exchange", which emerged in the 1990s, was seen as the result of social and political processes, and required effective relationships. Finally, "knowledge interaction", which has emerged in more recent years, has been regarded as embedded in systems and cultures, and which requires effective integration within organisations. Kamelarczyk and Gamborg (2014) distinguished between knowledge transfer models and knowledge interaction models: for them, knowledge transfer was characterised as a linear and one-way model, clearly separating the worlds of science and policymaking, and working under the assumption that knowledge was produced as a clearly defined, ready-to-use product. In contrast, knowledge transaction sees science and policymaking as an undivided whole, and characterised by blurry boundaries. Here, knowledge is coproduced in cycles. Kamelarczyk's and Gamborg's (2014) concept of knowledge transaction thus corresponds to the concepts of knowledge exchange and knowledge interaction proposed by Stewart et al. (2014).

A review of the literature on knowledge transfer and knowledge exchange in science/policy or science/practice interaction reveals that many scholars agree on these two basic concepts of "knowledge transfer" and "knowledge exchange" (Fazey et al. 2014). Firstly, there is a consensus that concepts of knowledge transfer that have emerged since the 1950s (e.g. Lasswell 1956) have understood transfer as a linear, one-way or unidirectional activity of knowledge production, from science to policy or practice (see Stone 2001; Pregering 2004; Birkland 2005; Böcher and Krott 2014; Kamelarczyk and Gamborg 2014; Stewart et al. 2014; Linke et al. 2014). Second, there is also a consensus that knowledge transfer may be complemented or substituted by more recent concepts of knowledge exchange or related terms, such as knowledge interaction or transaction (see Stone 2001; Keeley and Scoones 2003; Jasanoff 2004; Roux et al. 2006; Turnout et al. 2007; Kamelarczyk and Gamborg 2014; Böcher and Krott 2014; Prager and McKee 2015).

Theories and concepts of knowledge transfer and exchange draw on wider debates on science/policy or science/practice interaction. Following Sybille van den Hove, we define science/policy interaction

> *as social processes which encompass relations between scientists and other actors in the policy process, and which allow for exchanges, co-evolution, and joint construction of knowledge with the aim of enriching decision-making.* (van den Hove, 2007: 807).

In fact, van Enst et al. (2014: 13–16) used this same definition to develop a typology of science/policy interaction, specifying different ways of handling knowledge.

1. The first type, "individual science-policy mediation", focuses on individual scientists and experts that link science and policymaking by mediating between the two groups and thus help to make knowledge available and utilisable.
2. The second type concentrates on the process of "participatory knowledge development", in particular transdisciplinary and participatory coproduction of knowledge and common understandings. It includes all actors—from scientists, policymakers and other professionals to laypeople (van Enst et al. 2014: 16).
3. The third type includes "boundary organisations", which van Enst et al. described as "formal institutions, often having a legal basis, which serve as an institutional bridge between the worlds of science and policy" (van Enst et al. 2014: ibid.).

For our article, we interviewed actors representing one or more of these three types of science/policy interfaces in an effort to demonstrate examples of who can provide knowledge and who needs it.

9.3 Methods

We conducted 16 qualitative expert interviews with actors representing science/policy interfaces in Switzerland, including individual science/policy mediators and boundary organisations. The interviews, which took between 50 minutes and 2 hours, were generally conducted at the actors' work. We recorded these interviews, took minutes, and prepared transcripts. In the questionnaire, we asked the actors to describe how they exchange knowledge themselves, and how knowledge is exchanged in spatial planning, land use and soil management in Switzerland more generally. We thus used "knowledge exchange" as a superordinate concept. Insofar as this concept characterises a two- or multiple-path exchange process, we made clear to the interviewees that knowledge exchange might be understood in many different ways, and thus also asked what other kinds of knowledge exchange they had experienced.

To analyse the transcribed interviews, we performed a qualitative content analysis following Mayring's method (2015). We coded the data with MaxQDA coding software. We started the coding process with concepts from the literature that we expected to appear in the data ("knowledge transfer", "knowledge exchange" or "coproduction of knowledge"). Over the course of the coding process, we complemented this first set of codes with inductively created new codes. Drawing from the coded interviews, we identified barriers and opportunities, and developed a typology of knowledge exchange, including knowledge transfer. We complemented this typology with elements that had not been explicitly expressed in the interviews, but that represented patterns that had appeared in the data. In the following sections, we do not present the interview results in detail. Rather, we focus on the summarised results to present a typology of knowledge exchange, and an overview of the barriers to and opportunities for knowledge exchange.

9.4   A Typology of Knowledge Exchange in Spatial Planning, Land Use and Soil Management in Switzerland

Our analysis of how actors exchange knowledge in Swiss contexts of spatial planning, land use and soil management resulted in a typology of knowledge exchange. Based on the interviews, we were able to identify six types: (1) knowledge transfer, (2) knowledge transfer support, (3) knowledge exchange, (4) knowledge exchange support, (5) participatory knowledge development and use, (6) formal and informal knowledge exchange. These six types exhibit a total of 21 subtypes, characterising different ways of exchanging knowledge (Fig. 9.1).

In the typology, we have distinguished between transfer and exchange processes. We understand knowledge transfer (Type 1) as one-way processes, where knowledge is imparted from one actor to another actor unidirectionally. In contrast, we define knowledge exchange (Type 3) as a reciprocal process operating in both directions. We further differentiate between "knowledge transfer" and "knowledge transfer support" as well as between "knowledge exchange" and "knowledge exchange support". While knowledge transfer as such characterises processes where scientists and policymakers transfer knowledge they themselves have produced or possessed, knowledge transfer support (Type 2) describes processes where boundary organisations or individual knowledge advocates enable the transfer between science and policymaking. Similarly, we differentiate between knowledge exchange (Type 3) and knowledge exchange support (Type 4). Moreover, in some of the subtypes, we then distinguish between direct and indirect processes. Direct knowledge exchange takes place when actors are in direct contact with other actors. This direct contact may also be organised by either a boundary organisation or an individual knowledge advocate, or both. "Indirect knowledge exchange" characterises processes in which actors do not meet or communicate directly. "Participatory knowledge development

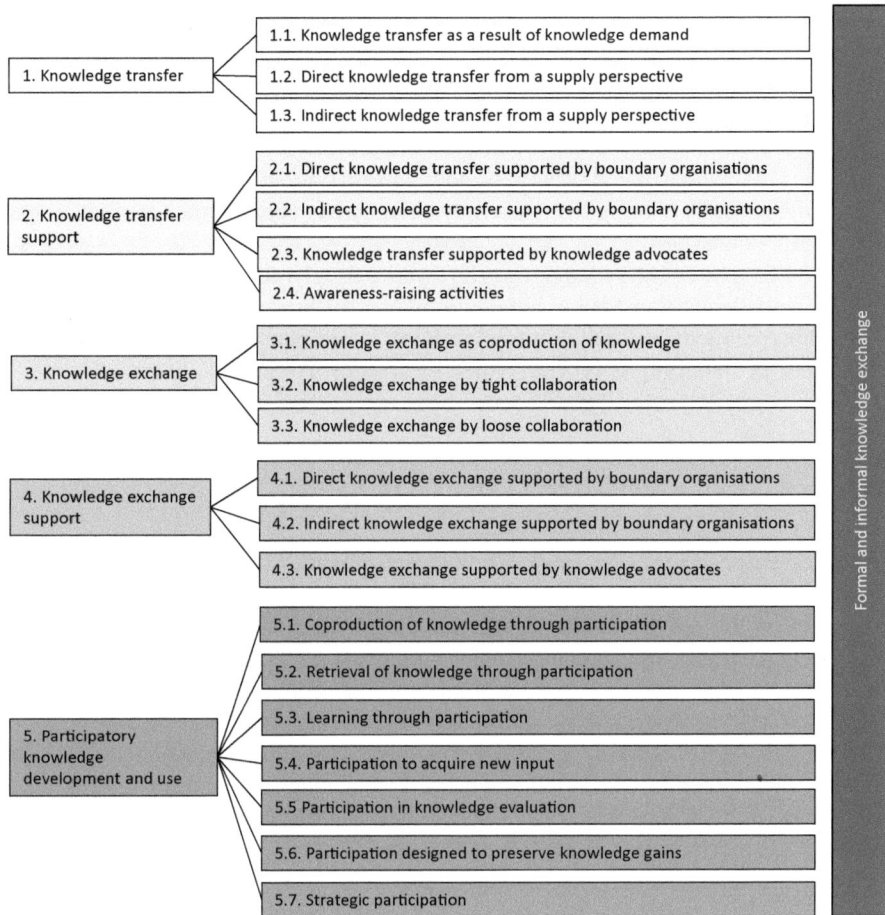

Fig. 9.1. Types and subtypes of knowledge exchange

and use" (Type 5) characterises how and why actors are involved in knowledge exchange processes. "Formal and informal knowledge exchange (Type 6) describes the formality of knowledge exchange processes. The following sections present these types and subtypes in greater detail.

9.4.1 Type 1: Knowledge Transfer

Knowledge exchange Type 1, "knowledge transfer", refers to knowledge exchange processes that are only one-way. We distinguish among three subtypes:

Type 1.1: Knowledge transfer as a result of knowledge demand. This describes knowledge transfer where either policymakers or scientists express a knowledge demand that can be met by a one-way transfer process. Knowledge demand, for example, occurs when policymakers approach scientists with a question the researchers can answer promptly (e.g. on the phone).

Type 1.2: Direct knowledge transfer from a supply perspective. This embraces knowledge transfer by researchers who disseminate their results not only via scientific journals, but in ways that also address policymakers directly. We understand this as knowledge transfer coming from a supply perspective, because the scientists impart their findings unasked.

Type 1.3: Indirect knowledge transfer from a supply perspective. This describes indirect knowledge transfer via publications that scientists or policymakers supply. This subtype includes, e.g. popular articles in newspapers, the trade press or factsheets, or scientific reports published by scientists or policymakers.

9.4.2 Type 2: Knowledge Transfer Support

The knowledge exchange type of "knowledge transfer support" is characterised by one-way transfer processes between science and policymaking enabled by boundary organisations or individual knowledge advocates. We established four subtypes.

Type 2.1: Direct knowledge transfer supported by boundary organisations. In this type, scientists and policymakers are in direct contact; knowledge transfer is supported, e.g. by universities within final presentations of research projects.

Type 2.2: Indirect knowledge transfer supported by boundary organisations. This indicates that scientists and policymakers do not meet or communicate directly; knowledge transfer is enabled by boundary organisations that are involved in science dissemination activities.

Type 2.3: Knowledge transfer supported by knowledge advocates. Here, individual advocates are understood to support knowledge transfer from science to policymaking or from policymaking to science in one-way processes, e.g. scientists who send publications directly to policymakers.

Type 2.4: Awareness-raising activities. These encompass events, excursions, marketing campaigns and publications to raise awareness regarding challenges in spatial planning, land use and soil management; they are organised by boundary organisations or individual knowledge advocates. This way, scientific and expert knowledge is transferred to policymakers and laypeople. This subtype shows features of all the other subtypes of "knowledge transfer support".

9.4.3 Type 3: Knowledge Exchange

"Knowledge exchange" describes knowledge exchange where dialogues or other interaction between actors from science and policymaking takes place. Here, knowledge is exchanged reciprocally. However, the kind of knowledge that science and policymaking exchange with each other are not necessarily the same. While science is understood to introduce scientific knowledge into the exchange process, policymakers impart knowledge on practical problems, on social networks or on administrative or legal processes. This type includes three subtypes:

Type 3.1: Knowledge exchange as coproduction of knowledge. This subtype is characterised by knowledge exchange between policymakers and scientists, where scientists carry out a research mandate for public authorities and subsequently coproduce knowledge. This does not mean working together constantly, but it does mean developing it in an iterative, collaborative way.

Type 3.2: Knowledge exchange by tight collaboration. Like Subtype 3.1, knowledge is exchanged between public authorities and scientists executing a research mandate. Here, policymakers only give input and feedback to accompany the research process closely.

Type 3.3: Knowledge exchange by loose collaboration. Again, like Subtypes 3.1 and 3.2, knowledge exchange takes place between public authorities and scientists executing a research mandate. Here, however, the contact between the two parties is only "loose". Policymakers provide feedback only occasionally, and do not follow the research process closely.

9.4.4   Type 4: Knowledge Exchange Support

The knowledge exchange type "knowledge exchange support" is characterised by reciprocal knowledge exchange between science and policymaking, but supported by boundary organisations or individual knowledge advocates. Once again, there are three subtypes.

Type 4.1: Direct knowledge exchange supported by boundary organisations. This is characterised by knowledge exchange between scientists and policymakers that occurs when boundary organisations arrange meetings, workshops, conferences, excursions and other events where actors meet directly.

Type 4.2: Indirect knowledge exchange supported by boundary organisations. This describes knowledge exchange processes where science and policymaking do not meet directly. Instead, knowledge is received from science and imparted to policymakers through boundary organisations.

Type 4.3: Knowledge exchange supported by knowledge advocates. In this subtype, individual knowledge advocates enable knowledge exchange between science and policymaking. This exchange is carried out, e.g. by interns in government agencies. Interns are often simultaneously students at a university, so that public authorities not only impart knowledge to interns, but also receive scientific knowledge back from them.

9.4.5 Type 5: Participatory Knowledge Development and Use

The "participatory knowledge development and use" knowledge exchange type describes actor participation in research projects and in policymaking. Who these actors are, how they are included in the process and for what purpose can vary. To show this variety more clearly, we have differentiated the subtypes of this knowledge exchange type into two groups: the first three subtypes focus on how participation is organised. The fourth to seventh subtypes describe the reasons participation is taking place, i.e. the purpose of introducing participation. Accordingly, the subtypes include two different aspects of participation (how and why) that overlap. There are seven subtypes in total:

Type 5.1: Coproduction of knowledge through participation. This is characterised by actor participation where knowledge is being coproduced by scientists, policymakers or actors from civil society. Coproduction means that actors can be involved, e.g. through stakeholder workshops, expert interviews or assessments. In transdisciplinary research projects, the coproduction of knowledge begins immediately, with different actors jointly defining the problem and setting project goals.

Type 5.2: Retrieval of knowledge through participation. This is focused on actor participation in workshops or surveys to retrieve knowledge. Here, participation does not imply a joint act of generating knowledge, but a demand-side act of requesting and receiving knowledge from actors.

Type 5.3: Learning through participation. This entails enabling individual learning or mutual learning between scientists and policymakers.

Type 5.4: Participation to acquire new input. The goal of this is to understand the variety of the actors' knowledge, opinions and interests; participating actors may be scientists or policymakers, as well as individuals from civil society.

Type 5.5: Participation in knowledge evaluation. This is characterised by actor participation with the goal of generating specific feedback on documents, prototypes, results, solutions or recommendations. Here, participation does not focus on acquiring new input, but on getting feedback and making evaluations.

Type 5.6: Participation designed to preserve knowledge gains. This concept refers to participation that incorporates crucial influential people who would need to be consulted before a policy could be applied.

Type 5.7: Strategic participation. "Strategic participation" refers to participation as a strategy to legitimise outcomes. Policymakers may ask for feedback by scientists, or scientists may be (formally) included in policymaking by public officials. Participation is only "pro forma", and designed to indicate that the actors have been involved, rather than actually exchanging knowledge as the primary goal.

9.4.6 Type 6: Formal and Informal Knowledge Exchange

The "formal and informal knowledge exchange" knowledge exchange type describes the formality of knowledge exchange. On the one hand, it describes in which setting (formal, informal) the other types of knowledge exchange in this typology take place. On the other hand, the subtypes can also represent themselves (e.g. informal knowledge exchange at conferences). It is important to note that we do not see "formal" and "informal" knowledge exchange as two completely separated types of knowledge exchange, but rather as extremes along a continuum. Knowledge exchange is a complex phenomenon, and often consists of multiple interactions between actors; for this reason, the formality of an exchange can change over time, e.g. informal exchange can lead to a project that is regulated by a formal agreement. This knowledge exchange type appears in two subtypes.

Type 6.1: Informal knowledge exchange. This occurs when actors exchange knowledge with other actors in an informal setting. This may include unofficial parts of meetings or conferences, as well as informal workshops organised by public agencies to stimulate creative and open exchange between policymakers and scientists in a confidential setting.

These networks are especially important before projects are formulated, when actors informally exchange information with their counterparts about problems they observe, about knowledge needs they have or about possible solutions they perceive.

The second subtype is 6.2: Formal knowledge exchange. This describes knowledge exchange that is embedded in formal structures and processes. For example, this subtype includes meetings organised by public agencies to gain insights on the opinions of different actors concerning a particular topic. Such meetings are designed in the formal style of a consultation, where the statements of the invited actors are written down and the course of the meeting is rather predetermined. However, this subtype also embraces formal calls for proposals and knowledge exchange that is regulated by agreements that define the tasks and duties of two contracting parties.

9.5 Barriers to and Opportunities for Knowledge Exchange in Spatial Planning, Land Use and Soil Management in Switzerland

Based on the interviews, we identified 11 "barriers" to and five "opportunities" for knowledge exchange (Tables 9.1 and 9.2). "Barriers" refer to factors that constrain knowledge exchange, while "opportunities" refer to factors that improve knowledge exchange. Basically, both barriers and opportunities are crucial factors influencing knowledge exchange. Obviously, barriers can turn into opportunities, and vice versa.

Table 9.1. Barriers to knowledge exchange

No	Barrier
1	The divide between science and policymaking
2	The different time frames of scientists and policymakers
3	Different "languages" of science and policymaking
4	A lack of science/policy interfaces
5	Difficult access to scientific knowledge
6	A lack of applicability of scientific findings
7	Information overload
8	Pressure to publish in scientific journals
9	A lack of financial resources and time
10	Conflicts of interest and power issues
11	A lack of links between soil and spatial planning experts

Table 9.2. Opportunities for knowledge exchange

No	Opportunity
1	Federalism and multilingualism
2	A small but well-connected expert community
3	The availability of informal and personal contacts
4	General openness to new developments
5	Motivation and a commitment to collaboration

9.5.1 Barriers to Knowledge Exchange

Some of the barriers to knowledge exchange that we identified can be classified into groups. One group of barriers refers to the interaction between science and policymaking (Barriers 1–4). A second group of barriers refers to issues around the use of knowledge (Barriers 4–8). Three other issues complement the list of barriers (Table 9.1).

Science and policymaking are often perceived as two distinct fields following different rules and demonstrating different characteristics (Barrier 1). They are seen as having a different "culture"—in terms of their ways of thinking, their methods, the knowledge they have and produce (scientific knowledge vs. practical knowledge), their spatial focus (international versus national), and/or their expectations regarding the outcomes of research. Scientists and policymakers also have different timeframes (Barrier 2). Science is perceived as long-term, policymaking as short-term.

Next, science and policymaking use different ways of expression (Barrier 3). Scientific knowledge has to be revised, edited and "translated" into language policymakers can understand and make use of. Moreover, the knowledge that has been produced needs to be calibrated for its target audience. Public agencies, consultants, farmers or civil society organisations all must be addressed differently, and need different degrees of detail. Policymakers look for solutions to practical problems, while scientists try to find a scientific gap in which they can perform their research and publish it.

To bridge the paradigmatic divide between science and policymaking, it is essential that scientists and policymakers have a common idea of the problem at hand, and that they exchange long enough to really understand what the other side knows and wants. To start with, science/policy interaction would immediately improve if the different expectations of science and policymaking regarding research were expressed and clarified, and if both sides would accept that these expectations will not always converge.

Finally, boundary organisations support science/policy interaction. They help bridge science/policymaking divides. However, it is obvious that a lack of science/policy interfaces represents a crucial barrier of knowledge exchange (Barrier 4). Clearly, this finding is strongly in line with the literature on transdisciplinarity (Zierhofer and Burger 2007; Zscheischler et al. 2018).

Knowledge exchange takes place when knowledge is used. Difficult access to scientific knowledge (Barrier 5), especially to research findings, results and guidance for practice hinders this usage of knowledge. For policymakers, researching knowledge can be an exhausting task. Furthermore, access to scientific publications and data is not always open access and free of charge.

However, even if knowledge is accessible, its usage is not guaranteed. Transferability, applicability and practicability of scientific knowledge is not self-evident (Barrier 6). Knowledge (e.g. a new method for land management) has to be tested and adapted to conditions in practice.

Another knowledge-specific barrier is the problem of selecting relevant information, and identifying new, innovative ideas in an age of information overload (Barrier 7).

Knowledge producers are often scientists. The reward system of scientists can in fact impede knowledge exchange at science/policy interfaces, because researchers are rewarded—and are mainly measured—by the frequency of their publications in high-ranking scientific journals (Barrier 8). This dissuades researchers from "wasting" time at the science/policy interface.

Afterall, it is obvious that knowledge exchange takes time and costs money. Accordingly, financial and time pressure in research projects result in fewer knowledge exchange activities organised by scientists or boundary organisations (Barrier 9).

Spatial planning, land use and soil management affect different sectors of the economy. Construction, tourism, agriculture and forestry, among others, all demand different uses for the land and the soil. Those with an interest in developing land could oppose those with an interest in preserving high-quality soil (Pütz 2011). Correspondingly, different interests make knowledge exchange difficult. Conflicts of interest and power issues might hinder the free flow of knowledge and its comprehensive exchange (Barrier 10). Related to the barrier concerning conflicting interests is the fact that experts in spatial planning, land use and soil management usually come from different disciplines and scientific communities as well as various fields of experience. This is especially true for the lack of links between soil and spatial planning experts, which typify the divide between natural sciences and social sciences.

9.5.2 Opportunities for Knowledge Exchange

Switzerland is a rather small country, with 8.3 million inhabitants. It is composed of 26 small cantons, within which are about 2,500 municipalities. Switzerland's federal structure and multilevel governance system, plus its four linguistic regions, result in a variety of ways to deal with spatial planning, land use and soil management. Correspondingly, a variety of different options exist to experiment and find distinct solutions. In light of both the smallness as well as the large number of entities, we identified five opportunities for knowledge exchange, summarised in Table 9.2.

Of course, the large number of small entities could potentially lead to either fragmentation or integration, and could thus be interpreted as either a barrier or an opportunity. According to the interviews, however, the advantages of the Swiss system and its practices outweigh the disadvantages. Therefore, we frame these factors as opportunities. The interviewees particularly noted the different linguistic areas and the necessity to coordinate across administrative levels and sectors as a challenge, but also as an opportunity. The German, the French and the Italian-speaking part of Switzerland have all taken the majority of their cues from concepts originating from the neighbouring countries with whom they share the same language, planning culture and ways of thinking.

9.6 Discussion

9.6.1 Types of Knowledge Exchange

Our typology of knowledge exchange shows that in Swiss spatial planning, land use and soil management actors exchange knowledge in six different ways. These six types do not always perfectly correspond to the concepts in the existing scholarly literature. The next few paragraphs highlight how our six types confirm or add to the debate.

Type 1, "knowledge transfer", describes one-way knowledge transfer processes. In the subtypes, we distinguished between knowledge transfer coming from either a demand or a supply perspective. This implies that knowledge is a thing that one actor has and another actor wants or lacks, and that can easily be conveyed from a holder to a receiver. This type clearly corresponds with Fazey et al. (2013: 20) who conceptualised knowledge transfer as a linear, one-way process, where knowledge is delivered and received, and is understood to be portable.

Type 2 (knowledge transfer support) and Type 4 (knowledge exchange support) characterise one-way transfer processes and reciprocal exchange processes, respectively. However, in these types of knowledge exchange, transfer and exchange are encouraged by boundary organisations or by individual knowledge advocates, rather than by the scientists and policymakers producing or possessing the knowledge. Boundary organisations and individual knowledge advocates supporting knowledge transfer and exchange in our typology correspond to the individual science/policy mediators and boundary organisations identified by van Enst et al. (2014: 13–16) in their typology of science/policy interaction, who conceptualised them as individuals or formal institutions operating as a bridge between science and policymaking (van Enst et al. 2014: 13–16). In our typology, individual knowledge advocates and boundary organisations also function as connecting elements between science and policymaking. However, we additionally distinguish between the knowledge transfer and the knowledge exchange that these third parties support.

Type 3, "knowledge exchange", characterises reciprocal exchange processes where interaction, dialogue or learning takes place. The exchanged knowledge does not necessarily have to be the same kind of knowledge: while scientists introduce scientific knowledge into the exchange, policymakers introduce practical knowledge and presumably also tactical knowledge for scientists (e.g. about funding opportunities). Fazey et al. (2013: 20) described knowledge exchange as reciprocal two-way or multiple-path processes, including learning, which is exactly how we conceptualise it.

Type 5, "participatory knowledge development and use", is characterised by the involvement of actors in policymaking processes. van Enst et al. (2014: 15) classified processes of participatory knowledge development as a third type in their typology of science-policy interfaces. However, they focused on participatory coproduction of knowledge, which only corresponds to one of our subtypes, 5.1—"the coproduction of knowledge through participation". We have demonstrated that participation can also be perceived in a more unidirectional way of simply retrieving knowledge (Subtype 5.2), where knowledge is not coproduced, but instead requested and received from actors; it may also be understood as mutual learning or individual learning within a collective body (Subtype 5.3). Apart from the Subtypes 5.1–5.3 that explain *how* participation operates, we also included four subtypes (5.4–5.7) that focus on *why* participation is organised. van Enst et al. (2014: 5–9) also discussed how knowledge is "strategically" used and produced. Our Subtype 5.7, "strategic participation" corresponds with that of van Enst et al. (2014) in terms of the strategic use of knowledge by policy of "knowledge being used selectively", which includes "politicians ask[ing] for advice only to legitimize their pre-formed decisions" (Hoppe 2005: 201 in van Enst et al. 2014: 7).

Type 6, "formal and informal knowledge exchange", focuses on the setting (formal, informal) in which knowledge exchange takes place. The Swiss respondents identified the informal component of knowledge exchange to be particularly important. Reed et al. (2013: 313) argued that "[k]nowledge exchange and transfer often take place through informal networks as well as through formalised and depersonalised forms of communication such as the mass media." This corresponds with our findings. However, they specified mass media as an example of formal knowledge exchange. Our data revealed other examples of formal knowledge exchange that emphasised the formality of the process more clearly, e.g. formal calls for proposals or bilateral agreements regulating knowledge exchange.

9.6.2 Barriers to and Opportunities for Knowledge Exchange

Our data showed that diverging differences in philosophy and priorities, as well as a lack of communication generally all impede knowledge exchange in Swiss spatial planning, land use and soil management. These include divides in "culture" and ways of thinking between scientists and policymakers, differences in the timeframes they work in, differences in the way they express themselves and a lack of institutionalised links between them. The two communities thesis, "[f]irst elaborated by Caplan (1979)... assumes that a fundamental gap exists between research and policy which is held to be the result of cultural differences between these two communities" (Nutley et al. 2014: 99). Referring to van Buuren and Edelenbos (2004); Wiltshire (2001) and Strydom et al. (2010); van Enst et al. (2014: 9 et seq.) also pointed out that differences in expression, timeframes and perceptions of reality between scientists and policymakers created problems for the science/policy interface. However, our data indicates that these divides, differences and lacking links that constrain knowledge exchange between science and policymaking are not limited to a gap between science and policymaking as such. They also include different "ways of thinking" and different concepts dominating in the different linguistic areas in Switzerland, administrative differences originating from Swiss federalism and a lack of communication paths between soil and spatial planning experts.

Interviewees mentioned several barriers to knowledge exchange between science and policymaking concerning the handling of scientific results. They criticised the results for sometimes being in a language policymakers did not understand or for being inapplicable or inaccessible. Hence, the interviewees called for adapting the communication of scientific results to the language of the addressees, investing in efforts to render the results applicable and providing open-access publications. The first two of these aspects correspond to the findings of Reed et al. (2014: 341 et seq.) whose interview partners suggested that an "effective" knowledge exchange would provide tangible and useful outputs, and would include outside actors in the formulation of policy implications so that the communication of results would actually reach the target audience. Referring to Sarawitz and Pielke (2007), van Enst et al. (2014: 10) also identified "insufficient access to knowledge" as a problem of the science/policy interface, which is in line with our findings.

Other barriers to knowledge exchange expressed in the interviews included the reward system of scientists, which encourages them to concentrate more on writing scientific publications, and less on providing guidance and ready-to-use tools; a lack of financial resources and time for knowledge exchange; an overload of information; power struggles and competing interests constraining knowledge exchange;

and informal contacts that may render knowledge exchange non-transparent. Reed et al. (2014: 340 et seq.) have indicated that adequate funding resources and time for knowledge exchange have to be included directly in the research design if the exchange is to be effective; this corresponds to our findings.

Aside from the barriers, the interviews revealed several opportunities for knowledge exchange. Some of these related explicitly to the Swiss situation. Interviewees considered the small expert community in Switzerland, where people know each other and communication paths are short, to be an opportunity for knowledge exchange. Similarly, the Swiss system of federalism, which produces a wide range of spatial planning solutions, facilitates learning about other cantons or even other language areas. Informal contacts and networks were also identified as opportunities for knowledge exchange. This corroborates findings by Reed et al. (2014: 341), who found that informal exchange between scientists and other actors were crucial for effective knowledge exchange. Another important factor in the facilitation of knowledge exchange was motivation and openness to new, different experiences among the actors involved. However, it also became clear that the Swiss system of federalism, the different language areas and informal networks could also serve as barriers to knowledge exchange.

9.7 Conclusions

Swiss actors in spatial planning, land use and soil management exchange knowledge in many different ways, corresponding to a variety of different concepts from the literature. Most of the actors use more than one way of exchanging knowledge, but they nevertheless exchange knowledge differently depending on the decision situation, the context or person they work with. A given actor might facilitate knowledge transfer as a knowledge advocate, but may also organise participatory and transdisciplinary processes of knowledge development and use.

In practice, multiple types of knowledge exchange occur at the same time and in parallel. Moreover, different situations, contexts and actors require different types of exchanging knowledge. Associated with these overlaps is the fact that actors have different expectations about knowledge exchange and bring in different understandings of what knowledge exchange is. Our article clearly confirms that knowledge exchange can be understood differently (e.g. to coproduce knowledge, to accompany research projects, to retrieve knowledge) and thus functions differently as well. Balancing different expectations and types of knowledge exchange is the crucial challenge to successfully exchanging knowledge.

We have shown that there are specific factors that impede knowledge exchange, including philosophical divides, differences in perspectives, a lack of links between actors and a lack of resources. Working to break down these barriers as well as making use of the opportunities for knowledge exchange will help Swiss scientists and policymakers improve knowledge exchange in the fields of spatial planning, land use and soil management.

References

Bielak, A. T., Campbell, A., Pope, S., Schaefer, K., & Shaxson, L. (2008). From science communication to knowledge brokering: The shift from "science push" to "policy pull." In D. Cheng, M. Claessens, N. R. J. Gascoigne, J. Metcalfe, B. Schiele, & S. Shi (Eds.), *Communicating Science in Social Contexts: New Models, New Practices* (pp. 201–226). New York: Springer.

Birkland, T. A. (2005). *An Introduction to the policy process: Theories, concepts and models of public policy making* (2nd ed.). Armonk: M.E. Sharpe.

Böcher, M., & Krott, M. (2014). The RIU model as an analytical framework for scientific knowledge transfer: the case of the "decision support system forest and climate change." *Biodiversity and Conservation, 23*(14), 3641–3656. https://doi.org/10.1007/s10531-014-0820-5.

Brils, J., Maring, L., Minixhofer, P., Zechmeister-Boltenstern, S., Stangl, R., Baumgarten, A., et al. (2016). National reports with a review and synthesis of the collated information. Final version as of 01.03.2016 of deliverable 2.5 of the HORIZON 2020 project INSPIRATION. (EC Grant agreement no: 642372), Dessau-Roßlau, Germany: Umweltbundesamt (UBA). Retrieved from https://www.inspiration-h2020.eu/sites/default/files/upload/documents/20160301_inspiration_d2.5_0.pdf.

Caplan, N. (1979). The two-communities theory and knowledge utilization. *American Behavioral Scientist, 22*(3), 459–470. https://doi.org/10.1177/000276427902200308.

Fazey, I., Bunse, L., Msika, J., Pinke, M., Preedy, K., Evely, A. C., et al. (2014). Evaluating knowledge exchange in interdisciplinary and multi-stakeholder research. *Global Environmental Change,25*, 204–220. https://doi.org/10.1016/j.gloenvcha.2013.12.012.

Fazey, I., Evely, A. C., Reed, M. S., Stringer, L. C., Kruijsen, J., White, P. C. L., et al. (2013). Knowledge exchange: A review and research agenda for environmental management. *Environmental Conservation, 40*(1), 19–36. https://doi.org/10.1017/S037689291200029X.

Funtowicz, S. O., & Ravetz, J. R. (1993). Science for the post-normal age. *Futures, 25*(7), 739–755. https://doi.org/10.1016/0016-3287(93)90022-L.

Gibbons, M., Limoges, C., Nowotny, H., Schwartzman, S., Scott, P., & Trow, M. (1994). *The new production of knowledge. The dynamics of science and research in contemporaty societies.* London, Thousand Oaks, New Dehli: SAGE Publications.

Hoppe, R. (2005). Rethinking the science-policy nexus: From knowledge utilization and science technology studies to types of boundary arrangements. *Poièsis and Praxis, 3*(3), 199–215. https://doi.org/10.1007/s10202-005-0074-0.

Jasanoff, S. (2004). States of knowledge: The Co-production of science and social order.Routledge, London.

Kamelarczyk, K. B. F., & Gamborg, C. (2014). Spanning boundaries: Science-policy interaction in developing countries—The Zambian REDD+ process case. *Environmental Development, 10,* 1–15. https://doi.org/10.1016/j.envdev.2014.01.001.

Keeley, J., & Scoones, I. (2003). *Understanding environmental policy processes: Cases from Africa.* London: Earthscan Publications.

Lasswell, H. D. (1956). *The decision process: Seven categories of functional analysis.* College Park: University of Maryland.

Linke, S., Gilek, M., Karlsson, M., & Udovyk, O. (2014). Unravelling science-policy interactions in environmental risk governance of the Baltic Sea: Comparing fisheries and eutrophication. *Journal of Risk Research, 17*(4), 505–523. https://doi.org/10.1080/13669877.2013.794154.

Mauser, W., Klepper, G., Rice, M., Schmalzbauer, B. S., Hackmann, H., Leemans, R., & Moore, H. (2013). Transdisciplinary global change research: the co-creation of knowledge for sustainability. *Current Opinion in Environmental Sustainability, 5*(3–4), 420–431.

Mayring, Ph. (2015). Qualitative Inhaltsanalyse. Grundlagen und Techniken. 12., überarbeitete Auflage (Qualitative Content Analysis. Basics and techniques). Weinheim, Basel: Beltz Verlag.

Nutley, S. M., Walter, I., & Davies, H. T. O. (2014). *Using evidence: How research can inform public services.* Bristol, UK: Policy Press.

Plieninger, T., Draux, H., Fagerholm, N., Bieling, C., Bürgi, M., Kizos, T., et al. (2016). The driving forces of landscape change in Europe: A systematic review of the evidence. *Land Use Policy, 57,* 204–214.

Prager, K., & McKee, A. (2015). Co-production of knowledge in soils governance. *International Journal of Rural Law and Policy Soil Governance, 1,* 1–17. https://doi.org/10.5130/ijrlp.i1.2015. 4352.

Pregering, M. (2004). Linking knowledge and action: The role of science in NFP processes. In P. Glück and J. Voitleithner (Eds.), (Hrsg.). NFP Research: Its retrospect and outlook. Proceedings of the Seminar of COST Action E19 "National Forest Programmes in a European Context", September, 2003, Vienna Vienna: Institute of Forest Sector Policy and Economics: 195–215.

Pretty, J., & Chambers, R. (1993). *Towards a learning paradigm: New professionalism and institutions for sustainable agriculture, IDS Discussion Paper DP334.* Brighton, UK: Institute for Development Studies.

Pütz, M. (2011). Power, scale and Ikea: analysing urban sprawl and land use planning in the metropolitan region of Munich, Germany. Regional Environmental Governance: Interdisciplinary Perspectives, Theoretical Issues, Comparative Designs (REGov). *Procedia Social and Behavioral Sciences, 14,* 177–185.

Ravetz, J. R. (1999). What is Post-Normal Science? *Futures, 31*(7), 647–653. https://doi.org/10. 1016/S0016-3287(99)00024-5.

Reed, M. S., Fazey, I., Stringer, L. C., Raymond, C. M., Akhtar-Schuster, M., Begni, G., et al. (2013). Knowledge management for land degradation monitoring and assessment: An analysis of contemporary thinking. *Land Degradation and Development, 24*(4), 307–322. https://doi.org/ 10.1002/ldr.1124.

Reed, M. S., Stringer, L. C., Fazey, I., Evely, A. C., & Kruijsen, J. H. J. (2014). Five principles for the practice of knowledge exchange in environmental management. *Journal of Environmental Management, 146,* 337–345. https://doi.org/10.1016/j.jenvman.2014.07.021.

Rogga, S., Weith, T., Aenis, T., Müller, K., Köhler, T., Härtel, L. & Kaiser, D. B. (2014). Wissenschaft-Praxis-Transfer jenseits der „Verladerampe". Zum Verständnis von Implementation und Transfer im Nachhaltigen Landmanagement. Nachhaltiges Landmanagement, Diskussionspapier Nr. 8., July 2014. Müncheberg.

Roux, D. J., Rogers, K. H., Biggs, H. C., Ashton, P. J. & Sergeant, A. (2006). Bridging the science-management divide: Moving from unidirectional knowledge transfer to knowledge interfacing and sharing. *Ecology and Society, 11*(1). https://doi.org/10.5751/ES-01643-110104.

Saarela, S. R., & Söderman, T. (2015). The challenge of knowledge exchange in national policy impact assessment—A case of Finnish climate policy. *Environmental Science & Policy, 54,* 340–348. https://doi.org/10.1016/j.envsci.2015.07.029.

Sarawitz, D., & Pielke, R. A. (2007). The neglected heart of science policy: Reconciling supply of and demand for science. *Environmental Science and Policy, 10*(1), 5–16. https://doi.org/10.1016/ j.envsci.2006.10.001.

Schweizerischer Bundesrat, KdK, BPUK, SSV, & SGV. (2012). Raumkonzept Schweiz. Überarbeitete Fassung (Spatial Strategy for Switzerland. Revised Version). Bern: Schweizerischer Bundesrat, Konferenz der Kantonsregierungen (KdK), Bau-, Planungs- und Umweltdirektoren-Konferenz (BPUK), Schweizerischer Städteverband (SSV), Schweizerischer Gemeindeverband (SGV). Retrieved from https://www.are.admin.ch/dam/are/de/dokumente/raumplanung/publik ationen/raumkonzept_schweiz.pdf.download.pdf/raumkonzept_schweiz.pdf.

Stewart, A., Edwards, D., & Lawrence, A. (2014). Improving the science–policy–practice interface: Decision support system uptake and use in the forestry sector in Great Britain. *Scandinavian Journal of Forest Research, 29*(sup1), 144–153. https://doi.org/10.1080/02827581.2013.849358.

Stone, D. (2001). Learning lessons, policy transfer and the international diffusion of policy ideas. Centre for the Study of Globalisation and Regionalisation Working Paper, No.69/01 University of Warwick.

Strydom, W. F., Funke, N., Nienaber, S., Nortje, K. & Steyn, M. (2010). Evidence-based policy making: A review. *South African Journal of Science, 106*, 5–6. https://doi.org/10.4102/sajs.v10 6i5/6.249.

Turnout, E., Hisschemöller, M., & Eijsackers, H. (2007). Ecological indicators: Between the two fires of science and policy. *Ecological Indicators, 7,* 215–228. https://doi.org/10.1016/j.ecolind. 2005.12.003.

van Buuren, A., & Edelenbos, J. (2004). Conflicting knowledge: Why is joint knowledge production such a problem? *Science and Public Policy, 31*(4), 289–299. https://doi.org/10.3152/147154304 781779967.

van den Hove, S. (2007). A rationale for science-policy interfaces. *Futures, 39*(7), 807–826. https:// doi.org/10.1016/j.futures.2006.12.004.

van Enst, W. I., Driessen, P. P. J. & Runhaar, H. A. C. (2014). Towards productive science-policy interfaces: A research agenda. *Journal of Environmental Assessment Policy and Management, 16*(1). https://doi.org/10.1142/S1464333214500070.

Wiltshire, K. (2001). Scientists and policy-makers: Towards a new partnership. *International Social Science Journal, 53*(170), 621–635. https://doi.org/10.1111/1468-2451.00349.

Zierhofer, W. & Burger, P. (2007). Transdisciplinary research—A distinct mode of knowledge production? Problem-orientation, knowledge integration and participation in transdisciplinary research projects. *GAIA—Ecology Perspectives for Science & Society 16*(1), 29–34.

Zscheischler, J., Rogga, S., & Lange, A. (2018). The success of transdisciplinary research for sustainable land use: Individual perceptions and assessments. *Sustainability Science, 13*(4), 1061–1074.

Chapter 10
Serious Games in Sustainable Land Management

Jacqueline Maaß

Abstract Over the last few decades, Germany has experienced a trend towards increased suburbanisation and urban sprawl, accompanied by growing distances between residential and commercial areas and growing numbers of the commuting population. These phenomena were made possible in part by cheap energy prices for fossil fuels and have accordingly determined our land use and settlement structures to a major extent. As energy prices for fossil energy rise, we feel the effects in the structures society has built, and these have an impact on land use and sustainable land management.

Keywords €LAN serious game · Land use and transport (LuT) model · Energy price · Land use · Political decision-makers

10.1 Introduction

The correlation between cheap fossil energy and land use has been the subject of several studies worldwide focusing on different aspects and using different methods. The €LAN—energy prices and land use project, funded by the German Sustainable Land Management funding programme, analysed this correlation for the German context. By simulating the complexity of sustainable land management as well as incorporating learning aspects for decision-makers, the project combined a serious game with a land use and transport (LuT) model, thus taking an integrated approach to analysing the effects of high prices for fossil fuels on land use. The €LAN project asked local and regional decision-makers to:

- Identify the effects of rising energy prices for their respective municipalities
- Generate possible responses about how to cope with rising energy prices, and
- Develop their own strategies to counter the negative effects of rising energy prices.

J. Maaß (✉)
Institute for Transport Planning and Logistics, Technische Universität Hamburg (TUHH), Hamburg, Germany
e-mail: jacqueline.maass@tuhh.de

© The Author(s) 2021   185
T. Weith et al. (eds.), *Sustainable Land Management in a European Context*,
Human-Environment Interactions 8, https://doi.org/10.1007/978-3-030-50841-8_10

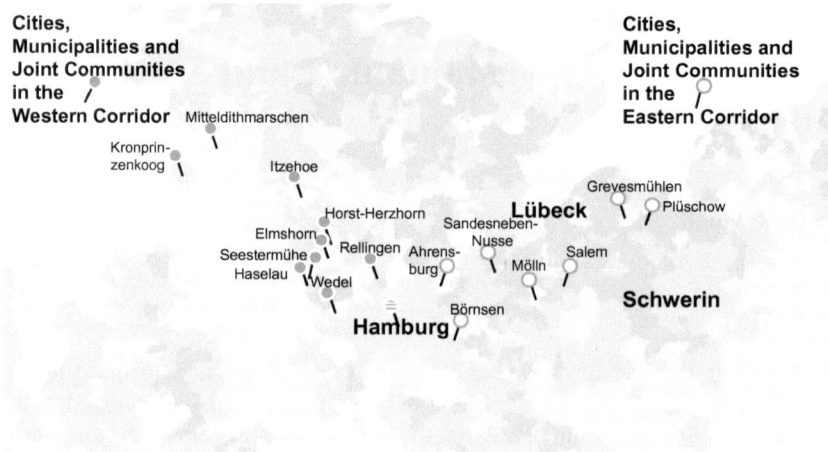

Fig. 10.1 Participating cities, municipalities and joint communities of both corridors (based on Gertz et al. 2015, p. 125, author's translation)

Participants in the serious game came from two different participating corridors in the Hamburg Metropolitan Region (HMR). The decision-makers of both corridors (northwest from Hamburg to the district of Dithmarschen, and northeast from Hamburg to the district of Northwest Mecklenburg) attended in several meetings to develop their individual strategies on how to cope with rising energy prices (Figure 10.1).

10.2 Post-Fossil Energy in the Twenty-First Century?

€LAN started in late 2010 and ended in 2014, coinciding with a phase of high crude oil prices between 2012 and 2014. Much of the debate on the post-fossil-energy era has been based on the assumption that the global fossil fuel supply was reaching (or had already reached) its peak. Over the course of the project, the debate as well as efforts to move into a post-fossil energy era became much more visible than they had been before. There were several reasons for the strong upwards trend in oil prices on the commodity markets (see also Carollo 2011; EWI and Prognos 2006; Newman 2008); prices went over USD 100 per barrel. This meant a sudden doubling in prices, and many market commentators considered the market to have crossed a psychological barrier.

Decades of cheap fossil energy have influenced our settlement structures: phenomena such as suburbanisation, accompanied by growing spatial distances between relevant destinations and a growing commuting population, is one major example of cheap and abundant fossil energy resources fostering such developments. The questions €LAN posed therefore were: What happens if high energy prices persist over a longer period of time? How would political decision-makers respond when areas such as transport, traffic and housing come under clear pressure from long-term rising oil prices? What effects might that have on land use?

Even in the 21st century, fossil energy is what keeps our society running in its current form. Germany imports most of its required energy in the form of coal, oil and gas from different countries (Bundesministerium für Wirtschaft und Energie (BMWI) August 2018). Being dependent on foreign energy sources leads to a dependency on the global structure of energy markets. Those energy markets are shaped by various factors, with global political developments playing an important role. The transport sector in Germany in particular depends on fossil energy, as more than 90% of the fossil energy used in this sector is imported (largely petroleum) (Bundesministerium für Wirtschaft und Energie (BMWI) August 2018). This means it is especially vulnerable to price fluctuations on the global crude oil market. Besides being a very car-dependent and car-centred society,[1] heating is commonly considered a major factor for private households in Germany in terms of fossil energy consumption (Deutsche Energie-Agentur GmbH (dena) September 2012).

10.3   How High Energy Prices Affect Communities

Rising energy prices also affect public budgets in Germany in multidimensional ways. This include immediate versus delayed effects, as well as direct and indirect effects.

As for the present trends, oil prices dropped remarkably quickly, and reached the level of 2003 (less than USD 30 per barrel) in January 2016 (International Energy Agency (IEA) 2016). Both upward and downward movements have fed discussions on the wide range of impacts that energy prices may have on economic activities and social life in general. One of the areas of concern is the resilience of spatial configurations of cities and regions to the effects of high energy prices. Newman and Kenworthy Jr. (1989) for example, were among the first to make a large-scale analysis of the reasons for the statistical connection between higher levels of fuel consumption per capita and lower population densities in urban areas of three continents. Their findings suggested a range of policy strategies for lowering the dependency on fossil fuels, e.g. by increasing urban density, enhancing the attractiveness of centres, and providing attractive public transport options as an alternative to travelling by car (Newman and Kenworthy Jr., 1989). Larson and Yezer (2015) used models to simulate the effects of energy on land consumption. A major policy concern has therefore been to make sustainable urban and regional development less vulnerable to energy price variations.

---

[1] More than 92% of all households with two or more persons have access to at least one car (Frondel et al. 2015); the car density in Germany as of January 2017 was as high as 684 motor vehicles per 1000 inhabitants (Kraftfahrtbundesamt 2017).

Such research and its results lead to questions about how much space sealed by infrastructure or settlement areas is desirable for our society, and how rising energy prices affect these structures. The answers to these questions should be given by planners and political decision-makers, but may to a certain degree also depend on the responses of private households in the sense of supply and demand.

The immediate and direct effects of rising energy prices on public budgets are to a degree comparable to those of private households. However, the delayed and indirect effects will affect municipalities the most if high energy prices persist over a longer period of time. Land management or questions concerning sustainable planning will therefore become more difficult to anticipate. Forecasting the possibly broad variety of private households' responses to rising energy prices seems necessary. Insofar as planning tries to anticipate and match the future needs of private households with the services and infrastructure provided, planning becomes more difficult as soon as outside influences disturb the current system. Linear projections of known development paths may therefore fail to grasp the effects.

Private households may be able to cover short peaks of higher fuel prices by adjusting their budgets and cutting down on other types of expenses, as shown in Fig. 10.2. Households with a higher income may be able to absorb rises in fuel prices and not change their habits in terms of housing and transport at all. However, households on a tighter budget might be forced to think sooner in more lifestyle-changing terms if energy prices start climbing and remain at a high level. The choices they previously made regarding residential or work locations or their daily travel routines, the mode of transport, etc. might become subject to efforts to minimise rising energy costs. It must be pointed out that rising oil prices affecting fuel prices for car use may lead to higher prices for car travel than for public transport, given

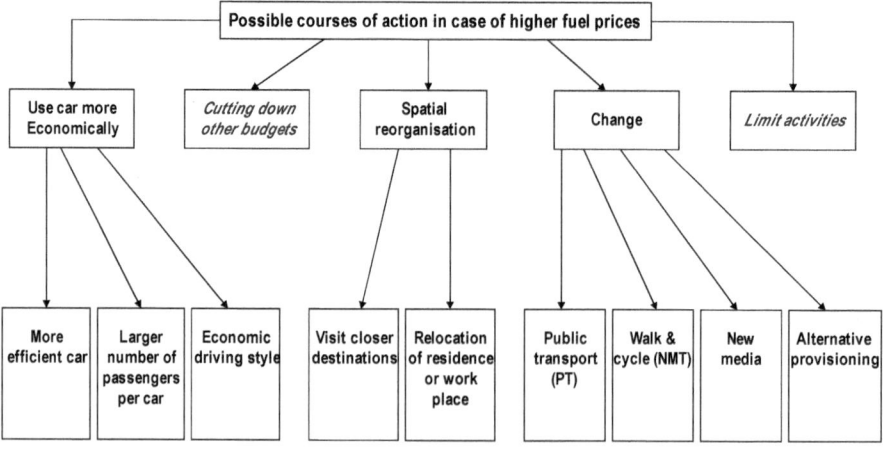

Fig. 10.2 Possible courses of action in case of higher fuel prices (BMVBS/BBSR March 2009, p. 60)

that only a fraction of the ticket price of public transport is directly related to the fuel price itself.

If several private households start changing their common routines or behaviour where the mode of transport is concerned, location choices might also be revised, as well as budgets for other activities. These effects might affect municipalities in various ways.

The two fields of "spatial reorganisation" and "change" (Fig. 10.2) will also lead to land use effects that municipalities must consider.

People might *change* from car transport to public transport or cycle more often for shorter distances. Current infrastructure might not be able to accommodate a rapid increase in public transport users where capacity limits have already been reached. Bike lanes might suddenly be necessary or insufficient in their current state of expansion. People might also tend to buy more items and services online and use delivery services more often, or try to use home office options (where possible) to cut down on (car) trips. Public infrastructures might not be able to accommodate sudden and numerous changes in demand.

Households deciding on *spatial reorganisation* to minimise commuting trips or change their workplaces might intensify current trends for municipalities or influence tax income. Municipalities already suffering population losses due to various causes—e.g. rural areas with fewer possibilities for higher education or work places have already suffered from a drain of younger inhabitants—might experience even faster outflows. Other centres experiencing a rise in the number of inhabitants might experience a stronger influx due to rising energy prices and households' relocation choices.

Due to all these fields of possible variations, differentiations and changes, municipalities may experience shifts in modal splits, tax revenue, social outlays, infrastructure, etc., which will make planning and hence land use decisions altogether more difficult in the light of rising energy prices.

10.4 Rising Energy Prices and Land Use—A New Research Focus

The stated combination of effects on global energy markets for fossil fuels, societies' dependency on fossil fuels, the responses of private households, and the physical and topographic municipal landscape as well as existing settlement layouts and transport-related infrastructure, have opened up new avenues for research. Few publications so far have focused on rising energy prices and land management in terms of a combined effect analysis of settlement structure, infrastructure and accessibility. Research in this context has so far focused on the following aspects:

- How to use resources more efficiently and manage changes in land management for a more resilient and sustainable outcome (i.e. Bundesministerium für Bildung und Forschung (BMBF), April 2013; Beckmann, 2000; Bundesamt für Bauwesen und Raumordnung 1999).

- The vulnerability of low-income households to rising energy costs, known as "fuel poverty" or "energy poverty" in general (e.g. Boardman 1991; Dubois and Meier 2016; González-Eguino 2015; Brunner et al. 2012), or with a supplemental focus incorporating aspects of accessibility, transport, mobility and vulnerability (e.g. Berry et al., 2016; Legendre and Ricci 2015; BMVBS/BBSR, March 2009; Dodson and Sipe 2007; or Roberts et al. 2015).
- Spatial aspects or accessibility questions in combination with scarce energy supplies or rising energy prices (e.g. Büttner 2017; Wegener 2009a; Dodson and Sipe 2008; Fiorello et al. 2006; or Shepherd et al. 2008).
- The social dimension of equity or (in)equality existing in the context of rising energy prices and transport, accessibility or car dependency (e.g. Dubois and Meier 2016; Mattioli 2014; Reames 2016; Walker and Day 2012).

Experience with and knowledge of the resilience of urban settlement areas (city layout) or of accessibility aspects has often been derived from software-based approaches. Using models or GIS tools to visualise or account for effects has been a major approach in the context of land use research (e.g. Fiorello et al. 2006; Wegener 2009b; Buettner et al. 2013).

However, little research so far has focused on administrative levels, the impact of rising energy prices on social-political structures, and the potential effects caused by decision-makers' responses. There is no research literature on how decision-makers at various levels of government cope with the complex, multidimensional problem of rising energy prices in a society dominated by car transport and the land use strategies associated with it. Timeframes for oil crises and rising energy prices have been rather short. But as the oil crisis in the 1970s showed, drastic effects and responses are possible. Further research in that field is therefore required.

10.5 Serious Games—A Different Approach to Sustainable Land Management

Usually, mathematical models are used to map future trends or evaluate political decisions based on numbers. However, those numeric models cannot include future (political) planning decisions that might arise in the light of high energy prices or (sudden) scarcity of fossil energy.

With fossil energy a major factor in transport, traffic and heating, high energy prices over a longer period of time increase pressure on decision-makers to find more sustainable ways of planning. Policy-makers and administrative decision-makers might therefore be in need of new or additional approaches to cope with such complex phenomena, assuming that fossil energy might (again) become more cost-intensive in the future. To explore which path future developments will take and to shape strategic decisions for the coming decades, a complex, integrative approach is necessary. €LAN achieved this by simulating the impacts of rising energy prices and capturing the responses of decision-makers through participative methods. The insights obtained during the project suggested this "serious game" as the most suitable approach for simulating sustainable land management decision-making.

The term *"serious game"* has not been precisely defined; different contexts have conceptualised it differently (e.g. Hitzler et al. 2010: 218–219, Rebmann 2001: 9). Definitions and understandings may also vary between the German and the Anglo-American contexts, where the expression "Planspiel" and the terms "simulation game" or "serious game" can be found (e.g. Hitzler et al. 2010: 218–219, Rebmann 2001: 9).

For the €LAN context, the closest definition of the "serious game" is the following: *"The serious game is an activity-oriented method, in which complex economic or social-political functional relationships are simulated within an illustrative modelled gaming scenario."* (Author's translation). "Das Planspiel ist eine handlungsorientierte Methode, bei der komplexe ökonomische oder politisch-soziale Funktionszusammenhänge in einem modellhaften Spielszenario simuliert werden." (Fischer 2008: 137).

This makes a serious game a good methodological fit for the project's purposes— a combination of modelled scenario in combination with an activity-oriented basic approach.

In the context of €LAN, the term *"Planspiel"* has been used in German publications and *"serious game"* as the direct equivalent to describe the research approach in English publications.

Serious games are not a new tool, and have been used in teaching, learning and evaluation contexts for several decades (e.g. Ebert 1992; Hitzler et al. 2010; Korte and Lehmbrock 2009). For political consulting, serious games have also been known to be a valuable tool for testing the effects of new laws or administrative regulations (e.g. Herz and Blätte 2000; Korte and Lehmbrock 2009; Hitzler et al. 2010; Böhret and Wordelmann 2000; Vissers and van der Meer 2000; Joldersma and Geurts 2000). As shown by way of example in Korte and Lehmbrock (2009), serious games have also a long history in political contexts for urban and transport planning (Korte and Lehmbrock 2009: 11–12). However, as Hitzler et al. (2010) noted, there is no homogeneity of serious games within Germany. Serious games have been designed by many different institutions, companies, universities and individuals for a multitude of different purposes (e.g. Hitzler et al. 2010: 220–221). Nevertheless, the combination of serious games with computer-supported elements is still very rare in the sector of planning or decision support for politicians and planners; one exception is the serious game TAU (Böhret and Wordelmann 2000).

Researchers on the €LAN project also found that, in spite of the potential usefulness of a serious game, there was nothing available to cover all the aspects of the project's research interest. €LAN researchers therefore chose to illuminate this less-explored object of research by developing a new serious game, supported by scenarios generated by an integrated land use and transport model (LuT). This LuT was developed within the project as the second pillar of research, and thus was an excellent fit for the serious game.

The serious game developed by €LAN focused on setting a scenario for the participants in which they would be able to react to rising energy prices according to their actual roles and responsibilities in local government.

10.6 The Serious Game Developed by €LAN—Methodological Outline

To get both elements—the serious game and the LuT model—to work seamlessly together, a translation process between both sides was a very important aspect of the project (Fig. 10.3).

The first step of the iterative process entailed selecting participants (decision-makers) and using various media to present them with the basic trends and developments according to a created scenario (Fig. 10.3, at the top). The scenario was created using an LuT model, which cast the effects of rising energy prices three years into the future[2] (Fig. 10.3, left-hand side). The scenario was based on the latest

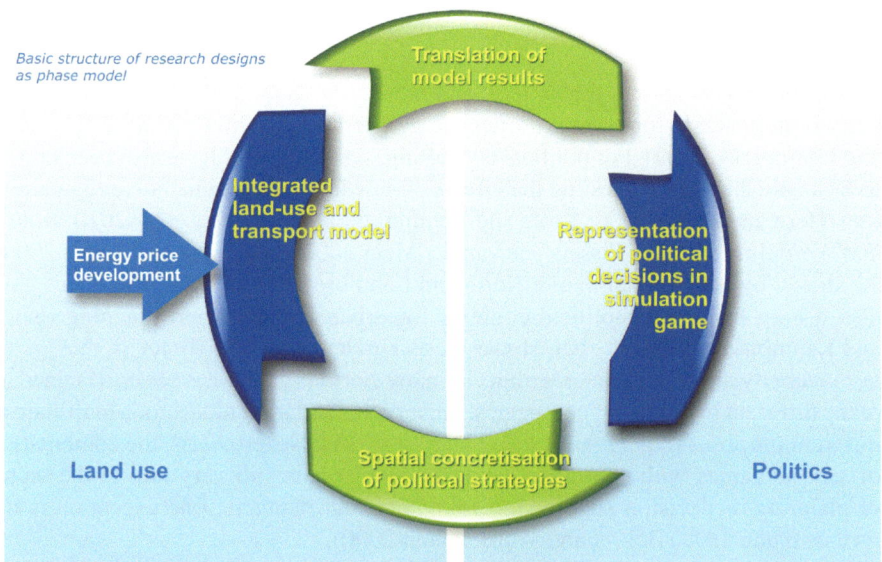

Fig. 10.3 Iterative phase model; €LAN 2014 (also in Guimarães et al. 2014)

---

[2]Since the first round took place in 2012, the scenario for the serious game was 2015—close enough in time to prevent a possible response in the direction of "technological evolution will solve the problem", but also far enough into the future to relate to substantially higher energy prices.

data available on demography, employment patterns, transport infrastructure, residential locations, taxes, household incomes, education and other important aspects (for further information about the model, see Gertz et al. 2015).

The second step of the project attempted to communicate this data via different channels in an effort to make simulation as realistic as possible for the participants. For the professional level, for example, the project created a regional planning report with common facts and figures about commuter data, demographic developments, etc. typically used in such media. Participants were also addressed on a more personal level by media clips and short video clips designed as news reports. The articles for those media clips were produced (based on scenario data) by a professional journalist and imitated the styles and characteristics of layouts of current print media products. The goal was to sensitise the participants to the topic and to enable them to relate to the scenario in different ways on a broad professional and personal basis.

These visualisations were used in the third step to prompt the participants to identify the effects of rising energy prices for their respective municipality and to develop measures, ideas and strategies to respond to such cost developments as decision-makers for their specialisation and/or municipality (Fig. 10.3, right-hand side).

During the fourth step, the participants' ideas were reformulated as far as possible into numeric values according to previously identified parameters, such as the time frame for implementation and financial effort.

The output generated by the serious game was translated and fed back into the model (Fig. 10.3, bottom) to generate a scenario for 2025. The new scenario was then translated into different media and presented to the decision-makers. Confronted with these outcomes, they were asked to respond once again. This created an iterative cycle between the model and decision-makers, and the project simulated a span of 20 years (2010–2030).

10.7 Investigating Energy-Price Effects in the Hamburg Metropolitan Region—How to Integrate Regional Decision-Makers

The study area of the €LAN project is the—still growing—Hamburg Metropolitan Region (HMR). Currently, it is made up of 1177 municipalities in four federal states in Northern Germany,[3] with about 5 million inhabitants. It comprises an area of around 26,000 km^2. The city of Hamburg, with around 1.7 million inhabitants, is the dominant centre in the region and is a hub for key economic activities. The region also contains sub-centres with universities, entertainment facilities, hospitals and

---

[3]In €LAN, the Hamburg Metropolitan Region encompasses the city-state of Hamburg, the western part of Mecklenburg-West Pomerania and the southern part of Schleswig–Holstein. The Hamburg Metropolitan Region also includes parts of Lower Saxony, but this area was omitted from the study.

health services, and displays a relatively diverse economic structure with employment opportunities in a variety of sectors.

Complex problems in planning require a comprehensive approach. This is why the project undertook a policy analysis to identify which stakeholders played which roles in the processes of decision-making and developing strategies pertaining to land use and regional development. The logic for coming to decisions was based on an analysis of the incentives or constraints, the relationships between stakeholders and the administrative-institutional system.

This led to identifying two central dimensions in that context: the "*role*" and the "*region*". The former contained three scopes, which were important as both descriptive and selective factors for potential participants and were called "*selection dimensions*" (Projekt €LAN, 30 June 2011, p. 11):

- Political power (in the sense of the separation of powers)
- Policy
- Administration (Projekt €LAN 30 June 2011, p. 11); author's translation).

The latter dimension, "region", led to the classification of five types of affectedness within the HMR (Fig. 10.4).

The dimensions of "*role*" and "*region*" were then cross-referenced to identify which stakeholders were relevant for the serious game at the municipal level. The project then approached these stakeholders to participate themselves or recommend suitable representatives to participate in the serious game.

The feedback to the first contact through the project was very positive overall. The actual circumstances of rising energy prices during the project might have also pushed stakeholders' interest in the project's favour. The mix of participants covered a wide range of knowledge relevant for planning (especially concerning transport, housing, energy and economics) at all three political levels (municipal, federal state (*Länder*) and federal government). The decision-makers participating

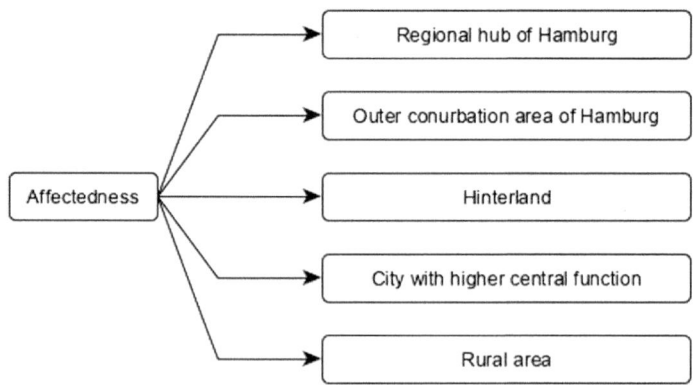

Fig. 10.4 Five communal classes of affectedness by rising energy prices within the HMR (according to Projekt €LAN, 30 June 2011, p. 11, author's translation)

in €LAN represented their own local, state or federal level and were chosen from different types of areas in the HMR, including parts of the city-state of Hamburg, the western part of Mecklenburg-West Pomerania and the southern part of Schleswig-Holstein. Thus the selected representatives of communal classes varied according to population, economic structures and degree of vulnerability to fluctuations in (fossil) energy costs. Unfortunately, no secondary city as a counterpart to Hamburg was able to participate, which led to one classification to be omitted in the project.

Since Mecklenburg-West Pomerania and the southern part of Schleswig-Holstein are very different in their settlement as well as administrative structure, the project constructed two corridors for the serious game. The Eastern Corridor (Hamburg to Mecklenburg-West Pomerania) and the Western Corridor (the southern part of Schleswig-Holstein to Hamburg) addressed these differences and made it possible to compare different structures for the serious game.

Continuity was a central factor for the project during the different parts of the serious game. The sessions were held over a period of one-and-a-half years, between April 2012 and October 2013. The long duration of the intervals between sessions was in part necessary for the project to be able to additionally incorporate the federal level. This was relevant for regional stakeholders' decisions, as the federal government sets the overall framework for possible responses at the regional level.[4] In order to most realistically simulate regional decision-making, the regional level actors and stakeholders therefore had to interact with the federal level to explore the relationship between both levels, and to develop an exchange between them.

The result is a two-tier serious game as shown in Fig. 10.5.

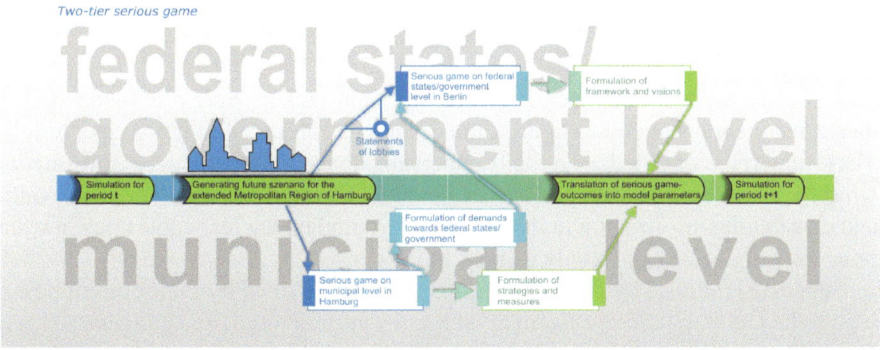

Fig. 10.5 The two-tier serious game of €LAN (€LAN 2014, also in Guimarães et al. 2014)

---

[4]I.e. The German Renewable Energy Sources Act—*Das Erneuerbare-Energien-Gesetz (EEG)*—which sets the frame for electricity to be preferentially produced by regenerative energy sources and refunded by certain mechanisms. For the German federal states, important also for electrical vehicles and transport modes.

10.8   How to Play the €LAN Serious Game

To implement the serious game, a series of moderated workshops was held in which key stakeholders from the case study area were asked to react to scenarios of rising energy prices. They were asked to act according to their best intentions and knowledge as they would in reality, and thus to decide on measures and strategies to counter the negative effects of rising energy prices identified for their municipality or their respective sphere of action.

Several serious game sessions were held during the €LAN project between 2012 and 2013. Decision-makers from different administrative levels reacted to a scenario of constantly high energy prices, which would stress the public purse through costs for providing transport, social services and energy for heating, street lights, etc. If energy prices consistently rose and were not a peak event that ceased after a few days or weeks (like in the 1970s or in 2008), stakeholders would be more likely to be forced to take action to counter the noticeable effects of increased fuel and heating costs. One possible option was to directly subsidise fuel costs for consumers. Another option would be a systemic option for hardship cases to improve their financial situation, strained by rising energy prices. Those were two possible strategies decision-makers might decide on. These and other possible answers were anticipated by the project as possible outcomes of the serious game.

For the project's starting scenario, "2015", a price of USD 200 per barrel of crude oil (Brent) was set. This was calculated to result in a pump price of EUR 2.20 per litre. The average additional cost per household paid for energy (heating, transport and electricity) in the scenario was subsequently calculated to be EUR 130 per month compared to prices in 2010. For the later scenario, "2025", the prices were set at USD 400 per barrel of crude oil (Brent), which amounted to an additional EUR 340 compared to 2010 prices.

To formulate decisions, measures, etc., decision-makers were provided with LuT model results translated into figures, tables and maps, as well as newspaper articles. These materials consisted of an easily understandable multifaceted set of information, which gave tangible meaning to abstract model numbers. The aim behind providing and distributing information in this way was to simulate the real conditions in which policymaking occurs as closely as possible.

Suggested policies by decision-makers were constrained by their current powers within the German federal system as well as by current municipal budgetary restrictions. The restrictions were defined to be as realistic as possible to prevent participants from pursuing "unrealistic" responses, e.g. in the sense of spending more on measures than they would be able to in reality. The project also attempted to prevent the inclusion of more measures or ideas in their policy portfolio than they would if the scenario was true.

The outcomes of the sessions were then quantified, translated into parameters for the LuT model and fed back into it. For a more detailed explanation of the serious game design, see Gertz et al. (2015).

During the first session in the Eastern Corridor, the participating decision-makers were presented with the results generated from the integrated land use and transport model as a first step, which was intended to aid them in relating to the scenario on a more personal basis. Through (custom-produced) newspaper articles, they got an impression of how rising energy prices might influence daily life. A report presenting statistics relating to the development scenario—a regional monitor—presented them with key facts and figures in a familiar format to relate to the scenario on a more professional and planning-oriented level. In the second step, the decision-makers identified likely effects on their respective municipality, and were then asked to devise actions and measures to counter the (adverse) consequences of rising energy prices.

On the regional level, participating decision-makers were split into two groups according to their regional affiliation or corridor. One group represented the Eastern Corridor; the other the Western Corridor. The project took different methodological approaches to each region in the serious game, varying in the number of sessions and therefore also in the depth of analysis of the effects for each participating municipality. This paper concentrates on explanations about the serious game within the "East" group, which had more sessions and a longer and therefore deeper phase of analysis. This gave participants more time to analyse the effects of rising energy prices on their own municipality and identify the resulting areas in which there would be a need to act. In the Western Corridor, the first phase of identifying such areas as well as the depth of analysis and time for developing appropriate policies was shorter. The results from both corridors are summarised in the following section.

10.9   The Resulting Policy Agenda

After two iterative cycles of the serious game, each decision-maker came up with proposals on how to minimise the negative effects of rising energy prices by 2030. The actions and policies developed by each decision-maker were clustered and labelled according to the general direction the measures implied. Depending on the municipal classification (e.g. small town with a rural context or larger city) for which each decision-maker acted, there was no consensus among stakeholders about the general direction that the proposed measures took, and in some cases they were even contradictory. But in some areas, they also complemented each other. An overview of the direction of measures and strategies the municipal representatives came up with is given in Table 10.1.

When taking a closer look at the various measures that decision-makers chose, it is remarkable that they largely come from five policy fields:

- Housing
- Transport
- Energy
- Technical and social infrastructure
- Comprehensive approaches.

Table 10.1 Policy measures suggested by decision-makers within the €LAN serious game

Direction of measures	Description	Examples of measures
More efficient, car-based society	(Small) Municipalities highly dependent on structures for private car use do not see any way of replacing the car. They attempt to increase car use efficiency through technological improvements as well as a higher degree of occupancy per car	• Promotion/development of electric transport • Online ride-sharing services
More efficient housing	Almost all municipalities see the housing sector as one with a high potential to save energy	• Energy-efficient renovation of houses • Energy advice for tenants and owners • Construction of new housing • Setting higher energy standards for urban land use planning
Public transport as the saviour of mobility ("Stepping out of the comfort zone")	Municipalities close to main public transport corridors see public transport as the saviour of mobility. Significant improvements can make public transport more attractive, but these investments are only worthwhile if their inhabitants discover sympathy for them	• Capacity improvements on existing public transport routes • Traffic management and combination of transport modes • Park & Ride options • Regional network expansion of public transport (including rail)
New happiness on two wheels	Many municipalities view electric bicycles and especially the potential connection with public transport as a major opportunity	• Mobility management • Combination of transport modes • Promotion and development of electric mobility (suggested by participants in the Western corridor)
Self-sufficiency on various levels	As climbing energy prices increase spatial resistance, some municipalities strive for self-sufficiency on multiple levels: energy supply (wind power stations, bio-energy village, …), attractiveness to businesses (of their own municipality), infrastructure, binding purchasing power	• Combined heat and power plants as the standard for urban land use planning • Promotion of bio-energy villages • Cooperation between heat producers and heat users through a "heat stock market"

(continued)

Table 10.1 (continued)

Direction of measures	Description	Examples of measures
Compensation of energy price—disadvantages of the location	Some (more peripheral) municipalities try to compensate for their location disadvantage (worsened through increased energy prices) by strengthening their local advantages (subsidised land for building, affordable and high-quality child care options, good reputation, green spaces, …)	• Attractive day care (nursery school) • Increased promotion of attractive housing areas • Citizen survey: What do we want? • Attracting businesses (to their own municipality)
Maintain short distances, coupled with the topic of "adaptation to demographic change"	Strategies for adaptation to demographic change. Especially when talking about retaining shops and services (social infrastructure, retail) through reorganisation, subsidies or restructuring	• Establishment of a health centre • Establishment of a cooperative town hall • Local supply of goods through a "marketplace shop"
E-Everything with local ties	Many municipalities acknowledge the potential of digital media in enabling more energy-efficient mobility for their residents This idea is tied together by purpose-built components (e.g. online ride sharing platforms in combination with safe and well-designed commuter parking spots) as well as social elements (e.g. municipal initiatives to use ride sharing, mentor programmes for using online services, etc.)	• Broadband and telecommunications • "E-Everything" (local supplies and all municipal services online) • Telemedicine (online medical services)
Services on wheels—services come to us	In very rural areas, mobile services and supplies are experiencing renewed popularity. Thus another thread of the discussion and strategies linked to demographic change debate has continued. The reality of implementation so far has yet to culminate to a significant amount	• Local supply of goods through a "marketplace truck", mobile medical services, Combi-Bus (bus taking on passengers as well as mail and goods)

(continued)

Table 10.1 (continued)

Direction of measures	Description	Examples of measures
Make space and avoid overheating	Due to rising energy prices, central urban areas enjoy additional popularity, adding another element to trends of reurbanisation. At the same time, communities fear an overheating of housing markets and look for possibilities to handle the sometimes significant needs for infrastructure expansion (e.g. along the main axes and main nodes of public transport). The remaining question of how those expansions are to be paid for still needs to be answered	• Capacity improvements on existing public transport routes, advertising new housing sites and inner-city development
There is still life in old regional centres	Model outcomes suggest that the attractiveness of medium-sized regional centres will rise along with increased energy prices. This advantage should be exploited by regional centres	• Capacity improvements on existing public transport routes • Advertising new housing sites and inner-city development

Interestingly, none of the participants suggested that subsidising energy costs directly should be part of their respective strategies or policies.

The majority of the measures suggested by decision-makers belonged to either the housing or transport field. One reason is certainly that decision-makers at all municipal levels believe the housing sector to have great potential for saving energy, e.g. by renovating existing buildings to save energy, and/or by setting higher energy standards for new buildings. The large number of measures suggested in the field of mobility is mainly caused by the assumption that travel as such should still be possible for their citizens or residents, but perhaps by a different mode of transport than before—such as switching from, e.g. car to public transport. Many measures tended to improve public transport services or promote a switch of transport modes. Where travelling by car is the only viable option, participants mentioned ridesharing programmes or related ideas as ways forward. The overall aim is to keep the level of mobility for their residents at least constant, or improve the existing options of transport means for necessary trips.

The results of the €LAN model reveal that the current social disparities within the Hamburg Metropolitan Region will deepen in line with rising energy prices. This result is in part a consequence of a higher car dependency in rural areas and a reasonably well developed mass public transport system in urbanised areas. Not only are different regions differently affected by rising energy prices, but different types of municipalities also experience different effects. Regional centres will face an over-heating housing market, as increasing energy prices create additional demand along with the noticeable trend of reurbanisation. One consequence of this development will be that existing mass transport infrastructure will experience increased demand along the main axes and at the main connecting nodes.

At the other end of the scale, rural areas will struggle to compensate for their car dependency by using various strategies to offer other benefits. Developing additional independent structures for energy production or even attracting more jobs are some of the measures municipalities are looking for, depending on their own requirements and capacities.

The effects of increasing energy prices simulated for distances travelled in €LAN show that while the City of Hamburg will experience an increase in traffic, the average distance travelled per day and per person in rural and low-income municipalities will drop by approximately 20 km in comparison to 2010. This sharp reduction in distances travelled affects whole regions, such as Harburg, Storman and Segeberg. Interestingly, inhabitants of some municipalities (mostly in the east) would be less affected or would even drive more than in the reference year. This would occur due to a lack of employment, because high energy prices (among other factors) help to contribute to turning part-time or small-income jobs unprofitable, if commuting costs consume a large fraction of that income. Coupled with the relatively low overall income levels in these areas already, these effects would deepen social inequities in the region.

The results indicate that it is not possible for all municipalities to compensate for the effects of rising energy prices in full. Moreover, some effects might cause already existing trends (demographic change, migration trends, etc.) to increase or to progress at a faster pace.

10.10   Future Options for Using New Modes of Integration

The participants agreed that the biggest value of the project came from the opportunity to discuss the effects of rising energy costs and possible responses in an unusual setting and with an unprecedented combination of decision-makers. They stated that the views from all regional classifications and different federal bodies enabled them to investigate the problem from multiple angles, and that they rarely had the opportunity to meet in such constellations in their professional setting or field of responsibility.

They also considered the presentation of their strategies incorporated into a model to be valuable, because it allowed them to gain an overview of possible developments and effects of policies, even when there was no numeric output to quantify the effects of individual policies or measures.

The serious game created an interface in which the decision-makers were able to freely discuss a complex problem with other stakeholders that was not (yet) on the agenda but deemed very important. The value added by the serious game to the discussion about new ways for the co-design and co-production of knowledge for land use governance can be seen as an approach for taking a broader view on a complex subject. Serious games can be used as a tool to come to better informed conclusions; the idea of co-design in the context of land use as a scarce resource and in meeting sustainability goals was especially fruitful. This was true not only within various administrative levels—to broaden the view on one specific problem—but also in a context, e.g. with local stakeholders, to use serious games as a participatory co-creation tool. As an example to actively involve additionally practical experience for a broader view on a complex subject, serious games can involve farmers in the process of agroecosystem design and identify factors and patterns of local-level decision-making to openly discuss and share their ideas (Speelman et al. 2014).

This is just one option how serious games can add value to sustainable land management due to their discursive design and the wide selection of participants. The serious game enabled the participants as a group to come to a better understanding of the problem by combining views from different (occupational) backgrounds, public institutions, panels, or other more formalised structures that would not meet in the traditional course of planning and policy; this thus enhanced the formal process at an early stage.

A broader understanding of the problem via serious games does not necessarily lead to a "better" solution for a problem in the sense of increased efficiency, but it may enable decision-makers to consider more facets and thus raise the group's level of knowledge through the inherently communicative design of a serious game.

To determine all important aspects of a complex problem, planning in such a broad context is necessary, but it depends on many variables. Exploring such problems with tools such as serious games, which are not standard procedure or formalised, can add value to the whole process. Nevertheless, the non-standard design requires individual commitment. To embed such commitment into more formal structures or procedures might in some sense be counterproductive, but should be understood as acknowledged evidence for the usefulness of serious games as a tool for improving administrative and political processes.

The complex context of sustainable land use also lends itself to serious games. Many serious games utilise dynamic computer-based elements to create scenarios for the participants or to demonstrate decision results or the outcomes of the session. Participants of €LAN stressed the need for more exchange about complex problems. Another example of exchange and format on complex land use problems was LandYOUs, where participants explored dimensions of sustainability with respect to economic, social and environmental conditions, while being continuously threatened by global trade fluctuations and limited resources as described in Schulze et al. (2015) and Seppelt et al. (2014).

Different formats for exchange and exploration besides the standard methods and boards of professional planning must still be discussed further to facilitate cross-border solutions adequate for the problem. The developments of the past ten years have shown that serious games are increasingly seen as collaborative tools to enhance planning decisions and sustainable land management. Research by den Haan and van der Voort (2018) has offered a glimpse of the possibilities, citing examples of more than 40 serious games around the world in the field of sustainable land management.

These results reinforce the idea of serious games as a platform for this necessary and desired exchange as part of the preparatory process for policymaking. This is especially the case in the complex fields of planning and policymaking, where planners and politicians need a combination of the opportunities to process complex information and respond to the possible effects of their decisions. In such circumstances, an integrated computer-supported serious game like €LAN can be a valuable tool of support at all levels of government and administration.

References

Beckmann K. J. (2000). Anforderungen einer nachhaltigen Verkehrsentwicklung: Chancen einer Integration von Raum- und Verkehrsplanung. In Forschungsgesellschaft für Straßen- und Verkehrswesen (FGSV) (ed.) *Zukunftsfähige Mobilität in Stadt und Region: FSGV-Kolloquium am 31. Mai und 1. Juni 1999 in Bonn.* Köln: FGSV, pp. 5–22.

Berry A., Jouffe Y., Coulombel N., & Guivarch C. (2016). Investigating fuel poverty in the transport sector: Toward a composite indicator of vulnerability. *Energy Research & Social Science 18,* 7–20. Accessed on January 10, 2017.

Boardman, B. (1991). *Fuel poverty: from cold homes to affordable warmth.* London, New York: Belhaven Press.

Böhret, C., & Wordelmann, P. (2000). Planspiele als Methode der Fortbildung und als Entscheidungshilfe: Das computergestützte Planspiel TAU. In D. Herz & A. Blätte (Eds.), *Simulation und Planspiel in den Sozialwissenschaften: Eine Bestandsaufnahme der internationalen Diskussion* (pp. 207–230). Lit: Münster.

Brunner, K.-M., Spitzer, M., & Christanell, A. (2012). Experiencing fuel poverty. Coping strategies of low-income households in Vienna/Austria. *Energy Policy, 49,* 53–59.

Buettner B., Wulfhorst G., Ji, C., Crozet, I., Mercier, A., & Ovtracht N. (2013). The impact of sharp increases in mobility costs analysed by means of the Vulnerability Assessment. In World conference on transport research society (ed.) *The 13th World Conference on Transport Research (WCTR).*

BMVBS/BBSR (März 2009) *Chancen und Risiken steigender Verkehrskosten für die Stadt- und Siedlungsentwicklung unter Beachtung der Aspekte der postfossilen Mobilität.* https://www.bbsr.bund.de/BBSR/DE/Veroeffentlichungen/BBSROnline/2009/DL_ON0620 09N.pdf?__blob=publicationFile&v=2. Accessed on April 12, 2016.

Bundesamt für Bauwesen und Raumordnung. (1999). *Steuerung der Flächennutzung.* Bonn: Selbstverlag des Bundesamts für Bauwesen und Raumordnung.

Bundesministerium für Bildung und Forschung. (2013). *Nachhaltiges Landmanagement; Eine Herausforderung für alle.* https://www.fona.de/mediathek/pdf/Broschuere_Landmanage ment_bf.pdf. Accessed on July 12, 2016.

Bundesministerium für Wirtschaft und Energie. (2018). *Gesamtausgabe der Grafiken zu Energiedaten.* BMWI. https://www.bmwi.de/Redaktion/DE/Downloads/Energiedaten/energieda ten-gesamt-pdf-grafiken.html. Accessed on February 21, 2019.

Büttner, B. (2017). *Consequences of sharp increases in mobility costs on accessibility: Suggestions for individual and public development strategies.* München.

Carollo, S. (2011). *Understanding oil prices: A guide to what drives the price of oil in today's markets.* Chichester, West Sussex, UK: John Wiley & Sons Ltd.

den Haan, R.-J., & van der Voort, M. (2018). On evaluating social learning outcomes of serious games to collaboratively address sustainability problems. *A Literature Review. Sustainability, 10,* 4529. https://doi.org/10.3390/su10124529

Deutsche Energie-Agentur GmbH. (2012). *Der dena-Gebäudereport 2012; Statistiken und Analysen zur Energieeffizienz im Gebäudebestand*
Dodson, J., & Sipe, N. (2007). Oil vulnerability in the Australian city: Assessing socioeconomic risks from higher urban fuel prices. *Urban Studies, 44*(1), 37–62.
Dodson, J. & Sipe, N. (2008). Shocking the suburbs: Urban location, homeownership and oil vulnerability in the Australian city. *Housing Studies, 23*(3), 377–401. Accessed on July 13, 2016.
Dubois, U., & Meier, H. (2016). Energy affordability and energy inequality in Europe: Implications for policymaking. *Energy Research & Social Science, 18,* 21–35 Accessed on January 10, 2017.
Ebert, G. (1992). Planspiel—eine aktive und attraktive Lehrmethode. In H. Keim & W. Buddensiek (Eds.), *Planspiel, Rollenspiel, Fallstudie: Zur Praxis und Theorie lernaktiver Methoden* (pp. 25–42). Wirtschaftsverl. Bachem: Köln.
EWI and Prognos. (2006). *Auswirkungen höherer Ölpreise auf Energieangebot und-nachfrage: Ölpreisvariante der Energiewirtschaftlichen Referenzprognose 2030.* Untersuchung im Auftrag des Bundesministeriums für Wirtschaft und Technologie, Berlin: Köln, Basel.
Fiorello, D., Huismans, G., López, E., Marques, C., Steenberghen T., & Wegener M. (2006). *Transport strategies under the scarcity of energy supply: STEPs Final Report.* The Hague: Buck Consultants International.
Fischer, C. (2008). Planspiel Planwirtschaft. *Gesellschaft, Wirtschaft, Politik(GWP) Sozialwissenschaften für politische Bildung 57*(1), 137–146.
Frondel, M., Andor, M., Ritter, N., Sommer, S., Vance, C., & Matuschek, P. (2015). *Erhebung des Energieverbrauchs der privaten Haushalte für die Jahre 2011–2013: Bericht für das Projekt Erhebung des Energieverbrauchs der privaten Haushalte für die Jahre 2006–2013 Forschungsprojekt Nr. 54/09 des Bundesministeriums für Wirtschaft und Technologie, BMWi.*
Gertz, C., Maaß, J., & Guimarães, T. (2015). *Auswirkungen von steigenden Energiepreisen auf die Mobilität und Landnutzung in der Metropolregion Hamburg: Ergebnisse des Projekts €LAN— Energiepreisentwicklung und Landnutzung.* Münster, Hannover: Monsenstein und Vannerdat; Technische Informationsbibliothek u. Universitätsbibliothek.
González-Eguino, M. (2015). Energy poverty: An overview. *Renewable and Sustainable Energy Reviews, 47,* 377–385. Accessed on January 11, 2017.
Guimarães, T., Maaß, J., & Gertz, C. (2014). Integrating a land use transport model with a serious game for supporting planning decisions under rising energy prices. *Transportation Research Procedia, 4,* 241–254. https://doi.org/10.1016/j.trpro.2014.11.019
Herz, D., & Blätte, A. (Eds.). (2000). *Simulation und Planspiel in den Sozialwissenschaften.* Lit, Münster: Eine Bestandsaufnahme der internationalen Diskussion.
Hitzler, S., Zürn, B., & Trautwein, F. (2010). Status Quo der europäischen Planspielszene. In F. Trautwein, S. Hitzler, & B. Zürn (Eds.), *Planspiele—Entwicklungen und Perspektiven: Rückblick auf den deutschen Planspielpreis 2010* (pp. 217–230). Norderstedt: Books on Demand.
IEA. (2016). *Oil Market Report.* https://www.iea.org/oilmarketreport/omrpublic/
Joldersma, C., & Geurts, J. L. A. (2000). Simulation/ Gaming for participatory policy-making. In D. Herz & A. Blätte (Eds.), *Simulation und Planspiel in den Sozialwissenschaften: Eine Bestandsaufnahme der internationalen Diskussion* (pp. 259–273). Lit: Münster.
Korte, L., & Lehmbrock, M. (2009). *Planspiel interkommunale Verkehrsentwicklungsplanung in der Region München: Dokumentation.* Berlin: Difu.
Kraftfahrtbundesamt (KBA). (2017). *Zahlen zum 1. Januar 2017 im Überblick.* Available at: https://www.kba.de/DE/Statistik/Fahrzeuge/Bestand/bestand_node.html. Accessed on November 2, 2017.
Larson, W., & Yezer, A. (2015). The energy implications of city size and density. *Journal of Urban Economics, 90,* 35–49. Accessed on February 17, 2016.
Legendre, B. & Ricci, O. (2015). Measuring fuel poverty in France: Which households are the most fuel vulnerable? *Energy Economics, 49,* 620–628. Accessed on January 4, 2017.
Mattioli, G. (2014). Where sustainable transport and social exclusion meet: Households without cars and car dependence in great Britain. *Journal of Environmental Policy & Planning, 16*(3), 379–400.

Newman, N. (2008). The high price of oil. *Engineering & Technology, 3*(8), 46–49.

Newman, P., & Kenworthy, J. R. (1989). Gasoline consumption and cities—a comparison of United-States cities with a global Survey. *Journal of the American Planning Association, 55*(1), 24–37.

Projekt €LAN. (2011). *Gesonderte Berichterstattung zum erreichten Arbeitsstand im Arbeitspaket C: Meilenstein 3.*

Reames, T. G. (2016). Targeting energy justice: Exploring spatial, racial/ethnic and socioeconomic disparities in urban residential heating energy efficiency. *Energy Policy 97*, 549–558. Accessed on January 5, 2017.

Rebmann, K. (2001). *Planspiel und Planspieleinsatz: Theoretische und empirische Explorationen zu einer konstruktivistischen Planspieldidaktik.* Hamburg: Kovac.

Roberts, D., Vera-Toscano, E. & Phimister, E. (2015). Fuel poverty in the UK: Is there a difference between rural and urban areas? *Energy Policy 87.* 216–223. Accessed on January 13, 2017.

Schulze, J., Martin, R., Finger, A., Henzen, C., Lindner, M., Pietzsch, K., et al. (2015). Design, implementation and test of a serious online game for exploring complex relationships of sustainable land management and human well-being. *Environmental Modelling & Software, 65,* 58–66. https://doi.org/10.1016/j.envsoft.2014.11.029

Seppelt, R., Martin, R., Finger, A., Henzen, C., Lindner, M., Pietzsch, K., et al. (2014). Experiences with a serious online game for exploring complex relationships of sustainable land management and human well-being: LandYOUs. In International Environmental Modelling and Software Society (ed.) Visions for Environmental Modeling, pp. 1121–1128

Shepherd, S. P., Pfaffenbichler, P., Martino, A., Fiorello, D., & Christidis, P. (2008). The effect of oil prices on transport policies for Europe. *International Journal of Sustainable Transportation, 2*(1), 19–40.

Speelman, E. N., García-Barrios, L. E., Groot, J. C. J., & Tittonell, P. (2014). Gaming for smallholder participation in the design of more sustainable agricultural landscapes. *Agricultural Systems, 126,* 62–75. https://doi.org/10.1016/j.agsy.2013.09.002

Vissers, G., & van der Meer, F.-B. (2000). Social simulation of polycentric policy making: Ex ante assessment of administrative reform in the region of Rotterdam. In D. Herz & A. Blätte (Eds.), *Simulation und Planspiel in den Sozialwissenschaften: Eine Bestandsaufnahme der internationalen Diskussion* (pp. 231–257). Lit: Münster.

Walker G. and Day R. (2012) Fuel poverty as injustice: Integrating distribution, recognition and procedure in the struggle for affordable warmth. *Energy Policy* 49: 69–75. Accessed on January 9, 2017.

Wegener, M. (2009). Energie, Raum und Verkehr: Auswirkungen hoher Energiepreise auf Stadtentwicklung und Mobilität. *Wissenschaft & Umwelt Interdisziplinär, 12,* 67–75.

Wegener, M. (2009b). Energy scarcity and climate change: the challenge for urban models. In University of Hong Kong (ed.) *Proceedings of the 11th International Conference on Computers in Urban Planning and Urban Management (CUPUM 2009).* Hong Kong.

Chapter 11
Real-World Laboratories Initiated by Practitioner Stakeholders for Sustainable Land Management—Characteristics and Challenges Using the Example of Energieavantgarde Anhalt

Helga Kanning, Bianca Richter-Harm, Babette Scurrell, and Özgür Yildiz

Abstract Real-world laboratories have gained substantially in importance as a format in sustainability and transformation research in recent years in Germany. This increase in significance is associated with the expectation of fostering and experimentally investigating transformations towards sustainability under real-world conditions in a bid to gain knowledge of their dynamics, to identify characteristics of successful transformation processes, and to be able to transfer this knowledge to other cases. Real-world laboratories are usually managed by a scientific partner, enabling use to be made of established procedures and methods in areas such as knowledge integration. Where responsibility for coordinating real-world laboratories lies with practitioner stakeholders, there is promising potential for their deployment. However, it also gives rise to situations, processes and challenges that are new to all parties involved and that have yet to be explored. In principle, experimental approaches that are characteristic of real-world laboratories are not new in the field of sustainable land management and spatial development. However, they are not traditionally alluded to as the real-world

---

For reasons of readability and simplicity, no use is made of the masculine and feminine form when referring to people or job titles in this publication. However, all genders are implied at all times.

H. Kanning (✉) · B. Richter-Harm
Sustainify GmbH, Institut für Nachhaltige Forschung, Bildung, Innovation, Große Düwelstr. 28, 30171 Hannover, Germany
e-mail: kanning@sustainify.de

B. Richter-Harm
e-mail: post@sustainify.de

B. Scurrell
Schefferweg 2, 12249 Berlin, Germany
e-mail: babette.scurell@bund.net

Ö. Yildiz
Department of Environmental Economics, Technische Universität Berlin, Straße des 17. Juni, 10623 Berlin, Germany
e-mail: oezguer.yildiz@campus.tu-berlin.de

© The Author(s) 2021   207
T. Weith et al. (eds.), *Sustainable Land Management in a European Context*,
Human-Environment Interactions 8, https://doi.org/10.1007/978-3-030-50841-8_11

laboratory format. The two desiderata above provide the starting point for the present article. The aim of this article is to classify and reflect on the possibilities generated by real-world laboratories that have been initiated by practitioner stakeholders. A prime example of such real-world laboratories are those developed by Energieavantgarde Anhalt. This registered association wishes to contribute to sustainable land management in the context of the energy transition in rural areas, featuring small and medium-sized towns. A comparative analysis of these real-world laboratories is conducted using core characteristics from the scientific debate on real-world laboratories. As a result, the insight gained from this analysis can be used for future development and research.

Keywords Regional energy transition · Real-world laboratories as a whole, within a project · Practitioner stakeholder's initiative · Participation · Co-design, Co-production

11.1 Introduction

Real-world laboratories are gaining importance in the field of sustainable land management and spatial development (see, for example, Augenstein et al. 2016; Hahne and Kegler 2016). In Germany, the discussion on these innovative formats was primarily triggered by the flagship report by the German Advisory Council on Global Change (WBGU) "World in Transition: A Social Contract for Sustainability" (WBGU 2011). In its report, WBGU recommends a new kind of interaction between politics, society, science and the economy (ibid.: 26). In this context, transdisciplinary and transformative sustainability research is encouraged (Schneidewind and Singer-Brodowski 2013: 2015) and, in particular, the format of real-world laboratories promoted (Schneidewind 2014).

The popularity of real-world laboratories has increased dramatically recently in Germany, due not least to the initiation of several support programmes at the federal and state level.[1] As such, both the variety of publications and publication density are developing dynamically and at a disproportionately high rate, as pointed out clearly in the analysis by Schäpke et al. (2018a). At the same time, a debate has emerged that is controversial to some extent. While some see little new in the real-world laboratory format, others pin their hopes on the (potentially) innovative power of real-world laboratories (Beecroft and Parodi 2016: 4). In other words, the debate on real-world laboratories and the potential they have for social transformations and

---

[1]The state government of Baden-Württemberg in particular was quick to embrace the WBGU recommendations, establishing a "Science for Sustainability" group of experts (MWK 2013) and initiating two support programmes for real-world laboratories (BaWü Labs); for an overview, see: https://mwk.baden-wuerttemberg.de/de/forschung/forschungspolitik/wissenschaft-fuer-nachhaltigkeit/reallabore/ (last accessed on 16 July 2019). The Federal Ministry of Education and Research (BMBF) also promotes real-world laboratories, for example in the context of the "City of the Future" funding initiative (Schmidt 2017), addressing not only universities in the process, but also communities or municipalities as practitioner stakeholders.

for transformation research has only just begun.[2] Real-world laboratories are "in their infancy" (Beecroft and Parodi 2016: 4), and a detailed methodological and theoretical concept has yet to be developed (Grunwald 2016: 204f). Although it can be assumed that real-world laboratories initiated by practitioner stakeholders have high deployment potential on account of the practical approach they take, they have attracted little attention in the scientific debate to date (Engels and Rogge 2018; Menny et al. 2018).

Real-world experiments are generally considered to be the core of the real-world laboratory approach (Schäpke et al. 2017: 3). This idea is not new in the field of sustainable land management. In fact, the approach of experimentally investigating social change processes at the urban level goes back to the sociological Chicago School of the 1920s (Gross et al. 2005: 65ff; Schneidewind 2014: 3). In land management, for example, there are many projects and model projects of an experimental nature that are not, however, termed real-world laboratories. Examples include the International Building Exhibitions (IBA), state and national flower shows, the regional structural aid measures in North Rhine-Westphalia called the REGIONALE, and various regional development processes in the context of European regional assistance (see De Flander et al. 2014: 285; Hohn et al. 2014). As yet, the experience gained in these measures is largely detached from the real-world laboratory debate, meaning that a great deal of research is needed to bring together these aspects (De Flander et al. 2014: 285).

These two desiderata provide the starting point for the present article. The objective is to integrate into the scientific debate real-world laboratories that have been initiated and coordinated by practitioner stakeholders for the purpose of sustainable land management, and to reflect on the possibilities and limitations of those approaches. In the process, questions that have not been addressed in the scientific discussion to date are of importance: What challenges arise when real-world laboratories are initiated by practitioner stakeholders? Are those challenges similar to those arising in real-world laboratories designed by scientists? How do they differ? What opportunities do they offer, and what added value can be expected from the real-world laboratory format? To answer these questions, Sect. 11.2 provides an account of the real-world laboratories initiated by the Energieavantgarde Anhalt (EAA) association, which include the urban laboratories undertaken within the joint research project "The re-productive town"[3] funded by the Federal Ministry of Education and Research (BMBF). The core characteristics of real-world laboratories are then identified from the scientific literature. These characteristics provide the theoretical basis for discussing the special features of real-world laboratories that have been initiated by practitioner stakeholders (Sect. 11.3). This discussion is based not

---

[2]Concerning classification in transdisciplinary research and transformation research, see, e.g. Wittmayer and Hölscher (2016), Rogga et al. (2018).

[3]The project entitled "The re-productive town. Changing towns for achieving the energy and sustainability transition" [original in German: Die re-produktive Stadt. Die Stadt verändern, um die Energie- und Nachhaltigkeitswende zu schaffen] receives BMBF funding under the FONA/Social-Ecological Research: "Sustainable Transformation of Urban Areas" funding line from August 2016 to July 2019; see https://re-produktive-stadt.energieavantgarde.de.

only on the experience gained from the BMBF-funded project "The re-productive town", but also on insights from three workshops held with EAA members and other interested participants (business representatives, especially from the utilities sector; representatives from local government and politics and from science) in 2017. The article concludes with a critical analysis and an outlook for future developments (Sect. 11.4).

11.2   Real-World Laboratories Initiated by the Practitioner Stakeholder Energieavantgarde Anhalt e.V.

Energieavantgarde Anhalt (EAA) is an association that acts as a network of stakeholders comprising civil activists, municipalities and rural districts, companies and other institutions in the Anhalt-Bitterfeld-Wittenberg region. This network is committed to accelerate the energy transition in the region in cooperation with national and European partners. The approach was developed in the context of the profound, multiple socio-economic transformation processes that pose huge challenges to towns and regions, such as the closure of businesses and the loss of livelihoods, demographic change, and the energy transition. The direct impact of these processes is very much apparent in the Anhalt-Bitterfeld-Wittenberg region: high cost pressure relating to infrastructure, a sharp fall in property prices, the demolition of entire neighbourhoods. A wide range of technical, economic and socio-cultural innovations are needed to meet the challenges associated with these developments in this region dominated by lignite mining and the chemical industry. These innovations radically change land uses, creating new decentralised, interconnected and energy systems based on renewables, as well as new urban-rural relations. In the process, EAA places particular emphasis on the regionalisation of energy production and energy use and on sector coupling. To achieve this, developments in the area of prosumer models and demand-side management measures should encourage not only resource efficiency, but also system-supporting, flexible energy consumption behaviour, and enable as many citizens as possible to participate in the regional energy transition through regional added value and democratic processes.

Since there is no ready guidance on how to meet these challenges and since a wide range of individual issues need to be resolved, the association calls this large-scale regional experiment the "*Anhalt Real-World Laboratory*" (www.energieavant garde.de). In this laboratory, partners engaged as practitioners in the region and scientists join forces to design a variety of sub-laboratories and experimental setups. With this in mind, the association brings together within its framework not only local authorities, public utility companies and technology companies from the renewables sector, but also civil society interest groups. The projects in the region initiated by EAA are generally based on the experience and issues raised by association members within their everyday operations and on collaboration with research institutions in other projects. In their role as project initiators and project coordinators in the region, members of the association explicitly represent the interests of the association and of the regional stakeholders it represents. As a result, the focus is on searching for workable approaches for promoting sustainable development by using renewable energies and achieving high resource efficiency. Considering contributions from the current scientific real-world laboratory debate, this regional institution could also be characterised as a "real-world laboratory as a whole" [own translation] (Beecroft et al. 2018: 80) where various transdisciplinary sustainability projects are implemented.

The joint research project entitled "The re-productive town", which has received BMBF funding for three years, is one of the outcomes of EAA's activities in the Anhalt region. Initiated by EAA, the research alliance comprises EAA and Bitterfeld-Wolfen Town Council as its practice partners, and Brandenburg University of Technology (BTU) Cottbus-Senftenberg/Chair of Urban Technical Infrastructure, the Fraunhofer Institute for Microstructure of Materials and Systems (IMWS) and inter 3 Institute for Resource Management as its science partners. The project is accompanied scientifically by sustainify Institut für nachhaltige Forschung, Bildung, Innovation. In the research project, the town of Bitterfeld-Wolfen is taken as an example of the urban planning challenges associated with transformation. This town seems to be particularly suited to develop and test new approaches for social-ecological urban development. The starting point of the project is the energy sector, from which inroads are made into agriculture and forestry, architecture and building services, industry and finance, citizenry, the urban economy and the urban landscape. Possibilities are systematically sought to consider unexploited resources such as brownfield sites, sun, wind and green waste as well as secondary resources such as waste heat and refuse as a starting point for something new. These innovations are then reutilised for the benefit of the town and its inhabitants or the processes that generate them are directly changed. Conceptually, the approach refers back to the concept of (re)productivity proposed by Biesecker and Hofmeister (2006). According to this concept, (urban) production and consumption processes must be designed in such a way that the town maintains or even improves its material/energy and economic/social reproductive capability in order to remain sustainable or to ensure its long-term survival. The aim is to use the systematic improvement of the material/energy and economic/social reproductive capability of Bitterfeld-Wolfen Town to develop a blueprint for a possible transformation path for a new, yet very common type of town as a result of territorial reforms: an extensive, medium-sized, polycentric town that can be expected to offer new starting points for energy and sustainability transition and, as a result, new townscapes and urban landscapes.

Urban laboratories are a core format. *Urban laboratories* are site-specific participatory and communication platforms that map ongoing local transformation processes and enable broad participation. They provide the experimental basis for developing, negotiating and implementing into urban design solutions for the use of secondary resources in urban spaces in cooperation with the population, companies and the administration. This is undertaken in work phases of living labs or real-world laboratories such as co-design, co-creation, co-exploration, co-experimentation/testing and co-evaluation steps. More specifically, four *urban laboratories* representing neighbourhoods typical of medium-sized towns were selected in consideration of characteristics such as resource potentials and stakeholder constellations. These neighbourhoods are

- A central, inner-city area, characterised by a combination of brownfield and industrial areas (neighbourhood type—inner-city brownfield in a central location: "Am Plan" urban laboratory)
- A detached housing estate, including listed buildings, that faces extensive changes in ownership structure (neighbourhood type—existing housing estate with a garden city character: "Gartenstadt" urban laboratory)
- A new housing estate with detached houses and multiple dwellings on an urban open space (neighbourhood type—new residential area: "Krondorfer Wiesen" urban laboratory)
- A multiple dwelling demolition area characterised by industrial housing construction as well as demographic and socio-economic challenges (neighbourhood type—industrial prefabricated large housing estate: "Wohnkomplex 4/4" urban laboratory).

11.3   A Comparison of Core Characteristics

As outlined in the introduction, there is as yet no uniform theoretical and detailed methodological concept of real-world laboratories (Grunwald 2016: 204f), and therefore no uniform definition either. However, several scientific institutions are currently performing further groundwork, especially also in the context of research in support of the real-world laboratories (BaWü Labs) funded in Baden-Württemberg.[4] On the international arena, there are also a multitude of other approaches that are similar to the real-world laboratory format or that were used as its basis. These include living labs, sustainable living labs and urban transition/living labs (for a comparative overview, see Schäpke et al. 2017: 28ff; Schäpke et al. 2018b). Furthermore, an almost inflationary (and simultaneously unspecific) use is currently being made of the term "lab" in other fields.

According to a definition originally introduced by Schneidewind (2014), a real-world laboratory generally describes "… a societal context in which researchers carry out interventions in the sense of 'real-life experiments' in order to learn about social dynamics and processes" [own translation] (Schneidewind 2014: 3). Real-life experiments are considered to be the core of the real-world laboratory approach (Schäpke et al. 2017: 3 with reference to WBGU 2014, 2016; Schneidewind 2014; De Flander et al. 2014; MWK 2013; Wagner and Grunwald 2015). The idea is to transfer the term "laboratory", as used in the natural sciences context, to the analysis of social and

---

[4]These include, in particular, the Wuppertal Institute (WI), the Institute for Technology Assessment and Systems Analysis (ITAS) at the Karlsruhe Institute of Technology (KIT), the Institute for Social-Ecological Research (ISOE), as well as Leuphana University and the University of Basel, especially in the context of research in support of BaWü Labs. BaWü Labs are supported by two teams of researchers: (1) the "ForReal" team (WI, ISOE, Leuphana University), (2) BF-Uni Basel (University of Basel).

political processes in concrete social contexts (Schäpke et al. 2017: 4). According to the definition coined by Gross et al. (2005), a hybrid form of the experiment is associated with this term, ranging between the production and application of knowledge and situation-specific and controlled boundary conditions (Schneidewind 2014: 2). Conceptually, real-world laboratories therefore build on the 'experimental turn' in the social, economic and sustainability sciences, and are similar to other transdisciplinary and participatory research approaches such as transdisciplinary case studies, participatory action research, fieldwork, intervention research or transition research (Schäpke et al. 2017: 4, referring in each case to prominent representatives of the approaches named).

In the scientific landscape, the concept of the real-world laboratory is therefore easily expandable and currently formative. In practice, however, the term is rejected by some because it evokes associations that experiments are being performed on the participants (Grießhammer and Brohmann 2015: 22). The term "urban laboratory" proved to be useful for work in the "The re-productive town" project. This is because the term is used in the field of urban development, albeit with diverse and different meanings, such as for educational institutions with an experimental laboratory character.

In order to shed light on what characterises real-world laboratories that have been initiated and largely shaped by practitioner stakeholders, we refer below to the core characteristics listed by Parodi et al. (2016): research orientation, normativity, transdisciplinarity, transformativity, civil society orientation/participation, long-term nature and laboratory character (see Table 11.1). These core characteristics largely correspond to or overlap with characterisations proposed by other authors such as WBGU (2016), Schäpke et al. (2017), Defila and Di Giulio (2018a). We add another core feature—continuous processes of reflection and learning with regard to one's own research practice and social effect; these characterise the research process (e.g. Schäpke et al. 2018b; Schneidewind and Singer-Brodowski 2015).

Based on these core characteristics, an outline is given below of how real-world laboratories initiated and coordinated by EAA can be characterised, whether and how they differ from those real-world laboratories that are initiated and coordinated by stakeholders from science, whether they face challenges and, if so, what those challenges are. The definition of the relevant core characteristics is given in Table 11.1.

11.3.1 Regarding Research Orientation

In the *Anhalt Real-World Laboratory* 'as a whole', the EAA association offers interested researchers the region's problems concerning energy design, energy policy and energy management, some of which have already been formulated and structured; its contacts with regional practitioner stakeholders; and its expertise in the development of a sustainable regional energy system for transdisciplinary research. In this sense, EAA serves as an institution for sustainability and transformation research;

application-oriented research is explicitly mentioned in the association's statutory objectives. However, whether or to what extent these institutions for real-world laboratories must always have a scientific character is not deemed absolutely necessary in the previous discussion. Instead, the association focuses primarily on practical orientation and on the search for practicable approaches towards sustainable development in its areas of interest; in the process, analytical work and the consideration of internal scientific interests are accepted as prerequisites for joint research and

Table 11.1 A comparison of the core characteristics and real-world laboratories undertaken by Energieavantgarde Anhalt (authors' compilation)

Core characteristics	Real-world laboratories …	Anhalt real-world laboratory as a whole	Urban laboratories in the project "The re-productive town"
Research orientation	… act as scientific facilities in sustainability and transformation research[1]	Not necessarily, focuses on practical orientation	✓ Research closely oriented to practice
Normativity	… are oriented towards the principles of sustainable development[1]	✓	✓
Transdisciplinarity	… function in a transdisciplinary way They directly connect science and society (practitioner stakeholders) and use forms and methods of transdisciplinary research in their experiments[1]	✓	✓
Transformativity	… conduct transformative research. They are hybrid undertakings that aim to concurrently achieve scientific findings and social design. They facilitate sustainability research and simultaneously make experimental contributions to sustainable development[1]	✓	✓

(continued)

Table 11.1   (continued)

Core characteristics	Real-world laboratories …	Anhalt real-world laboratory as a whole	Urban laboratories in the project "The re-productive town"
Civil society orientation, participation[3]	… integrate citizens and/or civil society in particular as strong partners and decision-makers into their work from the beginning … embrace participation, from information and consultation to cooperation and empowerment, and develop their transdisciplinary experiments in co-design[1]	✓ But opposite direction of activity: drive comes from civil society, which involves science	✓ But opposite direction of activity: drive comes from civil society, which involves science
Long-term nature	… are long-term research facilities spanning (many) decades[1]	✓	–
Laboratory character	… are laboratories. They are a transdisciplinary infrastructure in order to ensure the best and most stable conditions possible for experimental research and observation in complex real-world contexts[2]	✓	✓
Continuous reflection and learning process[4]	Research in real-world laboratories is devised and understood as a continuous reflection and learning process with regard to one's own research practice and social effect	✓	✓

[1]Own translation of Parodi et al. (2016: 16)
[2]Own translation of Parodi et al. (2016: 16f)
[3]Own addition to description of characteristics
[4]Own addition based on Schäpke et al. (2017: 5), Schneidewind and Singer-Brodowski (2015)

development activities. This principle of strong practical orientation can be adapted accordingly to current circumstances and needs at the specific project level.

One example are the *urban laboratories* in the BMBF joint research project entitled "The re-productive town"; these feature an explicit research orientation. The research questions were defined jointly by the scientific partners and the practice partners (co-design); they are accessible for scientific analysis and for practical changes for transformation towards sustainability.

11.3.2   Regarding Normativity

The normative orientation towards sustainability is one of the association's implicit statutory objectives and, as such, of the *Anhalt Real-World Laboratory*. This orientation is specified in the association's statutes on objectives such as to contribute to environmental protection and climate action, and the objective to conserve the natural basis of life.

In *urban laboratories*, the normative assumptions, principles and objectives regarding the reference to the concept of re-productivity by Biesecker and Hofmeister (2006) are made explicit. In this context, the following insight was gained from ongoing work: scientific partners proceed in accordance with elaborated sustainability concepts in real-world laboratories, and it would be helpful for companies, civil society organisations, other institutions and local authorities to use or develop concrete tools in the practical implementation of real-world laboratories. Examples that would make the integrative concept of sustainability tangible for practitioners in the process include environmental protection concepts, corporate social responsibility standards, a local climate action plan, the European Energy Award and other quality management systems. It may also be helpful in this context to adapt for practical use scientific-theoretical sustainability approaches such as the concept of re-productivity in an intermediate step, and to prepare such approaches for practical implementation in the real world (Yildiz et al. 2012).

11.3.3   Regarding Transformativity

In the *Anhalt Real-World Laboratory*, the EAA association focuses primarily on the shaping of society in terms of the energy transition, mainly by way of local experimental contributions. To do this, EAA draws on findings resulting from sustainability research. The work of EEA backs two aspects of real-world laboratories: first, the *Anhalt Real-World Laboratory* sees itself as an element of various niche experiments embedded in structuring processes somewhere between the niche level and the regime level. Second, in line with its objective, the work performed by the association should help further develop transformative sciences by portraying and investigating the abstract format of transformation in the real-world laboratory as a physical environment.

Owing to their origins, *urban laboratories* are likewise primarily practice-oriented. Potential changes in the practices of resource utilisation by municipal stakeholders play a key role in the selection of neighbourhoods, and therefore also in the constitution of the problem, the institutions and people involved, and the methods and intensity of participation. This framework also yields a wide range of options for sustainability research and scientific evidence (e.g. the methodological operational-isation of the characteristics of re-productivity in criteria for assessing technical and socio-economic solutions) (Schön et al. 2013).

The strong practical orientation in the EAA real-world laboratories necessitates a careful reflection and evaluation of the approaches taken so as to be able to make statements on the effect of interventions and on the course of transformation processes in real-world laboratories. The small scale and reach of the measures that can be implemented concerning sustainable urban development generally make it difficult to formulate transferable results, which is currently being hotly discussed as a general phenomenon of real-world laboratories.[5] One cannot help but suspect that a specific contribution to resource efficiency or a viable use of renewable energies in a certain neighbourhood arises more by accident than by design due to a certain constellation of problems and stakeholders. It is then impossible to repeat such a success at other locations. The result is that strong practical orientation represents a restriction, especially for scientific partners. Then again, precisely these small-scale changes can occur and be documented in the *Anhalt Real-World Laboratory*. These are the small steps that represent the details of social transformation, which is ultimately of greater significance from a practical point of view.

11.3.4   Regarding Civil Society Orientation, Participation

This characteristic exhibits the biggest difference between the state of the scientific discussion and the approaches taken by the EAA real-world laboratories. The drive to initiate the *Anhalt Real-World Laboratory* came from civil society, which involved scientists in the project as strong partners. The initiative for the BMBF joint research project and its *urban laboratories* also came from EAA. As such, the direction of activity is opposite to the characteristic portrayed in the literature. The idea for a scientifically supported, experimental transformation of the regional energy system arose from the realisation that the special constellation of stakeholders seeking change and the decisive issue of regional energy supply involving the broader shaping and economic participation of the population became apparent as an opportunity for innovative action. Although science, lobby institutions and financial backers were then involved in the subsequent establishment of the *Anhalt Real-World Laboratory* at the

---

[5]Concerning this, see also, e.g. Krohn et al. (2017) and www.td-academy.de (last accessed on 16 July 2019).

very beginning, the format can be described as a laboratory initiated by practitioner stakeholders, because:

- Practitioner stakeholders from the Anhalt region raised the issue of the regional energy transition, and had already addressed this issue with their own commitment using the resources available to them for more than three years,
- The establishment of the real-world laboratory was only conceivable and feasible due to the active work of key regional stakeholders seeking to change the existing system, and
- It was only possible to address additional practitioner stakeholders with success because of the trusting relationships among regional stakeholders that had been in existence for several years.

Against this backdrop, an important finding that is a compelling case for the establishment and long-term operation of real-world laboratories by practitioner stakeholders is the fact that it takes a long time to establish successful participatory constellations. This longer-term option is missing in *urban laboratories* (see the long-term nature criterion). To achieve effective cooperation in the transformation of society, all stakeholders must also act proactively so as to position their issues and other concerns. After all, a form of cooperation that always expects the drive and organisation to come from the same partner will soon show signs of fatigue. It is clear that the involvement of local stakeholders remains a challenge, even if the real-world laboratory is initiated by practice partners. Even if a region has activists who are interested in transformative research, this does not mean that all of the stakeholders needed to tackle a specific issue are willing to get involved. At best, the initiating practice partner will be powerful, influential and well networked, enabling it to organise the constructive participation of the necessary stakeholders.

11.3.5   Regarding the Long-Term Nature and Laboratory Character

The *Anhalt Real-World Laboratory* is designed for the long term. The association seeks to establish transdisciplinary infrastructure with adequate physical and personnel resources (criterion laboratory character) to be able to ensure the best, most stable possible conditions for experimental research and observation in complex real-world contexts (see Parodi et al. 2016). In contrast, *urban laboratories* are based on a three-year time frame and are project-related, despite ideally desiring their longer-term and autonomous establishment.

The availability of sufficient resources is a prerequisite for this. However, the non-profit association has very limited resources. One possibility would be to raise funds by providing services, but this would imply a change of role to that of a market economy stakeholder like an energy agency, planning office or consulting agency. However, if EAA represents its own business interests, it runs the risk of losing credibility with regard to the handling of sensitive data. This is likely to affect the trust required to acquire regional cooperation partners for transformative research, and the quasi-public role of mediating between possibly competing partners could only be played to a very limited extent in the best case (Yildiz and Schön 2014).

As a result, in order to maintain transdisciplinary infrastructure in the long run, other ways of obtaining sustainable funding must be found by EAA in this specific case and by other practitioner stakeholders seeking to establish real-world laboratories. One option could be a system of mixed financing, comprising continuous funding from state and local resources, together with the acquisition of external funding for research to ensure the independence and impartiality of the real-world laboratory. In this way, the *Anhalt Real-World Laboratory* could be stabilised as an independent sponsor of transformative research and regional development, akin to an (economic) development agency. With regard to ensuring sustainable infrastructure, the challenge is principally to ensure continuous work processes. This is not possible in the case of project funding alone. After all, funding shortfalls will inevitably occur between a funding project and the next funding projects, ideally following straight on from the first. Unlike research institutions, which are equipped with basic funding, practitioner stakeholders are particularly affected by such shortfalls. Moreover, subsequent funding is uncertain, and there are limitations to the capability of the content to tie in with previous funding, due to the fact that support programmes are usually themed. Besides the (political) will to establish experimental spaces and to actively co-create them, the funding issue therefore becomes a key issue for the establishment of longer-term, viable infrastructures for real-world laboratories (see Kanning and Scurrell 2018).

11.3.6   Regarding Continuous Processes of Reflection and Learning

To assess the processes of reflection and learning, emphasis is placed below on the level of interdisciplinary and transdisciplinary cooperation (Singer-Brodowski et al. 2018) as well as the associated role of accompanying research (Defila and Di Giulio 2018b).

When it established the *Anhalt Real-World Laboratory*, EAA already made provision for accompanying research. The discussions about the topic proved to be difficult. This was because partners with previous experience in transdisciplinary research expected to be closely involved in the real-world laboratory, while the majority of partners assumed that traditional observational research would be conducted. The accompanying research was thus established in the context of a Ph.D. project at the Berlin Social Science Center (WZB), financed by the real-world laboratory. In the light of the findings on relations between researchers, accompanying researchers and financial backers presented in the meantime by Defila and Di Giulio (2018b), it is now possible to make a more detailed assessment of this issue.

The existing accompanying research in the *Anhalt Real-World Laboratory* is indeed geared towards producing knowledge on the processes that take place in the real-world laboratory. According to Defila and Di Giulio (2018b), however, the relation to individual projects in the real-world laboratory can be described as a relation to the "object of research" that is characterised by dependence and an unequal distribution of power. This relation to the object is very much apparent in the real-world laboratory. The strong substantive involvement of the association's main financial backer gives it access to information about the individual projects. What is more, in addition to the geographical proximity of the accompanying research to the association's sponsors (both from outside the region), the financial backer's interests are close to those of the research institution, namely the effectiveness of energy policy and the national recognition of achievements. As such, the tensions described by Defila and Di Giulio (2018b) do not occur in the sponsor's relation to accompanying research, but there are tensions in both their relations to regional stakeholders. There is a realisation that the establishment of the *Anhalt Real-World Laboratory*, driven by practitioner stakeholders, could well have benefited from accompaniment experienced in transdisciplinary research in order to cope with integrating the different bases and forms of knowledge.

In the BMBF research project entitled "The re-productive town", within which *urban laboratories* are initiated and developed, this was achieved by contracting out support in the experience process, the knowledge process and the process of transferring results—although it was not possible to describe this that clearly at the time of the application. The knowledge of experienced transdisciplinary researchers is necessary to integrate knowledge bases from practice and science; to produce transferable knowledge; and, not least, to ensure the continuous self-reflection of different, sometimes changing roles in the transdisciplinary learning process. Ideally, such knowledge should be involved as early as at the stage of conceptually designing real-world laboratories. In this case, scientific accompaniment by a neutral moderator such as sustainify GmbH proved to be successful in the joint research project, ensuring the integration of different bases and forms of knowledge as well as the self-reflection of the practice and scientific partners involved. This insight is consistent with the recommendations already made by Parodi et al. to ensure "co-created accompaniment that supports real-world laboratories in a cooperative, advisory manner" [own translation] (2018: 179).

11.4   A Summary Critical Appraisal and Outlook

Real-world laboratories are a relatively young and yet highly diverse format that is interpreted and shaped in a strongly divergent manner by practitioners in some cases. The debate has only just begun and is still being shaped. In principle, many of the characteristics of real-world laboratories discussed are not new for sustainable land management, such as the development of common problem definitions and solutions (co-design, co-production), as is often the case especially in informal processes of sustainable urban and regional developments. What is more, knowledge on partic-ipation in planning processes also virtually serves as a role model for real-world laboratories (Eckart et al. 2018: 131ff; Kanning 2018). As such, real-world labora-tory formats are compatible with sustainable land management, and also offer added value. Especially real-world laboratory formats that are initiated and coordinated by practitioner stakeholders offer specific implementation potential and, at the same time, are faced with particular challenges.

In our opinion, the direct and explicit integration of objectives for practice and research associated with the real-world laboratory format (Defila and Di Giulio 2018c: 40) represents particular added value over common participatory land management processes. In real-world laboratories, all of the stakeholders involved, whether practitioners or scientists, are considered to be "researchers" [own transla-tion] (Eckart et al. 2018: 105f) who jointly define the solution to the problem and produce new knowledge (co-design, co-production), integrating different specialist disciplines as well as science and practice. In contrast to the planning and develop-ment approaches established in land management, an extended self-conception can be identified that could help bridge the oft-criticised gap between theory and prac-tice (e.g. Lamker et al. 2017). In real-world laboratories, the transformation approach is oriented to radical innovations and change processes towards sustainability in a much more proactive manner than is often the case to date in sustainable urban and regional development processes (see Heyen et al. 2018: 26). Where the principle of sustainable development is reflected critically and understood integratively in corre-spondence with the state of scientific knowledge in real-world laboratories, this goes beyond the current prevailing understanding of sustainability in land management. The latter focuses primarily on the safeguarding or creation of ecological qualities, and pays little attention to the original core of the idea of sustainability, i.e. the trans-formation of social, economic action in a social-ecological direction (see Kanning 2005; Hofmeister 2014). In this connection, it would also be necessary to include in the discussion the generally inherent, unquestioned concept of material growth (see Fröhlich and Gerhard 2017: 28ff). As such, the real-world laboratory format—in line with the design currently featuring strongly in the scientific discourse—could help establish experimental spaces for radical innovations in which the various areas of expertise in transformation and planning (science) are brought together for sustain-able land management and, ideally, linked to educational objectives (see Beecroft et al. 2018: 78).

Real-world laboratories that are initiated and coordinated by practitioner stake-holders also offer special deployment potential due their practical approach, and at the same time face special challenges. On that point, a number of insights and hypotheses can be summarised from EAA's experiences and discussions for further scientific discourse and practical development:

Practitioner stakeholders must satisfy various conditions and have certain skills to be able to initiate real-world laboratories. Among other things, they must be capable of organising a research alliance; making their results publicly accessible; and participating in scientific discourse. They must also either have their own financial resources for conducting research or at least have a strong position in the relevant stakeholder network, enabling them to generate the financial resources needed to operate a real-world laboratory.

Scientific and practitioner stakeholders often face the same challenges when establishing the research process, because the interests of many stakeholders must be accommodated when it comes to complex transformation processes. There is always a need to formulate issues in a practically relevant as well as scientifically interesting and challenging manner at the constituent stage of the project, irrespective of the real-world laboratory initiator's institutional background. It is only the weighting of the practical and scientific relevance that may vary to a certain extent. The good position of a research-affine practitioner stakeholder may make it easier to develop stable stakeholder networks for the purpose of achieving cooperation among the relevant practitioner stakeholders, but the various aspects of effective participation must be borne in mind nonetheless. On the other hand, practitioner stakeholders find it particularly challenging to find scientific partners from several disciplines who go along with a joint problem definition and who are not only interested in obtaining data or conducting purely scientific experiments. In addition, social or economic practitioners who have initiated real-world laboratories tend to be challenged more by the need for experience in methods of knowledge integration and modelling for the purpose of transferring results. One solution for this may be to seek support from experienced transdisciplinary researchers and to involve these experts in the conceptual design stage of the real-world laboratory.

Based on these findings, real-world laboratories led by practitioner stakeholders offer particularly favourable conditions when they are backed by local authorities or public bodies. Strong local governments have excellent links; they know the stakeholders' interests; they have experience in planning participatory processes, which can be largely transferred to real-world laboratories (Eckart et al. 2018: 131ff); they can perpetuate transdisciplinary research, ensuring continuity and, on this basis, learning processes. However, small and medium-sized towns, and towns undergoing socio-economic structural change that feature disproportionate demographics are often under financial supervision and rarely have the human resources capacity to be able to undertake the research that is urgently required for their strategic realignment. Such local authorities therefore tend to be unable to support real-world laboratories, which means that they are only rarely able to incorporate their particular problems in research projects. Consequently, support structures are required to make real-world laboratories accessible to all local authority types.

Against this backdrop, real-world laboratories should not only be financed by research funding in the future, but at least in equal parts by structural support from the relevant ministries (e.g. the Ministry of Energy and/or the Ministry for Economic Affairs). After all, besides producing effects in research and science, real-world laboratories (are supposed to) actively drive forward transformation towards a sustainable society (see Kanning and Scurrell 2018).

Several recent changes in science and structural policy will improve the conditions for real-world laboratories initiated by practitioner stakeholders in future. These include a greater shift towards citizen science, including its transformation towards to more complex civic research beyond mere data collection. Citizens are more frequently involved in the formulation of research issues, and the definition of criteria for data collection and analysis. Various institutions besides universities and research institutes give citizens the possibility to participate in research. The "Green Paper Citizen Sciences Strategy 2020 for Germany", published in 2016, provides guidance on activating citizen science for the purpose of transformation that is appreciated, acknowledged and embraced by society and science alike (Bonn et al. 2016: 25). From the perspective of practitioner stakeholders, it is important to create stronger links in future between citizen science formats, ranging from data collection to active co-design and active co-production (ibid.), to the original real-world laboratory format developed by science, creating synergies. In addition, since the 2017 Bundestag elections at the latest, greater attention is being paid in structural policy to the development of rural regions and their small and medium-sized towns. If such attention can also be translated into supporting measures for the sustainable development of rural areas, there will be new financial leeway for real-world laboratories, which can be established and used as experimental spaces for sustainable land management.

References

Augenstein, K., Haake, H., Palzkill, A., Schneidewind, U., Singer-Brodowski, M., Stelzer, F., & Wanner, M. (2016). Von der Stadt zum urbanen Reallabor - eine Einführung am Beispiel des Reallabors Wuppertal. In U. Hahne & H. Kegler (Eds.), *Resilienz. Stadt und Region: Reallabore der resilienzorientierten Transformation* (Vol. 1, pp. 167–195). Frankfurt am Main, New York: PL Academic Research (Stadtentwicklung Urban development).

Beecroft, R., & Parodi, O. (2016). Reallabore als Orte der Nachhaltigkeitsforschung und Transformation. Einführung in den Schwerpunkt. *Technikfolgenabschätzung–Theorie und Praxis, 25*(3), 4–8.

Beecroft, R., Trenks, H., Rhodius, R., Benighaus, C., & Parodi, O. (2018). Reallabore als Rahmen transformativer und transdisziplinärer Forschung: Ziele und Designprinzipien. In A. Di Giulio & R. Defila (Eds.), *Transdisziplinär und transformativ forschen – Eine Methodensammlung* (pp. 75–100). Wiesbaden.

Biesecker, A., & Hofmeister, S. (2006). Die Neuerfindung des Ökonomischen. *Ein (re)produktionstheoretischer Beitrag zur Sozial-ökologischen Forschung*. München.

Bonn, A., Richter, A., Vohland, K., Pettibone, L., Brandt, M., Feldmann, R., et al. (2016). *Green Paper Citizen Science Strategy 2020 for Germany*. Helmholtz-Zentrum für Umweltforschung (UFZ), Deutsches Zentrum für integrative Biodiversitätsforschung (iDiv) Halle-Jena-Leipzig, Leipzig, Museum für Naturkunde Berlin, Leibniz-Institut für Evolutions- und Biodiversitätsforschung (MfN), Berlin-Brandenburgisches Institut für Biodiversitätsforschung (BBIB). Berlin.

De Flander, K., Hahne, U., Kegler, H., Lang, D., Lucas, R., Schneidewind, U., et al. (2014). Resilience and real-life laboratories as key concepts for urban transformation research. Resilienz und Reallabore als Schlüsselkonzepte urbaner Transformationsforschung. Zwölf Thesen. *GAIA—Ecological Perspectives for Science and Society, 23*(3), 284–286.

Defila, R., & Di Giulio, A. (2018a). Reallabore als Quelle für die Methodik transdisziplinären und transformativen Forschens – eine Einführung. In R. Defila, & A. Di Giulio (Eds.), *Transdisziplinär und transformativ forschen* (pp. 9–35). Eine Methodensammlung. Wiesbaden.

Defila, R., & Di Giulio, A. (2018b). What is it good for? Reflecting and systematizing accompanying research to research programs. *GAIA—Ecological Perspectives for Science and Society, 27*(S1), 97–104.

Defila, R., & Di Giulio, A. (2018c). Partizipative Wissenserzeugung und Wissenschaftlichkeit – ein methodologischer Beitrag. In R. Defila, & Di Giulio, A. (Eds.), *Transdisziplinär und transformativ forschen* (pp. 39–67). Eine Methodensammlung. Wiesbaden.

Eckart, J., Ley, A., Häußler, E., & Erl, T. (2018). Leitfragen für die Gestaltung von Partizipationsprozessen in Reallaboren. In A. Di Giulio & R. Defila (Ed.), *Transdisziplinär und transformativ forschen: Eine Methodensammlung* (pp. 105–135). Wiesbaden.

Engels, F., & Rogge, J. C. (2018). Tensions and trade-offs in real-world laboratories-the participants' perspective. *GAIA—Ecological Perspectives for Science and Society, 27*(S1), 28–31.

Fröhlich, K., & Gerhard, U. (2017). Wissensbasierte Stadtentwicklung – ein Erfolgskonzept auch für Nachhaltigkeit? Einblicke in die Entwicklung der Heidelberger Südstadt aus Reallaborperspektive. In Deutsche Akademie für Landeskunde e.V. und Leibniz Institut für Länderkunde (Eds.), Reallabore als Forschungsformat nachhaltiger Stadtentwicklung, Leipzig, Berichte (Vol. 91, Issue 1, pp. 13–33). Geographie und Landeskunde (BGL).

Grießhammer, R., & Brohmann, B. (2015). *Wie Transformationen und gesellschaftliche Innovationen gelingen können.* UFOPLAN-Vorhaben—FKZ 371211103. (Ed.: Umweltbundesamt). Dessau-Roßlau.

Gross, M., Hoffmann-Riem, H., & Krohn, W. (2005). Realexperimente. *Ökologische Gestaltungsprozesse in der Wissensgesellschaft.* Bielefeld.

Grunwald, A. (2016). *Nachhaltigkeit verstehen: Arbeiten an der Bedeutung nachhaltiger Entwicklung.* München.

Hahne, U., & Kegler, H. (Ed.). (2016). Resilienz. *Stadt und Region: Reallabore der resilienzorientierten Transformation* (p. 1). Frankfurt am Main, New York: PL Academic Research (Stadtentwicklung Urban development).

Heyen, D. A., Brohmann, B., Libbe, J., Riechel, R., & Trapp, J. H. (2018). *Stand der Transformationsforschung unter besonderer Berücksichtigung der kommunalen Ebene.* Paper within the "Vom Stadtumbau zur städtischen Transformationsstrategie project" as part of the "Experimenteller Wohnungs- und Städtebau" (ExWoSt) research programme. Sine loco.

Hofmeister, S. (2014). Das Leitbild Nachhaltigkeit – Anforderungen an die Raum- und Umweltplanung. In: H. Heinrichs & G. Michelsen (Eds.), *Nachhaltigkeitswissenschaften* (pp. 304–320). Berlin, Heidelberg.

Hohn, U., Kemming, H., & Reimer, M. (Eds.). (2014). Formate der Innovation in der Stadt- und Regionalentwicklung. Reflexionen aus Planungstheorie und Planungspraxis, Detmold = Reihe Metropolis und Region des Stadt- und regionalwissenschaftlichen Forschungsnetzwerks Ruhr (SURF) (p. 13).

Kanning, H. (2005). *Brücken zwischen Ökologie und Ökonomie – Umweltplanerisches und ökonomisches Wissen für ein nachhaltiges regionales Wirtschaften.* München.

Kanning, H. (2018). Reallabore aus planerischer Perspektive. sustainify Arbeits- und Diskussionspapier 3|2018, sustainify Institut für nachhaltige Forschung, Bildung, Innovation. Hannover.

Kanning, H., & Scurrell, B. (2018). Reallabore der Praxisakteure – Merkmale und methodische Herausforderungen. Sustainify Arbeits- und Diskussionspapier 1|2018, sustainify Institut für nachhaltige Forschung, Bildung, Innovation. Hannover.

Krohn, W., Grunwald, A., & Ukowitz, M. (2017). Transdisziplinäre Forschung revisited: Erkenntnisinteresse, Forschungsgegenstände, Wissensform und Methodologie. *GAIA—Ecological Perspectives for Science and Society, 26*(4), 341–347.

Lamker, C., Peer, C., & Sondermann, M. (2017). Zum Verhältnis von Planungswissenschaft und –praxis. *Nachrichten der ARL, 47*(1), 10–13.

Menny, M., Palgan, Y. V., & McCormick, K. (2018). Urban living labs and the role of users in co-creation. *GAIA—Ecological Perspectives for Science and Society, 27*(S1), 68–77.

MWK—Ministerium für Wissenschaft, Forschung und Kunst Baden-Württemberg (Ed.). (2013). Wissenschaft für Nachhaltigkeit. *Herausforderung und Chance für das baden-württembergische Wissenschaftssystem.* Stuttgart.

Parodi, O., Beecroft, R., Albiez, M., Quint, A., Seebacher, A., Tamm, K., & Waitz, C. (2016). Von "Aktionsforschung" bis "Zielkonflikte"–Schlüsselbegriffe der Reallaborforschung. *Technikfolgenabschätzung – Theorie und Praxis, 25*(3), 9–18.

Parodi, O., Ley, A., Fokdal, J., & Seebacher, A. (2018). Empfehlungen für die Förderung und Weiterentwicklung von Reallaboren: Erkenntnisse aus der Arbeit der BaWü-Labs. *GAIA—Ecological Perspectives for Science and Society, 27*(S1), 178–179.

Rogga, S., Zscheischler, J., & Gaasch, N. (2018). How much of the real-world laboratory is hidden in current transdisciplinary research? *GAIA—Ecological Perspectives for Science and Society, 27*(S1), 18–22.

Schäpke, N., et al. (2017). Reallabore im Kontext transformativer Forschung. Ansatzpunkte zur Konzeption und Einbettung in den internationalen Forschungsstand. ETSR Discussion papers in Transdisciplinary Sustainability Research, Leuphana Universität, Lüneburg.

Schäpke, N., Bergmann, M., Stelzer, F., & Lang, D. J. (2018a). Labs in the real world: Advancing transdisciplinary research and sustainability transformation: Mapping the field and emerging lines of inquiry. *GAIA—Ecological Perspectives for Science and Society, 27*(S1), 8–11.

Schäpke, N., Stelzer, F., Caniglia, G., Bergmann, M., Wanner, M., Singer-Brodowski, M., et al. (2018b). Jointly experimenting for transformation? Shaping real-world laboratories by comparing them. *GAIA—Ecological Perspectives for Science and Society, 27*(S1), 85–96.

Schmidt, A. (2017). Zukunftsstadt. Forschung für klimaresiliente, sozial-ökologisch gerechte und lebenswerte Städte. *GAIA—Ecological Perspectives for Science and Society, 26*(4), 355–356.

Schneidewind, U. (2014). Urbane Reallabore – ein Blick in die aktuelle Forschungswerkstatt. *Planung neu denken (pnd) online, 10*(3), 1–7.

Schneidewind, U., & Singer-Brodowski, M. (2013). *Transformative Wissenschaft: Klimawandel im deutschen Wissenschafts-und Hochschulsystem.* Marburg.

Schneidewind, U., & Singer-Brodowski, M. (2015). Vom experimentellen Lernen zum transformativen Experimentieren: Reallabore als Katalysator für eine lernende Gesellschaft auf dem Weg zu einer Nachhaltigen Entwicklung. *Zeitschrift für Wirtschafts- und Unternehmensethik (ZfWU), 16*(1), 10–23.

Schön, S., Biesecker, A., Hofmeister, S., & Scurrell, B. (2013). (Re)Produktives Wirtschaften im Dialog mit der Praxis. In Netzwerk Vorsorgendes Wirtschaften (Ed.), *Wege Vorsorgenden Wirtschaftens* (2nd ed., pp. 159–200). Marburg.

Singer-Brodowski, M., Beecroft, R., & Parodi, O. (2018). Learning in real-world laboratories: A systematic impulse for discussion. *GAIA—Ecological Perspectives for Science and Society, 27*(S1), 23–27.

Wagner, F., & Grunwald, A. (2015). Reallabore als Forschungs- und Transformationsinstrument: Die Quadratur des hermeneutischen Zirkels. *GAIA—Ecological Perspectives for Science and Society, 24*(1), 26–31.

WBGU—German Advisory Council on Global Change. (2011). *World in transition: A social contract for sustainability.* Summary for policy makers. Berlin.

WBGU—German Advisory Council on Global Change. (2014). *Special report climate protection as a world citizen movement.* Berlin.

WBGU—German Advisory Council on Global Change. (2016). *Humanity on the move: Unlocking the transformative power of cities.* Berlin.

Wittmayer, J. M., & Hölscher, K. (2016). *Transformation research: Goals, contents, methods.* Workshop Report. Drift 216.

Yildiz, Ö., Drießen, F., Pobloth, S., & Schön, S. (2012). Re-Produktionsketten als Ansatz koevolutionärer Regionalwirtschaft. *Ökologisches Wirtschaften, 27*(S1), 30–36.

Chapter 12
Knowledge Management for Sustainability: The Spatial Dimension of Higher Education as an Opportunity for Land Management

Jens Schulz, Thomas Köhler, and Thomas Weith

Abstract The production and dissemination of knowledge are seen as essential elements of sustainable land management (Salet 2014; Kaiser et al. 2016). Digitalisation creates a variety of opportunities for knowledge creation and communication. Knowledge cooperation and digitalisation for governance of land is a concept coined by the various and perhaps diverse interests of diverse stakeholders. In the era of digitalisation, new developments trigger further changes that may be of a challenging nature. However, digitalisation also offers innovative options for collaborative activity in land use, bringing together these diverse interests and eventually enabling new patterns of collaboration. The authors address principal patterns of collaboration in multi-stakeholder networks that have only recently been considered meaningful for research. Concepts from both domains—higher education and land management—are advantageously combined, allowing new interpretations of spatial and digital artefacts. Digitalisation is dramatically changing the knowledge-related domain itself at the end of the 2010s. Besides encouraging new teaching and learning activities, new technologies also have an impact on knowledge spaces and on the institutional and personnel knowledge carriers established therein. However, the institutional and sectoral development of academic education practice has rarely been addressed in the context of higher education research. To illustrate the new way in which knowledge management is directed under such conditions, the authors briefly present digital networks from both sectors. Obviously, different layers of

J. Schulz
Institute of Knowledge Transfer and Digital Transformation, University of Applied Sciences, Technikumplatz 17, Mittweida, Germany
e-mail: Schulz3@hs-mittweida.de

T. Köhler (✉)
Faculty of Education, TU Dresden, Weberplatz 5, Dresden, Germany
e-mail: Thomas.Koehler@tu-dresden.de

T. Weith
Institut für Umweltwissenschaften und Geographie, University of Potsdam, Campus-Golm, Karl-Liebknecht-Str. 24-25, Potsdam 14476, Germany
e-mail: Thomas.Weith@zalf.de

Working Group "Co-Design of Change and Innovation", Leibniz Centre for Agricultural Landscape Research, Eberswalder Str. 84, Müncheberg 15374, Germany

© The Author(s) 2021
T. Weith et al. (eds.), *Sustainable Land Management in a European Context*,
Human-Environment Interactions 8, https://doi.org/10.1007/978-3-030-50841-8_12

stakeholders (state ministries and other authorities; higher education institutions and their sub-units, and other societal actors, etc.) need to collaborate in order to define and run the processes prevailing in and for higher education. How do these spatially distributed institutions interrelate in order to co-design the higher education landscape across an entire federal state? Which structures and processes are applied in co-design practices? Consequently, this paper outlines these developments, and combines knowledge economics, educational geography and educational science approaches in the context of higher education research.

Keywords Land management · Higher education · Digitalisation · Knowledge management · Sustainability

12.1 Introduction

The production of new knowledge and the dissemination of knowledge are seen as essential elements of sustainable land management (Salet 2014; Kaiser et al. 2016). Digitalisation creates a variety of opportunities for knowledge creation and communication. Until now, new methods have mainly been discussed and implemented in research projects and programmes. But how can knowledge creation be combined with processes in the higher education system in the context of digitalisation? And what opportunities exist for sustainable land use and spatial development under these circumstances? Besides developing single curricula, the following chapter will reflect on the spatial conceptual models behind current practices in higher education, enabling the authors to discuss alternatives for the future in this field of knowledge creation and dissemination.

Digital media, as a core component and important output of digitalisation processes, differs significantly from conventional knowledge media, given that it emphasises cooperative action—whether in the form of human-machine interaction or as interpersonal mediated communication (Köhler 2006). Friedrich W. Hesse, the founding director of the Leibniz Institute for Knowledge Media in Tübingen, pointed out: "Through digital resources, a knowledge space is changing and knowledge work is changing" (Hesse 2017). Although in this case knowledge space stands for a mathematical structure that reflects the results of the learning process, the knowledge work clearly refers to a spatially effective size in the geographical understanding, since production, storage and transfer of knowledge require appropriate infrastructures or people.

Scientific discourse in spatial sciences (e.g. Kaiser et al. 2017) outlines how relevant opportunities for collaborative activity in land use occur, possibly bringing together diverse interests and eventually enabling new patterns of collaboration to be detected. Has research been able to identify principal structures of collaborative interrelations in multi-stakeholder networks? And how will governance models be affected? This aspect is especially relevant in the European context, where equal living conditions are quite often an underlying political target of spatial develop-ment, and where spatial justice is often seen as a basis for the distribution of both infrastructure and subsidies. Addressing these rather complex questions, the authors seek to open up the agenda for a theoretical discourse and subsequently an empirical discourse, to be documented in the present chapter. A possibly innovative momentum will be developed from the combination of theoretical domains—knowledge coop-eration and digitalisation in higher education for the governance of land, which have so far been rather independent.

An interdisciplinary analysis of the technological, social, (organisation) didactic and policy-related challenges is required so as not only to accompany, but also to shape developments in the concrete case. The link between digitalisation and the concept of space is addressed by introducing the concept of knowledge as an interface. Future research in digitised knowledge cooperation will be linked with research and explanatory approaches of higher education research and research on informational land use, applying aspects such as organisational research, the knowledge economy and educational geography.

First, the authors provide a brief introduction to the status quo of the two chosen domains—research on knowledge in land management (related to information and communication, and relevant technologies) and higher education research (focusing on e-learning, organisational development and digitalisation).

12.1.1 Generating and Disseminating Knowledge for Sustainable Land Use

Knowledge can generally be understood as the conscious processing and adoption of information. In the same vein, knowledge is also defined as the process of purposeful connection of information (North 2011: 36f). Change processes, also in land use, are characterised by making individual knowledge accessible to the actors involved (Nonaka and Takeuchi 1995). New knowledge is generated by the cyclical repetition of the internalisation of information and processing into knowledge (learning), and its externalisation (communication). Learning, feedback and communication thus serve the purpose of knowledge generation and knowledge dissemination, supported by the moderation of intermediaries. In recent years, considerable progress has been made with regard to the development and availability of codified knowledge stocks and new knowledge generation processes by digitalisation.

The handling of real sustainability problems such as challenges in land use depends on the production and dissemination of new knowledge produced in this way. This is often developed in transdisciplinary settings in a co-design approach. Such practice enables the integration of different forms of knowledge, resulting, in particular, from the diversity of practitioners involved in the research process. As a result, contributions to the solution of complex societal challenges are grouped around problems, rather than scientific disciplines (Pohl 2014).

With sustainability research in mind, several authors stress the need for this kind of research; the current lack of knowledge; and new methods of knowledge management (e.g. Miller et al. 2014: 244; Kajikawa et al. 2014). For several years, research efforts have focused more on producing the knowledge required to understand the challenges. Until now, the 'quality and validity of knowledge systems in the context of sustainability research' have been a major challenge (Cornell et al. 2013) within knowledge systems (ibid.: 61).

Scientific debates in this context are accompanied by science policy debates and political discourses. The Sustainable Development Goals (SDGs) document is an important political statement in this context. Knowledge management is addressed in three SDG objectives (Goal 4: quality education; Goal 16: peace, justice and strong institutions; and Goal 17: partnerships for the goals). It is simultaneously linked to education and lifelong learning. An explicit reference to digitalisation is also mentioned for the first time in the SDGs. Besides being relevant to many sustainability goals, digitalisation is explicitly addressed in sub-goal 9c (Industry, Innovation and Infrastructure): "To substantially improve access to information and communication technology and to ensure universal and affordable access to the Internet in at least developed countries by 2020".[1] The accompanying "Future Earth" research process is of great importance for the development and implementation of the SDGs. This global network focuses on the development and dissemination of new knowledge and the exchange of knowledge (www.futureearth.org). One of the network's explicit goals is to support the transformation towards sustainability. The associated research agenda states that this also requires a new form of knowledge generation across disciplinary boundaries, together with social partners.

While knowledge is a precondition for action, co-designed processes for the generation and management of knowledge are considered to be core elements for sustainable spatial development and land use (Salet 2014). Davoudi (2015) grasps the knowledge-action relationship in a manner that defines "planning as practice of knowing". In this context, knowledge is quite often based on a variety of experiences in practice, embedded in social networks and organisational structures (Zimmermann 2010: 118). In this way, knowledge production and dissemination is not a single task to be addressed by several experts, but is accompanied by regional and local knowledge from a variety of actors.

The processes must be complemented by actors who manage the processes of knowledge cooperation or learning. This task often involves moderating, comparable to the tasks of facilitators and, in some cases, mediating actors (cf. Stützer et al. 2013). Moderation implements the "cohesion function", i.e. guiding group work, keeping the group together, introducing rules, overseeing, and harmonising group members, as well as the "locomotion function", i.e. setting group work in motion, and ensuring effective and focused working methods (Ziegler 1993). On this basis, the mediating actor fulfils two functions: those of discussion leader and consultant. Another concept of facilitated interaction is that of intermediaries, which is close to the method of moderation. Millar and Choi (2003) define knowledge intermediaries

---

[1] https://www.un.org/development/desa/disabilities/envision2030.html

as organisations that "facilitate a recipient's measurement of the intangible value of knowledge received" (ibid.: 269). One function of the knowledge intermediary is to provide firms and/or knowledge producers with technology and knowledge transfer processes in the context of regional innovation systems (Parker and Hine 2013). The findings about intermediaries provide helpful information for determining who is responsible for knowledge management. Recent approaches often refer to the changing role of knowledge professionals. This role may even disappear completely, such as in the case where the question is asked: "Does a class need a teacher?" (Köhler and Kahnwald 2005). Indeed, greater emphasis is placed on sharing interests in practice than on following a predefined curriculum (Kahnwald and Köhler 2007).

Regarding the transfer of knowledge, the use of a unidirectional or one-way transfer from knowledge producers to knowledge consumers (also called "mode 1") to deal with information and knowledge flows cannot adequately reflect the challenges of a decision-making process or of a planning support process in a complex multi-stakeholder environment. This mode of science-policy interaction can be described as the "science push" and/or the "demand pull" model (Dilling and Lemos 2011). The new focus of knowledge transfer activities also considers the communication, translation and mediation of knowledge (Cash et al. 2003). Gibbons et al. (1994) call this focus "mode 2 knowledge" in the context of transdisciplinary research. The experiences of transdisciplinary research, primarily characterised by the cooperation of scientists and practitioner stakeholders, confirm the barriers and disadvantages of "mode-1" versus "mode-2" knowledge production (cf. Hirsch Hadorn et al. 2008). Thus, assembling different stakeholders in various groups throughout the planning process is one of the crucial conditions for a sustainable decision-making process and a key pillar of the currently consolidating concept of transdisciplinarity (Gibbons et al. 1994; Pohl 2011; Opdam et al. 2015). From this point of view, mutual collaboration of the stakeholders concerned is required in order to support knowledge production, transfer and implementation—this is the idea behind the concept of the co-production of knowledge (Pohl et al. 2010; Enengel et al. 2012). Lemos and Morehouse (2005) argue that an iterative and interactive model for the co-production of science and policy requires interdisciplinarity, stakeholder participation and the production of usable knowledge, which can be incorporated into all stakeholders' decision-making processes. They also acknowledge in this context that usable knowledge "not only must be tailored to fit stakeholders' needs and uses, but must also be made accessible to those users" (ibid.: 62).

Usability, accessibility and transferability depend in many ways on "producers" and "producing processes". Perspectives on challenges and problems define the starting points of agenda processes. Co-design processes are essential, followed by the co-production of knowledge, accompanied by exchange and interaction (Mauser et al. 2013: 427). In consequence, one of the most important questions at the beginning of the process is whom to define as an important actor. In addition, actor networks and organisational units are highly relevant, because single actors are embedded, and also develop and share knowledge within their organisations. In consequence, the use of transdisciplinary approaches also means exchanging knowledge between organisations that start bargaining processes, and developing framing narratives in institutional settings. This will raise awareness beyond projects to whole systems such as higher education.

12.1.2 New Ways of Knowledge Generation, Dissemination and Management: Higher Education Research with a Focus on E-Learning, Organisational Development and Digitalisation

The institutional and sectoral development of academic education practice has rarely been addressed in the context of higher education research. While a major interest of education research is linked to understanding specific didactic effects, the interdisciplinary approach is crucial to cope with the complexity of organisational relationships and the institutional contexts in which the digitalisation of higher education takes place. Decision-making and the forecasting of such processes appear to be challenging in this sector. Dimensions to be covered include the following aspects or areas of research and practice:

- Structures and actors,
- Administration, governance and organisational development,
- Didactics and competence development, and
- Educational technologies and technical infrastructures.

The professions represented in higher education development committees should ideally reflect these four points. But what does this mean for knowledge practices in a completely digitised knowledge society with more and more immaterial goods? Indeed, it is also possible to recruit expertise at any time and disseminate it through various exchange formats within research or practice communities, or to elaborate such expertise in projects via digital media and relevant formats of collaboration. As a result, higher education research must consider society in general in order to gain an appropriate understanding of how knowledge is generated, shared and transferred. Extending such understanding would also require ideas about spatial organisation beyond the embedding of this knowledge-related practice in the physical space of educational buildings.

12.2 Digitalisation of Higher Education

12.2.1 Classification into Megatrends of the Twenty-First Century

Two thematic areas are particularly striking when analysing current global mega-trends: there are demographic and social scientific considerations as well as economic interpretations. In many cases, however, the boundaries between these domains are fluid because there are interactions and links between them. In addition, each of these megatrends can be accompanied by current developments in the field of educa-tion, which require detailed analysis in terms of digitalisation processes. Interesting connections can be established, particularly for universities (see Table 12.1).

The challenges presented here are only examples, and are not consistent in the list. They provide an overview in the sense of impetus, seeking to reveal the connections that are currently important to the education system. With regard to conclusions and consequences, the table shows how important it is to combine new forms of knowledge management with higher education. This goes hand in hand with the changing framework conditions, particularly in higher education, from a greater focus on the labour market, to digitalisation, globalisation and new forms of governance.

Table 12.1 Global megatrends (Stang and Eigenbrodt 2014: 233ff) and exemplary challenges for higher education in the context of German approaches

Global megatrend	Exemplary challenges for higher education
Demographic change	Lifelong learning
Individualisation of the mass market	New (multi-site) study and learning opportunities
Social fragmentation	"Ascension through education"; "open access universities"
Urbanisation	Competition between university locations
Mobility	Online learning/studying versus "global villages"[a]
Digitalisation of all areas of life	Skills, competences, habits and preferences of teachers and students
Automation and flexibility in collaborative working environments	Learning analytics; artificial intelligence; big data
Knowledge-based economy	Knowledge transfer in knowledge regions and globally; new requirements for education and training
Significance of education	Weakening the prohibition of cooperation under Article 30 of the German Constitution; "Quality Pact Teaching" of the BMBF

[a]This term, first introduced by Marshall McLuhan in 1962, is a media-theoretical description of the world that grows together through electronic networking (in the sense of electronic interdependence) (McLuhan 2008)

12.2.2 Digital Media and the Resulting Changes in Higher Education Teaching and Learning

The digitalisation of (higher) education unfortunately remains an issue that fails to create an optimistic and positive image of the future among the public (Schulz 2018). This is certainly not unfounded and is sometimes important; but as a societal cross-cutting task, there should at least be a vision for action in the future. However, digital transformation in education is particularly evident when comparing courses requiring presence to online courses. The example of "seminar versus webinar" shows the extent to which change is generally understood. In universities, students often sit in rows facing the lecturer, with a notebook or tablet on their desks in front of them, while students interact with teachers or each other, depending on the nature of the course. The lecturer uses a digital projector to show PowerPoint slides. The difference to face-to-face courses 20 years ago is astounding: notebooks and tablets take the place of books and paper; PowerPoint presentations have ousted the blackboard. Features may, of course, depend on the teacher and the subject culture, but most courses, whether seminars or lectures, now take place in such a format. Indeed, recent forecasting using scenario and Delphi techniques suggests that there will no longer be any pure offline educational situations by 2030 (Köhler et al. 2018, 2019).

Webinars are purely online seminars. They dissolve the physical space of the course, as both lecturers and students can participate from anywhere. Citizens of the "global village" come together in a virtual space; link up with knowledge of others in real time; generate their own knowledge; and finally share that knowledge with others. On closer inspection, however, most offers are a 1:1 transfer of the seminar concept to a non-physical space: the lecturer determines the degree of interaction by assigning rights; students sit (or lie) in front of a computer with an image transfer of the lecturer and the PowerPoint slides, and note important information in virtual and/or physical exercise books. Although there has been significant progress with digitally enhanced courses in terms of participation in educational processes, the opportunities associated with digital technologies are rarely fully exploited.

As mentioned above, digital teaching and learning technologies also open up new opportunities for imparting knowledge and learning material. Media didactics differentiate between "traditional" and "new" media formats. Traditional media formats are understood to be mainly analogue texts, images, graphics, presentations, films, videos and audio files. In this case, information and communication technologies are used for unidirectional performance or instruction. Newer mediation formats allow students to apply knowledge, obtain feedback or interact with fellow students in a more connectivist, co-constructionist way. As a consequence, the spatial distance is no longer decisive, but online access and interaction possibilities are

Excursus: Online interactions include video, audio and chat conferences, and digitally based forums, as well as digital simulations and business games, e-tests, common media products, wikis and blogs. Riedel et al. (2017) found in a survey among all lecturers at universities in Saxony that the use of traditional media formats clearly predominates. More than 98 per cent of the respondents stated that they used texts, presentations, graphics and images in digital form, almost 77 per cent used films, videos and audio files. In contrast, the new media formats mentioned, which encourage interaction, are only part of the teaching portfolio for about a third of the teaching staff (e.g. forums: 37.3%; e-tests: 24.8%).

At first glance, the primary media-didactic question about the use of different media formats in teaching seems to have little spatial impact in the geographical sense. However, if one includes social and communicative aspects of virtual cooperation, changes in the room can be expected. Regardless of an increase in the degree of use of such new media formats, devaluation, or at least a reclassification of currently used learning spaces, may occur. In contrast, new learning spaces are being created in places that are not yet intended for such use. Examples include spaces used while commuting to university, and student services cafés.

12.3 Spatial Dimensions: Informal Spaces and New Learning Worlds in Adult and Continuing Education

12.3.1 Governmental Argumentation in Favour of Digital Continuous Education

Official education policy, however, seems to consider online elements of education as being rather relevant for postgraduate education. Yet educational processes also take place outside of schools and universities. In adult and continuing education in particular, far-reaching developments can be observed, and are expected in the years ahead. The Federal Ministry for Economic Affairs and Energy (BMWi), for example, helps companies to digitise their business processes or sales channels, or to develop personnel by way of digitised training and e-learning. The ministry argues for the latter as follows: "If employees can use digital media for further training, companies and employees benefit:

- A large number of employees can be continuously trained.
- Further training can take place regardless of time and location.
- Standardised content can be prepared interactively, quickly and in a way that promotes learning.
- Compulsory training can be demonstrated more effectively (compliance).

- Learning groups can be virtually networked across locations.
- Customer training (product training) can take place digitally.
- Costs can be saved by, e.g. shortening learning times or by eliminating travel costs" (BMWi 2016: 24).

In the context of equal spatial development through digitalisation in education, the key words in this list are "location and time independent", "networked across locations ..." and "no travel expenses". The interpretation can go so far that digitalised educational offers mean that traders and their employees in peripheral, rural areas no longer have to fear location disadvantages with regard to further and advanced training. All the necessary events and training courses, including the product training mentioned, would be ubiquitously available, contributing to equal opportunities between differently characterised areas and regions.

12.3.2 Changes in Organisational Learning and Communication Behaviour

What conclusions can be drawn concerning the domain of educational governance? On the one hand, political aims and targets stress the desire for equal living conditions and fair opportunities for development. In recent years, discourses about new modes of governance have stressed development policies away from simply state-driven hierarchical systems towards integrative approaches, adopting the variety of actors' positions and network organisation. The organisational and regulatory system coordinates interaction between state and non-state actors of all kinds. "It is ... about how we establish goals, how we define rules for reaching the defined goals, and finally how we control outcomes following from the use of these rules" (Vatn 2005). There will be fewer conflicts, and common goals will be achieved.

On the other hand, there are still differences between rural areas and urban regions concerning the use and acceptance of digital resources.[2] An additional aspect that has only been so far been given little consideration is the potential different patterns of information culture between urban versus rural regions and their inhabitants. Indeed, the research literature shows that this topic of organisational learning and communication behaviour is a relatively undiscovered territory when it comes to the question of implementing information and communications technologies (ICT).

Another contradictory observation in relation to the far-reaching potential of technologies is the observation that administrative advantages are often serve an easy entry point of ICT into any domain, before applications arrive in the sectoral practices. For example, in education learning was mainly adopting administrative moments before educational practice started to reform (Erber et al. 2004). However, recent studies on ICT-based management procedures in general education (Bergner

---

[2]https://www.spiegel.de/netzwelt/web/breitband-ausbau-auf-dem-land-hinkt-hinterher-a-130
2174.html

2019) confirm that virtually no schools in Germany apply enterprise resource planning (ERP) technologies, which are commonplace in industry, not even in a simple way.

As information and communication devices, and related applications, increasingly penetrate everyday private and professional life, behaviours also change. However, learning can also take place in informal arrangements, whilst formal settings can be configured and enriched in very different ways to one or two decades ago. This new learning behaviour is characterised by:

- Interactivity,
- Anonymity,
- Individualisation,
- Hyper mediality,
- Topicality and
- Globality (Semar 2014: 12f).

This has been described briefly in Sect. 12.2. But a second change can already be observed in addition to "new learning", because the new technical conditions are able to turn almost all learners into lecturers (whereby this term should be used with caution, since it is usually associated with a formally proven teaching qualification). Any individual who believes that they have specific knowledge or have solved a practical problem can create an explanatory video with the simplest of means; upload it onto an appropriate internet platform; and hope that interested parties will find it. It instantaneously becomes an educational offer, even though the material may have not been elaborated with didactics in mind, and can be integrated into formal learning arrangements without any barriers. A seventh point "Knowledge transfer through production and distribution" can therefore be added to the above description of learning behaviour, whereby "Changed communication behaviour during the learning process" would be a more precise description in many cases.

12.3.3   Economic Versus Educational Perspectives: A Sample Case

To illustrate this development in the educational versus economic domain, the authors decided to present a sample case in the context of one of the largest German university associations for digital learning, the Education Portal Saxony (*Bildungsportal Sachsen*).The goal of this collaborative project, established only in 2019–2020, focusing on "virtual teaching cooperation", is to pilot cross-university teaching processes, taking into account the didactics of collaborative teaching and learning, ideally in a specialist domain and possibly between different university types. It also seeks to develop solutions for cooperation with international partners. Furthermore, the project seeks to promote the subject-independent (interdisciplinary) qualification of educational staff to strengthen the effectiveness of the self-regulation of students in online or blended learning scenarios in such open collaborative networks, for example by acquiring online tutoring skills and abilities. Finally, individual Saxony-wide preliminary courses and online self-assessment tests are supported with the use and creation of open educational resources (OER).[3]

12.4 Interactions of Space and Education in the Context of Digitalisation

12.4.1 The Spatial Dimension

Digitalization goes far beyond purely economic applications, and therefore illustrates the specific relevance for spatial and land-related developments. The link between digitalisation and the concept of space is addressed at this point by introducing the concept of knowledge as an interface. An outline has already been given of how the acquisition of knowledge using new information and communication tools will change, and continue to change. The possibilities of the location-independent and time-independent design of educational processes reveal that digitalisation is not a point-source event, but one that takes place in an area. The concept of knowledge in its spatial dimension is analysed as the central point of the discourse, since all actors become effective as part of their complex spatial environments.

New dynamics subsequently arise following the detachment of the physical connection of knowledge and information towards individuals. The concept of knowledge regions, which, when understood traditionally, was based on geographical, almost demarcated areas and easily comprehensible actor interactions, needs to be readjusted in the light of digitalisation. It can be assumed that there will be developments towards differentiated knowledge networks or global knowledge clusters, i.e. the spatial dimension will decline in favour of the knowledge economy, meaning that planning processes for any education domain will take place in new spatial thinking patterns.

12.4.2 Relevant Knowledge Economics and Educational Geographic Assumptions

It makes sense to use other relevant knowledge-economic and educational geographic assumptions to understand the spatial dimension described above. Initially, these must be interpreted as being detached from parallel digitalisation. Yet it is obvious

---

[3] https://bildungsportal.sachsen.de/portal/parentpage/projekte/hochschulvorhaben/projekte-2019-2020/virtuelle-lehrkooperationen

that a process is already taking place on a national or global scale, receiving new dynamics through digitalisation, which will certainly be preserved.

The first assumption relates to the production and (re)use of knowledge. Innovation processes are driven accordingly by different knowledge bases (Kujath and Peiker 2014), i.e. by networking knowledge producers in specialist communities or the ubiquity of classic and new knowledge media. The emergence of innovations is basically conceivable anytime, anywhere. In these traditional knowledge networks, places and regions, institutions and organisations are interconnected, forming the common basis for processes that can then culminate in a result.

This is followed by a second assumption, which states that competition and cooperation within a common theme are not mutually exclusive, and take place regionally, supra-regionally and globally. There are fixed interaction mechanisms between actors, which may be both different and of the same institutional type, e.g.

- Colleges and universities, R&D institutions,
- Primary and secondary educational institutions,
- Actors in knowledge-intensive regional development,
- The population as a carrier of knowledge-related identity, and
- Established and newly founded knowledge-intensive companies.

This is likely to have the greatest possible impact on each individual network member (Fromhold-Eisebith 2010). In addition to this concept, Kujath (2010) offers a third assumption for the above-mentioned interaction processes, which are also not only a social element, but primarily a transfer of knowledge and know-how: specialised transaction service providers support the transfer of knowledge between people, network nodes, knowledge regions and organisations. These are groups of actors who collect, store and process knowledge, and do business with it. Information and knowledge have, in fact, become a global commodity. However, the economic pursuit of profit need not necessarily be in the foreground, because not only international conferences or—with the leap towards digitalisation—operators of digital research and open access, but also publication networks may act as transfer service providers. Such a process has been described by Köhler and Schilde (2003) for higher education institutions by adopting the information broker concept from industry. The main finding would be moving away from thinking about territories and places, to inter-relational spaces. This would have to be reconsidered in the higher education system, in the sense of theoretical analysis.

12.4.3   Exemplary Developments in the Context of Research on Knowledge Networks in Land Use in Germany

One example of web-based knowledge platforms is the "Sustainable Land Management" knowledge platform, which simultaneously supports and implements aspects of communication and learning (www.nachhaltiges-landmanagement.de/en/library/

documents). The platform also contains a discussion forum. The aim of the platform is primarily to enable the exchange of information and codified knowledge. An online forum is the basis for developing a community of practice (CoP) for all those involved in the funding measure provided by the German Federal Ministry of Education and Research, i.e. scientists, practitioners, representatives from authorities, and so on. The platform enables research results and products to be exchanged, analysed and discussed, also to jointly initiate learning processes for sustainable land management (Weith and Kaiser 2015).

The platform has a graphical user interface. In addition to full-text search, the knowledge base can also be accessed by category-based search. Knowledge resources are differentiated according to products, target group, topics, regions and projects in sustainable land management. On the next level, these five categories are subdivided again into group-specific headings, which also feature selected keywords.

The interactive knowledge platform enabled all registered users to upload their own contributions (texts, images, etc.) between 2013 and 2018. Indexing of the uploaded contributions in the upload area (input mask for uploading contributions) played a major role. Indexing of contributions by the knowledge provider ensures that knowledge stocks can be accessed by category. Category-based access enables knowledge seekers to control the search results. The uploaded contributions were simultaneously made available for discussion in the forum.

In addition to the creation of a knowledge platform, presentations for formalised educational units for higher education were also developed (*Weiterbildungsmodule*). These units, which focus on a specific topic (e.g. the connection of supply chain management and sustainable land management), are also included in the knowledge platform, besides being specifically disseminated to organisations that deliver vocational training.

In a recent study, Köhler et al. (2019) argue that, in 10 years' time, higher education will resemble augmented human-computer interaction that actively combines educational offers from different educational institutions in a virtual space. In their fourth scenario, entitled "Open educational resources + open science - informal learning using the offers of several universities on the basis of recommender systems", they state that "opening up of the university through digitalisation means that study programs or learning objects from various providers are becoming more and more widespread" when simultaneously "offers from the university are made accessible to external users – which, although technically possible, has barely been implemented." The authors primarily see the following characteristics in this scenario:

- Students' shared knowledge production in a connectivist learning setting;
- Learning objects that are not clearly assigned to a single university, but taken from a virtual information market;
- The implementation of 'trusted repositories' as comprehensive subject-related collections of digital teaching and knowledge objects in the form of OER, OAP, etc.

As a result, universities may enable the portfolio-based networking of learners, brought together on the basis of suitable portfolios, also when working on inter-disciplinary topics, known as "knowledge dating" (ibid.), which leave geographical limitations behind.

12.5 Discussion and Summary Remarks

12.5.1 A Solution? Multifaceted, Networked Digital Knowledge Construction as an Opportunity for Equal Spatial Development

What options are available for the geographic design of educational spaces? We note tendencies to dissociate the offerings of place and time dependencies, and note tremendous technological advances, which always seem to be one step ahead of the previously proven teaching/learning scenarios. However, if one seeks a creative futurist approach not just in the educational meaning, but also in the spatial planning sense, then we need a new interpretation of geographical/physical space.

In this sense, spatial planning will shift more from functional place-making towards the creation of a virtual network of knowledge co-creators and adopters. This may be easy to imagine using the example of "grassland becomes a land for building maintenance" but, of course, it is harder for us to shift from physical reality to a digital, mediated reality of digital products and environments. As a result, at least two perspectives of knowledge cooperation and digitalisation in the higher education sector for spatial governance can be identified:

1. So far, the integration of different knowledge bases in educational processes has been characterised by information systems and the physical participation of actors (lecturers, students, and to a lesser extent, administrative staff). This is the currently widespread approach of e-learning or online education. However, the practice of education, especially in the higher education sector, is still strongly related to the physical location; regional virtualisation is hardly practiced, if at all. Virtual university cooperation is also neglected. For Saxony, this is evident in the goal of a joint project, initiated only in 2019/2020, focusing on "vir-tual teaching cooperation" to pilot cross-university teaching networking, taking into account the didactics of collaborative teaching and learning, ideally in a specialist domain and possibly between different types of university. Surpris-ingly, the newly published coalition agreement of December 2019 states: "Dig-ital forms of teaching and learning as well as open learning materials should find their way into all subjects. We create incentives for this through an innovation fund and drive the digital networking of study programmes. In the long run, the

Virtual University of Saxony can act as a shared platform."[4] This could also be regarded as a solution for cooperation with international partners. Furthermore, the subject-independent (interdisciplinary) qualification of educational staff should be promoted to strengthen the effectiveness of the self-regulation of students in online or blended learning scenarios, for example by acquiring online tutoring skills and abilities. Finally, individual Saxony-wide preliminary courses and online self-assessments are supported with the use and creation of OER.[5]

2. In the future, an innovative approach could be the cross-location consideration of partly new forms of generating and disseminating knowledge and learning as a starting point. In this respect, the network—the meaning to be generated in co-construction—becomes the starting point and replaces the primarily local focus, which is usually only brought together in a few cases afterwards. Area-related co-existence is also practiced, especially in the higher education sector, overcoming the primary location reference. Indeed, examples exist of such common practices of operating and, above all, using digital educational infrastructures. These include the Saxony educational portal[6] (as mentioned above), the education servers of school administrations, and the nationwide network for open educational resources, represented by OER Info[7] at the German Institute for Educational Research (DIPF). Obviously, one may find examples of shared developments across individual institutions, leading to new institutional forms.

What are the lessons learned from and opportunities for equal spatial development? As already shown, digitalisation enables especially peripheral and rural areas to participate fully in educational processes. However, this aspiration has only been met to a certain degree as yet because there is no successful interplay of a nationally available, powerful and sustainable digital infrastructure, digital services, and the highest possible degree of innovation. The Digitalization Strategy of the Free State of Saxony calls for the facilitation of a "high quality and attractiveness of a digital offer for all areas of life" (cf. SMWA 2019: 15). It is important to note, however, that uninterrupted broadband access is an urgent requirement that had not yet been met by 2019, especially in the rural areas of Saxony. In this respect, great hopes are associated with the introduction of the new 5G standard in mobile communications, which is an essential element of the infrastructure for networked systems, such as required for autonomous driving on roads.

In his recent study on how to apply digitalisation in education in an effort to trigger opportunities for equal spatial development, Schulz (2018) proposes reorganising broadband expansion and related investments in the digital infrastructure in special-purpose (public or municipal) associations. Indeed, he suggests that such

---

[4]https://www.spd-sachsen.de/wp-content/uploads/2019/12/Koalitionsvertrag_2019-2024.pdf (cf. page 16)

[5]https://bildungsportal.sachsen.de/portal/parentpage/projekte/hochschulvorhaben/projekte-2019-2020/virtuelle-lehrkooperationen

[6]https://bildungsportal.sachsen.de

[7]https://open-educational-resources.de

a transformation should also take place in wastewater disposal and public transport. After all, the digital infrastructure, and to some extent water and wastewater disposal and public transport, are not only needed in five years' time, but are now required for today's sustainably planned public services. In this way, a political solution could be created in which public institutions or even the municipalities, serve as stakeholders and act independently or in place of commercial providers. The fact that the federal states and the federal government act as a driving force would certainly support this development, enabling rural communities to move into the digitised future at greater speed.

With regard to the education sector and, detached from the discussion about broadband expansion, Schulz (2018) suggests that the existing physical and digital function and area types in cities and municipalities be recorded and rededicated to new learning, information, knowledge and transfer spaces. These would be interlinked places for education, enriched with digitised offers, whereby no clear assignment is possible or necessary in every case. After all, not only classrooms, public libraries, cafés, city parks, or in fact any public space, as well as workplaces and living rooms at home can be used as places of learning. In this way, learning islands are created which, thanks to their equipment and characteristics, meet learners' individual needs. In a digitised world, the physical offer would feature interfaces, enabling these islands to be integrated and linked to multifaceted[8] educational spaces that grow together. Reallocations and functional extensions would end the need for a one and only educational infrastructure. The physical location is left behind to enable learning regardless of time and location, with the ideal scenario being a convergence of physical and digital space. Mobile learning already addresses such a media design—but what functions then remain for physical locations?

12.5.2 Three Assumptions

Finally, the question of equal spatial development should be asked again. The authors suggest that digitalisation offers considerable opportunities, such as compensating for location disadvantages or developing peripheral regions into attractive places to work and live. Since digitalisation reduces distances to educational offers, this takes place in two directions: a Berlin resident can attend an interesting course online with a provider in Paris, whilst a Parisian can take advantage of an educational provider's offer from Berlin.

With regard to land use and spatial development, governance should reflect actor-based functional interrelations in coupled spaces more broadly. Current scientific discussions have proposed initial options (Newig et al. 2019). To connect spatial planning with the development of higher education systems, the overall framing concept must first be revised as described above. In addition, actors' roles must be

---

[8]The term "multifaceted" is derived from the description of a transformation option for libraries by Eigenbrodt (2014)

redefined. Specialist areas that previously only held advisory roles, as well as civil society, with its needs and independent initiatives, must be involved in the planning process. Bottom-up and top-down actors then form a planning process that takes all aspects into account, contributing to the equal development of educational spaces.

As shown by Köhler and Schilde (2003) and Schulz (2018), many different factors affect the topic of digitalisation, education and space. On the one hand, there are users' behaviours arising from the technical possibilities. These are characterised by individual influences driven by society and peer groups. As associated changes occur in social behaviour, the methodological case of learners and their educational needs also change. Digital components support individuality and are simultaneously becoming increasingly important. Similar to the case of online mail order, the proximity of the offer also plays a largely subordinate role in this case. As the distance from the learner increases, the proportion of digitally depicted elements and processes increases, which is an important point for education providers, be it universities, schools, libraries or private further education companies. However, in the absence of local providers, the competitive situation may be intensified, particularly at local or regional borders. This is shown in exaggerated form in the following three hypotheses:

1. Despite new ICTs, knowledge, control and monitoring functions are spatially concentrated, and routine functions are generally performed decentral (Meusburger et al. 2011).
2. Traditional teaching/learning arrangements will be largely superseded by self-directed, flexible and individualised forms in the medium term (Köhler et al. 2018, 2019).
3. Physical learning spaces in schools, universities/colleges and vocational education and training institutions will be replaced by digital learning spaces, as learners can access the same resources anywhere, anytime (ibid.).

Obviously, the necessary research would have to address the greater integration of (A) spatial ideas about cooperation in the context of digitalisation and (B) the development of new visions and models based on this, including all domains involving the creation of knowledge and its dissemination into society.

12.5.3 *Methodological Critique and Future Research Needs*

Finally, it is necessary to briefly present a methodological critique of the presented research, and to address the need for future research. Due to the multi-perspectives and the high dynamics of the (socio-)technological development around digitalisation and the variety in its adoption (cf. Fischer 2012), there are certainly inaccuracies in the scientific findings, both in theoretical and methodological terms, which must be discussed briefly in the following.

I. *Dynamics of the research domain*:

As argued by Meusburger et al. (2011), research has to take into account the historicity of the artefact under investigation. This means that certain elements or configurations are no longer existent, and thus the research object is not always consistent in itself. Due to a learner potentially changing into a teacher or provider of an educational offer via social media channels, open platforms or research networks, it can be assumed that digitalisation will blur the boundaries between decentralised and central functional spaces.

II. *Development and dissemination of new knowledge*:

Overall, the influence of digitalisation on knowledge genesis in the context of sustainable, fair development should not be underestimated—and is relevant for achieving strategic development goals (SDGs; Weith and Köhler 2019). The focus here is on the development and dissemination of new knowledge and the exchange of knowledge (www.futureearth.org). Supporting a transformation towards sustainability is mentioned there as an explicit goal. The associated research agenda states that this also necessitates a new form of knowledge generation across disciplinary boundaries—together with social partners.

References

Bergner, C. (2019). Improving education through technology in school management. *Empirical evidence from schools in Germany*. Dresden: Technische Universität.

Bundesministerium für Wirtschaft und Energie (BMWi). (2016). Zukunftschance Digitalisierung. In *Gute Geschäfte, zufriedene Kunden, erfolgreicher Mittelstand. Ein Wegweiser* (2nd updated ed.). Berlin.

Cash, D. W., Clark, W. C., Alcock, F., Dickson, N. M., Eckley, N., Guston, D. H., et al. (2003). Knowledge systems for sustainable development. *PNAS, 100*(14), 8086–8091.

Cornell, S., Berkhout, F., Tuinstra, W., Tàbara, J. D., Jäger, J., Chabay, I., & De Wit, B. (2013). Opening up knowledge systems for better responses to global environmental change. *Environmental Science and Policy, 28*(C), 60–70.

Davoudi, S. (2015). Planning as practice of knowing. *Planning Theory, 14,* 316–331.

Dilling, L., & Lemos, M. C. (2011). Creating usable science: Opportunities and constraints for climate knowledge use and their implications for science policy. *Global Environmental Change, 21,* 680–689.

Eigenbrodt, O. (2014). Auf dem Weg zur fluiden Bibliothek: Formierung und Konvergenz in integrierten Wissensräumen, S.207-220. In Stang & Eigenbrodt: Formierungen von Wissensräumen. Optionen des Zugangs zu Information und Bildung. Berlin : de Gruyter Saur.

Enengel, B., Muhar, A., Penker, M., Freyer, B., Drlik, S., & Ritter, F. (2012). Co-production of knowledge in transdisciplinary doctoral theses on landscape development—An analysis of actor roles and knowledge types in different research phases. *Landscape and Urban Planning, 105*(1–2), 106–117.

Erber, G., Köhler, T., Lattemann, C., Preissl, B., & Rentmeister, J. (2004). *Rahmenbedingungen für eine Breitbandoffensive in Deutschland*. Berlin: Deutsches Institut für Wirtschaft.

Fischer, H. (2012). Know Your Types—Konstruktion eines Bezugs zur Analyse der Adoption von E-Learning-Innovation in der Hochschullehre. Dissertation in den Fachgebieten Digital Culture an der Universitetet i Bergen (NO) und Bildungstechnologie an der TU Dresden.

Fromhold-Eisebith, M. (2010). Von der Technologie- zur Wissensregion—Neukonzeption anhand des Beispiels der Region Aachen. In F. Roost (Ed.), *Metropolregionen in der Wissensökonomie* (pp. 61–81). Detmold: Verlag Dorothea Rohn.

Gibbons, M., Limoges, C., Nowotny, H., Schwartzman, S., Scott, P., & Trow, M. (1994). *The new production of knowledge—The dynamics of science and research in contemporary societies*. London: SAGE.

Hesse, F. (2017). *Denken und Handeln auf der Basis externen Wissens*. Online at: https://www.youtube.com/embed/EsU01gc9bAk?list=PLOw5l9nehTTxbP9G46-T2O0ppzcLmNQnF&autoplay=1. Accessed on June 20, 2020.

Hirsch Hadorn, G., Hoffmann-Riem, H., Biber-Klemm, S., Grossenbacher-Mansuy, W., Joye, D., Pohl, C., & Zemp, E. (Eds.). (2008). *Handbook of transdisciplinary research*. Heidelberg: Springer.

Kahnwald, N., & Köhler, T. (2007). Microlearning in virtual communities of practice? An explorative analysis of changing information behaviour. In: T. Hug, M. Lindner, & P. A. Bruck (Eds.), Micromedia & eLearning 2.0: Gaining the Big Picture. *Proceedings of Microlearning 06*. Innsbruck: University Press.

Kaiser, D. B., Köhler, T., & Weith, T. (2016). Knowledge management in sustainability research projects: Concepts, effective models, and examples in a multi-stakeholder environment. *Journal of Applied Environmental Education & Communication, 15*(1).

Kaiser, D. B., Gaasch, N., & Weith, Th. (2017). Co-production of knowledge: A conceptual approach for integrative knowledge management in planning. Trans Assoc Eur Schools Plann 1.

Kajikawa, Y., Tacoa, F., & Yamaguchi, K. (2014). Sustainability science: The changing landscape of sustainability research. *Sustainability Science, 9*, 431–438.

Köhler, T., & Kahnwald, N. (2005). Does a class need a teacher? New teaching and learning paradigms for virtual learning communities. In Online Communities and Social Computing. *Proceedings of the HCI 2005*. New York: Lawrence Erlbaum Associates.

Köhler, T., & Schilde, P. (2003). From project teams to a virtual organization: The case of the education portal Thuringia. *Frontiers of e-Business Research, 2*(2).

Köhler, T. (2006). Wissen oder Handeln? Neue Medien aus lerntheoretischer Sicht. In D. Gebert (Ed.), *Innovation aus Tradition*. Dokumentation der 23. Arbeitstagung des Arbeitskreises der Sprachenzentren, Sprachlehrinstitute und Fremdspracheninstitute (AKS) 2004. Bochum: AKS-Verlag.

Köhler, T., Igel, C., & Wollersheim, H.-W. (2018). Szenarien des technology enhanced learning (TEL) und technology enhanced teaching (TET) in der akademischen Bildung 2028. In: B. Getto, & M. Kerres (Eds.), *Digitalisierung: Motor der Hochschulentwicklung? Münster, Waxmann*. https://www.waxmann.com/buch3868. Accessed on June 20, 2020.

Köhler, T., Wollersheim, H.-W., & Igel, C. (2019). Scenarios of technology enhanced learning (TEL) and technology enhanced teaching (TET) in academic education. A forecast for the next decade and its consequences for teaching staff. In *IEEE Proceedings of the 8th International Conference on Learning Technologies and Learning Environments (LTLE2019)*. Toyama: IAA.

Kujath, H. J., & Peiker, W. (2014). Wandel des internationalen Städtesystems unter dem Einfluss der Wissensökonomie. *Geographische Rundschau, 12*(2014), 12–18.

Kujath, H. J. (2010). Institutionen und räumliche Organisation der Wissensökonomie. In H. J. Kujath & S. Zillmer (Eds.), *Räume der Wissensökonomie*. Implikationen für das deutsche Städtesystem (pp. 51–81). Berlin: LIT Verlag Dr. W. Hopf.

Lemos, M. C., & Morehouse, B. J. (2005). The co-production of science and policy in integrated climate assessments. *Global Environmental Change, 15*, 57–68.

Mauser, W., Klepper, G., Rice, M., Schmalzbauer, B. S., Hackmann, H., Lee-mans, R., & Moore, H. (2013). Transdisciplinary global change research: The co-creation of knowledge for sustainability. *Current Opinion in Environmental Sustainability, 5*(3–4), 420–431.

McLuhan, E. (2008). The source of the term 'Global Village'. In *McLuhan studies* (Issue 2). https://projects.chass.utoronto.ca/mcluhan-studies/v1_iss2/1_2art2.htm. Accessed on June 20, 2020.

Meusburger, P., Koch, G., Christmann, G. B., et al. (2011). Nähe- und Distanzpraktiken in der Wissenserzeugung – Zur Notwendigkeit einer kontextbezogenen Analyse. In O. Ibert & H. J. Kujath (Eds.), *Räume in der Wissensarbeit*. Zur Funktion von Nähe und Distanz in der Wissensökonomie (pp. 221–249). Wiesbaden: VS Verlag für Sozialwissenschaften.

Millar, C. C. J. M., & Choi, C. J. (2003). Advertising and knowledge intermediaries: Managing the ethical challenges of intangibles. *Journal of Business Ethics, 48,* 267–277.

Miller, T. R., Wiek, A., Sarewitz, D., Robinson, J., Olsson, L., Kriebel, D., & Loorbach, D. (2014). The future of sustainability science: A solutions-oriented research agenda. *Sustainability Science, 9,* 239–246.

Newig, J., Lenschow, A., Challies, E., Cotta, B., & Schilling-Vacaflor, A. (2019). What is governance in global telecoupling? *Ecology and Society, 24*(3), 26. https://doi.org/10.5751/ES-11178-240326.

Nonaka, I., & Takeuchi, H. (1995). *The knowledge-creating company: How Japanese companies create the dynamics of information*. New York: Oxford University Press.

North, K. (2011). Wissensorientierte Unternehmensführung: Wertschöpfung durch Wissen. 5. Wiesbaden: Gabler Verlag.

Opdam, P., Westerink, J., Vos, C., & De Vries, B. (2015). The role and evolution of boundary concepts in transdisciplinary landscape planning. *Planning Theory & Practice, 16,* 63–78.

Parker, R., & Hine, D. (2013). The role of knowledge intermediaries in developing firm learning capabilities. *European Planning Studies, 22,* 1048–1061.

Pohl, C. (2011). What is progress in transdisciplinary research? *Futures, 43,* 618–626.

Pohl, C., Rist, S., Zimmermann, A., Fry, P., Gurung, G. S., Schneider, F., et al. (2010). Researchers' roles in knowledge co-production: Experience from sustainability research in Kenya, Switzerland, Bolivia and Nepal. *Science and Public Policy, 37,* 267–281.

Pohl, C. (2014). Eine Theorie transdisziplinärer Forschung für wen? Reaktion auf M.Ukowitz.2014. Auf dem Weg zu einer Theorie transdisziplinärer Forschung. GAIA, 23 (1), 19–22.

Riedel, J., Dubrau, M., Möbius, K., & Berthold, S. (2017). Digitales lehren & lernen in der Hochschule. https://nbn-resolving.de/urn:nbn:de:bsz:14-qucosa-217606. Accessed on June 20, 2020.

Salet, W. (2014). The authenticity of spatial planning knowledge. *European Planning Studies, 22,* 293–305.

Schulz, J. (2018). Digitalisierung in der Bildung. Chancen für eine gleichwertige räumliche Entwicklung? Vortrag auf der LAG-Frühjahrstagung 2018 "Digitalisierung: Der Nordosten ist flach?" Digitale Transformation und gleichwertige räumliche Entwicklung" der Akademie für Raumforschung und Landesplanung (ARL), Landesarbeitsgemeinschaft Berlin/Brandenburg/Mecklenburg-Vorpommern. https://slub.qucosa.de/api/qucosa%3A2 3393/attachment/ATT-0. Accessed on June 20, 2020.

Semar, W. (2014). Digitale Veränderungsprozesse: Konsequenzen für das Lern- und Kommunikationsverhalten. In O. Eigenbrodt & R. Stang (Eds.), *Formierungen von Wissensräumen. Optionen des Zugangs zu Information und Bildung* (pp. 11–21). Berlin/Boston: De Gruyter Saur.

SMWA—Staatsministerium für Wirtschaft, Arbeit und Verkehr. (2019). Sachsen Digital—Digitalisierungsstrategie des Freistaates Sachsen, 3. (Fully revised ed.). Dresden. https://publikationen.sachsen.de/bdb/artikel/33501/documents/51221.

Stang, R., & Eigenbrodt, O. (2014). Informations- und Wissensräume der Zukunft: Von Hochgefühlen und lernenden Städten. In O. Eigenbrodt & R. Stang (Eds.), *Formierungen von Wissensräumen Optionen des Zugangs zu Information und Bildung* (pp. 232–244). Berlin/Boston: De Gruyter Saur.

Stützer, C. M., Köhler, T., Carley, K. M., & Thiem, G. (2013). "Brokering" behavior in collaborative learning systems. *Procedia—Social and Behavioral Sciences, 100.* https://doi.org/10.1016/j.sbs pro.2013.10.702.

Vatn, A. (2005). *Institutions and the environment*. Cheltenham, UK, Northampton, Massachusetts: Edward Elgar.

Weith, Th., & Kaiser, D. B. (2015). Wissensmanagement: die Plattform "Nachhaltiges Landmanagement". In G. Meinel, et al. (Hg.), *Flächennutzungsmonitoring VII*. Boden - Flächenmanagement – Analysen und Szenarien. Berlin (Rhombos) (pp. 61–66).

Weith, T., & Köhler, T. (2019). *Der Einfluss der Digitalisierung auf die Wissensgenese im Kontext einer nachhaltig-gerechten Entwicklung*. Synergie. Fachmagazin für Digitalisierung in der Lehre, Ausgabe #07. https://www.synergie.uni-hamburg.de. Accessed on June 20, 2020.

Weith, T. (2020). "Smart Countryside" im Osten? Zum Wandel ländlicher Räume und den Herausforderungen der Digitalisierung. In M. Naumann & S. Becker (Eds.), *Regionalentwicklung in Ostdeutschland - Dynamiken, Perspektiven und der Beitrag der Humangeographie*. Berlin: Springer.

Ziegler, A. (1993). Wer moderieren will, muß Maß nehmen und Maß geben (Who wants to moderate, has to take the measurements and to provide tailor-made solutions). In A. Wohlgemuth (Ed.), *Moderation in Organisationen (Facilitation in organisations)* (pp. 17–53). Stuttgart: Haupt.

Zimmermann, K. (2010). Der veränderte Stellenwert von Wissen in der Planung. *Raumforschung und Raumordnung, 68,* 115–125.

Chapter 13
Transcending the Loading Dock Paradigm—Rethinking Science-Practice Transfer and Implementation in Sustainable Land Management

Sebastian Rogga

Abstract Modern science is increasingly called on to produce societally relevant and usable knowledge to tackle global challenges. Academics respond by conducting research projects that transcend the boundaries of single disciplines and institutions by actively engaging with non-academic actors. Such institutional arrangements open up entirely new perspectives for science communication and the problem-solving of real-world issues. However, they also call for elaborate management tasks and demand learning processes on the part of all those involved. This chapter introduces transfer and implementation (T&I) as a conceptual pair of terms to grasp the challenges, without compromising the opportunities of this new research mode. In doing so, this chapter discusses contemporary approaches of science knowledge transfer, and promotes a notion that prioritises knowledge transfer over information transfer through artefacts. After reframing T&I as a management area in research projects, I present three strategic policy pathways for sustainable land management.

Keywords Transfer · Implementation · Science-practice collaboration · Transdisciplinarity · Sustainable land management · Societal impact · Conceptual framework · Strategy

13.1 Introduction

After the discovery of the "ozone hole", it took policymakers five years to adopt effective measures to reduce ozone-depleting substances induced by humans on a global scale. What appears to be a long haul is actually a spectacularly short time scale if the cause of the discovery—a scientific study by Joe Farman et al. published in Nature magazine in May 1985—is placed in relation to the action—a global ban on the production of ozone-depleting substances in June 1990. In fact, the Farman report is a rare, yet outstanding example of scientific knowledge being transferred

S. Rogga (✉)
Leibniz Centre for Agricultural Landscape Research, Eberswalder Str. 84, 15374 Müncheberg, Germany
e-mail: sebastian.rogga@zalf.de

© The Author(s) 2021

249

T. Weith et al. (eds.), *Sustainable Land Management in a European Context*, Human-Environment Interactions 8, https://doi.org/10.1007/978-3-030-50841-8_13

directly to society, i.e. scientific evidence that led to concrete action being taken by decision makers in line with the authors of the study.

The case of the ozone layer is often cited when addressing the potential societal impact induced by research. The science-practice transfer, which appears as a straight path in our example, is rather diffuse and acts in a permeable environment at the intersection of science and society in which information is shared, reassembled and re-evaluated. What appears to be an action arena that is too complex to steer, or even to fully understand, is an upcoming field of investigation for scientific action that aspires to bring about societal effects as a result of the research undertaken. To be clear from the outset: science-to-practice transfer cannot be conducted and planned to the full extent. This is especially true for contested policy arenas, within areas of high scientific uncertainty, and areas of application involving a variety of actors. It is nonetheless possible to design science-policy interfaces that increase the likelihood of science results being adopted by societal actors. But how should such interfaces be designed in the field of sustainable research?

In this chapter, I seek to reframe the science-society interface in sustainable land management (SLM). The result is not a "how-to manual", but rather a framework for initiating and guiding SLM processes with a focus on the transfer of knowledge. SLM encompasses a purposeful process of managing land use and development by integrating scientific and practical knowledge. From my perspective, SLM embraces project-based research processes in which actors from practice and academia form a project consortium for a certain period with the common goal of achieving the more sustainable use of land resources. As such, SLM project partners form an alliance that proactively seeks to intervene in the highly regulated field of land use. I refer to these proactive intervention measures as **transfer and implementation** (T&I), despite a variety of related terms, including diffusion, impact or outreach.

T&I is applied in a wide variety of contexts such as political science, information technology and other domains. Nevertheless, this chapter will show that T&I, as a conceptual pair of terms, is ideal for supporting the framework for that particular area of SLM practice that receives greater attention as research is increasingly asked to "perform" with a higher societal impact. Besides the societal performance aspect, I also apply T&I as *guiding principles* for transdisciplinary processes. In doing so, I take into account the fact that T&I activities in SLM encompass more than a sole compilation of artefacts (i.e. books, reports and software) that are gathered and compiled at the end of a project cycle. Thus, I seek to transcend the prevalent practice that Cash (2006) calls the "loading dock paradigm" of transfer.

The "loading dock" is a metaphor for a location or deposit where scientific results are stored for the one-way transfer of results from science to other areas of society. These results, often peppered with scientific jargon and offering limited opportunity for feedback, are indeed generally available to everybody (unless they are hidden behind paywalls), but do not specifically address non-scientific target audiences. In addition, the "loading dock" oversimplifies the complex intersectional network that emerges in the transition area between science and society, which leads to a linear, causal understanding of how knowledge transfer actually works (ibid.).

Even though the knowledge base about the question of "how to transfer knowledge" is growing annually, the "loading dock approach" is still pervasive in many academic areas. However, I consider it to be outdated for the purposes of research on sustainability issues, which places communication and social learning in the centre of research activities (Fazey et al. 2013).

Clearly, the discussion on transfer and implementation is played out against the background of a new understanding of a reflexive, engaged and process-minded form of science that faces unknown challenges and seeks to gather new forms of knowledge. Thus, in Sect. 13.2, I briefly describe this new mode of science and how it differs from other (more conventional) approaches. Against that backdrop, I argue in Sect. 13.3 how transfer and implementation can be understood in transdisciplinary science approaches in contrast to prevailing transfer models. In Sect. 13.4, I present a T&I framework for transdisciplinary research projects, and apply this framework to SLM in Sect. 13.5.

13.2 Science, Society and the Drive Towards Transformation

New demands have recently been placed on science as an institution. Science is increasingly called on to serve society and justify the resources invested in it ("return on investment"). Academia is to make a greater contribution to solving *real-world issues* that society expects from science. These real-world issues, such as climate change and demographic change, differ from conventional scientific questions. They entail normative dimensions (i.e. human values, norms and preferences); they are complex (i.e. factors influencing the systems under investigation are manifold and sometimes unknown); they must be solved within democratically legitimised processes; and they call for a synthesised approach that incorporates the knowledge of many perspectives. Consequently, this new mode of science is asked to "produce" different forms of knowledge (Pohl and Hirsch-Hadorn 2006; Zscheischler et al. 2017), namely:

- *systems knowledge* about systemic interrelations in a given research context (the real-world issue),
- *target knowledge* about desired future developments in society, and
- *transformative knowledge* about strategies and instruments on how to approach the desired future.

Whereas the production of *systems knowledge* may be provided by established research methods, the production of *target knowledge* and *transformative knowledge* calls for entirely new methods that incorporate multiple scientific disciplines as well as so-called practitioners or stakeholders.[1] This novel mode of science (often

---

[1]For easier understanding, I refer in this text to "practitioners" as non-academic project partners (i.e. they represent non-academic institutions, but are contractually bound to the project consortium)

referred to as "transdisciplinary research", "transformative science" or "intervention science") seeks to change the current (non-sustainable) state of our societies by different means (and intervention levels), backed up by scientific evidence.

Not only is it necessary to generate new sets of knowledge that—unlike disciplinary knowledge—are "different" and are fed from society to science. It is also necessary to adequately communicate and feed back knowledge, and vice versa, if it is to be effective. This is at least the inherent claim of many sustainability-oriented research activities (cf. Jong et al. 2016).

From a systems theory perspective, the trend towards transdisciplinary science can be summarised as follows: The boundaries between the scientific subsystem and the other subsystems of society become more permeable. At the intersections of these subsystems, new intra- and inter-directional communication patterns are established that work bi-directionally. Transdisciplinary research projects can be located at these intersections. They form boundary organisations (Guston 2001) in which knowledge is produced and transferred to multiple subsystems such as science, politics, economy and administration.

If we continue the notion that transdisciplinary research is accompanied by new patterns of communication, then it follows that research outputs change as well since they reflect the changing mode of communication. Modern science, as it has evolved since the nineteenth century, has developed elaborated media channels for communicating within its own system boundaries. However, these communication channels may have changed in the digitised world, yet the supposed output of science has remained the same: the production of truth. As the outcomes of sustainability research change, so does the adaptation of what science has previously considered the output of a research project. Thus, concrete questions arise for the areas of T&I in the context of changing research, since T&I forms the bridge between "output" and "outcome".

13.3   From the "Loading Dock" to Reflexive Discourses—A Short Anthology of T&I as Objects of Scientific Investigation

The term **implementation** is used primarily in the fields of computer science, data processing and, most prominently, in the fields of politics and administration (Nohlen 1998). In the innovation context, implementation refers to the phase in which innovation has reached "market maturity" and is distributed to anticipated customer groups accordingly (cf. Ibert 2003). This applies in modified form to non-economically exploitable innovations, too. The innovation will be linked to certain goals, but these may have unintended consequences.

---

and "stakeholders" when I mention actors or actor groups affected by the issue under investigation. Stakeholders may become part of the research process (via participatory methods, public hearings, exhibitions, etc.), but are not bound to the project by contract.

Despite a variety of usage contexts, **transfer** can be defined as a "transmission" of information or objects between a sender and a receiver, some of which operate in specific contexts. In application-oriented research, transfer is conventionally equated with the transfer of research results to potential users. In this context, transfer becomes an instrument that appears in the context of innovation and diffusion research (Schröder et al. 2011).

In applied science, aspects of T&I have been the focus of numerous empirical studies on the diffusion of innovations since the 1920s and 1930s. Since then, mainly application-related disciplines, such as engineering and technical sciences, medicine and geography have addressed questions relating to T&I research. The aims of research activities include the systematic analysis of transfer conditions and modelling approaches (Gräsel et al. 2006: 479).

From the perspective of research on T&I, different approaches have been developed. Of particular note here are *actor-centred approaches*, which operate with a typology of persons (groups) in the diffusion process. According to these approaches, innovations by persons or groups of people are adapted at different speeds (Hägerstrand 1952; Rogers 1995). Another approach, based on the findings of the *network theory*, considers actors as objects in a superordinate social network whose connections are of central importance (see, for example, Granovetter 1973).

A third model that has been widely adapted in German innovation policy (Blümel 2016) is the *linear model of technology push or science push* (based on Bush (1945)). According to this model, innovations go through a gradual evolution that ranges from basic research, through applied research, to product development and innovation. This model is also based on unidirectional knowledge transfer, which understands "society" as the addressee of scientific results.

The principle of unidirectional transfer was dominant in the past (between the 1960s and the 1990s), but increasingly came under criticism. Cash et al. (2006) used the metaphor of the "loading dock" of science transfer to describe its deficiencies. According to this term, scientists perform their research activities in a house (with an adjacent loading dock) that symbolises academia. At the end of the research cycle, the results are eventually stored as readymade "information packages" or "products" on a dock and made available to potential end users: "*You take it out there, and you leave it on the dock and you say, there it is. And then you walk away and go back inside*" (ibid. 484). This approach is based on the premise that the information that reaches the recipient triggers appropriate action, or that research results are adapted from the practical side.

In the field of science communication, this phenomenon is also known as the "information deficit model", which attributes public scepticism to science to a lack of understanding, resulting from a lack of information. In science-based consultancy, an artificial separation of scientific information from policymaking was promoted by its clients (Weith 2011) and thus manifested the inherent logic of the loading dock metaphor.

The loading dock approach underlines the notion of a purely knowledge-driven science which, as the sole knowledge producer, remained separate from the application-relevant areas of knowledge ("policy knowledge"). It not only leaves potential user groups undefined, it also vaguely describes the benefit of research. It is assumed that the public needs innovative products and scientific knowledge of solutions. In addition, numerous empirical studies have shown that the implementation of practice transfer has fallen short of expectations (Böcher and Krott 2013; Ascher et al. 2010; Fry et al. 2003; Pregernig 2000, etc.) (Table 13.1).

Contemporary transfer approaches that especially concentrate on knowledge transfer and advocate a move beyond the transfer of artefacts reflect much more on the fact that the generation of knowledge requires a co-constructed (learning) process (Fazey et al. 2013). This means that the mere provision of knowledge in different media forms of preparation is not the same as the successful transfer of that knowledge. The way in which actors tap into knowledge plays an increasingly important role.

Best and Holmes (2010) see two paradigms of knowledge transfer as an extension to the linear model. First, relationship models, in which knowledge producers and consumers come together in collaborative networks and exchange knowledge,

Table 13.1 Comparison of linear transfer models versus systemic transfer models by characteristics (own source)

Characteristic	Linear transfer model	Systemic transfer model
Transfer impulse	Science-pushed or demand-driven	Science-pushed and demand-driven
Direction of transfer	Unidirectional; from science to practice	Reflexive and iterative between science and practice
Dimensions of transfer	Transfer of research results (technological transfer, transfer of codified information)	Transfer of research results (technological transfer, transfer of codified information); knowledge transfer (exchange of individual, disciplinary and organisation-based knowledge)
Date of transfer	Ex-post	Continuous and ex-post
Content of transfer	Results of research	Results of research, perspectives of disciplines and stakeholders, experiential knowledge of practitioners and stakeholders, data, evaluation of complex issues
Who transfers?	Science (via intermediaries)	Science and practice
Instruments of transfer	Publications (artefacts and presentations), academic teaching	Publications (artefacts and presentations), academic teaching, methods of knowledge exchange and integration
Transfer objects	Artefacts, scientific publications and dissemination formats (written and oral)	Artefacts, scientific publications and dissemination formats (written and oral), tools, policy briefs, exchange forums, decision support systems, etc.

have increasingly been discussed since the 1990s. Transdisciplinary research projects represent a suitable form of organisation in this case.

Second, "system models" that assume transfer-relevant knowledge is embedded in organisations or other system units have increasingly been addressed for about ten years. These models reflect the priorities of an organisation and its "culture". In this approach, transfer extends to the aspects of the integration, translation and mobilisation of knowledge (Partidario and Sheate 2013, p. 28).

13.4 Reframing T&I for SLM

Based on an extended understanding of T&I, I advocate a **multi-dimensional concept of transfer and implementation**. This concept extends the hitherto dominating unidirectional, mostly technology-oriented transfer concept that is science-driven and focuses on practice diffusion.

Communication and cognitive science aspects of T&I are stressed as the notion shifts from the *transfer of information* through artefacts to so-called *knowledge transfer*. Knowledge transfer is defined as a process in which knowledge (or information) is externalised and communicated internally and intrapersonally by means of media (Beckers 2012, p. 95). Accordingly, transfer takes place not only between individuals in the form of interaction, but also involves intrapersonal processes of learning, describing and explaining. Transfer is no longer just a process in which "knowers" connect with "non-knowers". It rather reflects a more dynamic view of how knowledge is generated; and that knowledge is embedded in individual and organisational contexts, and circulates among the actors involved, (ideally) creating a meta-learning process (co-learning) that runs parallel to individual learning (cf. Kaiser et al. 2017).

Given the context of transdisciplinary research, T&I activities tie in with the inherent goal of stimulating change in both the academic world (as new modes of knowledge production are tested) and the real world.

T&I can be captured in a bi-dimensional, conceptual framework that includes an **internal and external dimension** (see Fig. 13.1). In the following section, the framework is projected onto the SLM case study, but can in principle be adapted for any transdisciplinary research activity.

Our unit of consideration is the transdisciplinary research project, i.e. a temporally, financially and staff-wise limited unit of activities in relation to one or more related research goals (Newig et al. 2019) that exists within a real-world research setting (see Fig. 13.2). Following the normative and programmatic nature of transdisciplinary research, the project involves different actors from academia and practice (i.e. the project consists of at least two different parties); it involves conducting research in a collaborative fashion and aspires to transcend its boundaries by tackling real-world issues. The boundaries of a research project are thus constituted not only by resources (namely time, workforce and money), but also by the affiliation of the

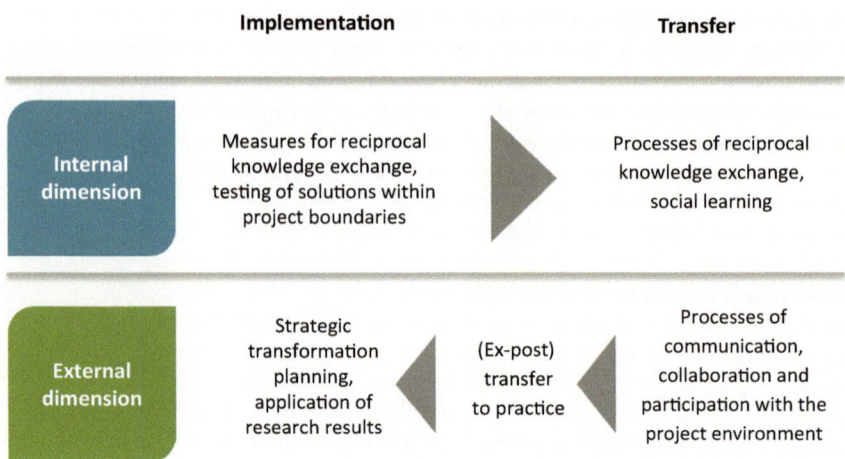

Fig. 13.1 Multidimensionality and causal links between T&I in transdisciplinary projects (own source)

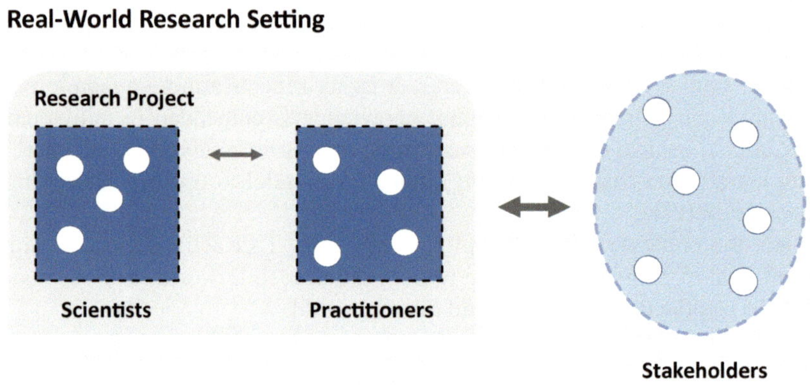

Real-World Research Setting

Fig. 13.2 Basic model of a transdisciplinary research project and its boundary (own source)

project members. In externally funded projects, project partners are often contract-bound and, thus, form the boundary of a project by the receipt of grants (as opposed to non-grantees).[2]

Either way, successful research activities, outputs and outcomes in those kinds of institutional arrangements cannot be taken for granted (Newig et al. 2019) as they

---

[2]In reality, however, project boundaries cannot be drawn in such a clear-cut way because project partners may also be bound simply by cooperation agreement without the awarding of grants.

form complex management tasks. I therefore stress the **internal dimension** of T&I as a key field of action that should be covered by project coordination.

In this context, **transfer** refers mainly to the realm of knowledge transfer among project team members. It includes processes that aim at informative exchange about knowledge stocks (expert knowledge, interests, agendas, norms, values, etc.) and at an integration of different epistemologies for the cause of the research question. Transfer evolves into a field of coordination action that must be continually considered and highlighted during the early stages of the project cycle in which fundamental research issues are framed that may be impossible to correct at later stages of the research process. Transfer in this context can also be understood as processes of knowledge internalisation (individual learning, group learning) and externalisation (i.e. making tacit knowledge explicit).

To foster internal transfer, it makes sense to **implement** a set of rules and regulations, as well as integration methods, i.e. methods that support the exchange of people and their knowledge. Research projects usually consist of relatively loose networks that are bound by their common research interest and that are not subject to any other regulations than those provided by the project funding agency and the research partners' respective affiliation. Thus, project rules such as control mechanisms and codes of conduct are ideally a result of negotiation processes within the project. In other words, internal implementation is a matter of transdisciplinary research design.

Another notion of implementation in this context refers to projects that form experimental spaces or work on technological innovations. In the course of research projects, prototypes or comparable artefacts may be implemented to gather data or observe their performance in a laboratory setting.

Whereas the internal dimension covers activities *within* the project boundary, the **external dimension** of T&I highlights action arenas *beyond* the boundary of the research project.

In this context, **transfer** invokes processes and activities of interaction with the project environment (e.g. public relations, participation, and feedback loops). The purpose and function of external transfer can range between simply providing information about research activities and empowering target audiences. Transfer is not solely considered as a punctual, one-way minded, communicative act, but rather understood as a reflexive mode of science that stimulates social learning and as a vehicle to close the application gap of scientific results.

In contrast, external **implementation** can be understood as a result of external transfer activities. It describes two realms: first, it is the strategic approach (or pathway) on how to stimulate change towards a preferred direction in a given research context. This requires a mental model acquired by the research team about how transformations actually (might) work in their project setting to be effective. Second (and in accordance with the conventional understanding of implementation in applied research), it describes the output(s) of a research project; the communicative format or artefact that is designed to suit the target audience's expectations.

13.5   T&I Strategic Policy Pathways in SLM

Up to this point, I have outlined the framework for all kinds of subjects in which transdisciplinary research can be performed. In the following section, this framework is applied to **land management settings** (see Fig. 13.2), seeking to achieve a strategic mental model of T&I. This mental model is suitable as a starting point for considering and planning T&I activities in land-related research projects that seek to stimulate change towards sustainability. To capture the rich nature of land-based issues, the model picks a rather high level of generalisation. My description of the elements of the model will include practical examples to facilitate understanding.

The model was inspired by theoretical assumptions from the literature on T&I using deduction (see Sect. 13.3) as well as by observed transfer strategies applied by SLM projects (from the funding measure under the same title 2010–2016; see also Weith et al. 2019), acquired by induction during project activities.

The *research project* is the starting point and central subject of the model (Fig. 13.3; left side). Even though the project assembles a group of individuals with very specific goals and interests, I assume that the central connective element is the mutual goal to make *land use more sustainable* (Fig. 13.3; green circle). This also includes scientists, because they demonstrate a high degree of commitment to the normative goal of SLM projects (Zscheischler et al. 2018). Thus, the research project is considered to be a unit that proactively pushes towards sustainability by means of various measures.

Since research projects (usually) lack the capacity to directly transform land use practices (cf. Lux et al. 2019), the mental model incorporates intermediate actors as transmitters that are proactive in both directions. Besides building a bridge from the research project to land use decision arenas or decision makers, they fuel and enrich the research process by feeding knowledge (e.g. factual knowledge, experiential knowledge) back to the research project. Thus, they may be a part of the research

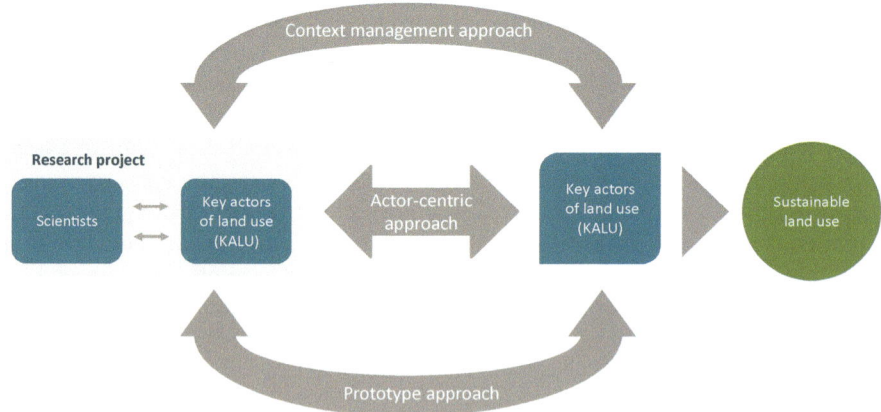

Fig. 13.3 Strategic conceptual model of T&I in SLM (own source)

project itself (bound by contract), as this increases the likelihood that research results will be adopted after the end of the project cycle (Lux et al. 2019). Numerous studies on transdisciplinary research indicate that an early and continuous involvement of practitioners in the research process (e.g. mutual problem definition, co-design of the research process) correlate with the higher societal impact of (transdisciplinary) research (Lux et al. 2019; Jong et al. 2016).

For SLM, such intermediaries are introduced as *"key actors of land use"* (KALU). In the model, they constitute both the main addressee of external T&I actions and the co-producers of knowledge in internal T&I activities. KALU can be individuals or (representatives of) institutions that play a decisive or influential role in the land use context such as.

(a) individuals or collective institutions that possess direct decisive power over land use options on a given plot (e.g. farmers, agricultural enterprises, forest owners, land owners in general).
(b) institutions that steer land use by setting the legislative and regulative framework of land use in a given spatial context (e.g. planning authorities, administrations).
(c) institutions that influence land use by setting decisive context conditions and/or incentives for land use decisions (e.g. markets, politics), and
(d) individuals or collective institutions that influence the societal discourse of land use in society. The latter covers the full range of institutions such as NGOs, political parties, professional organisations, the media, civil society actors, businesses, and science-policy interface organisations. Individuals can also be considered key actors, e.g. if they own the capacity and commitment to achieve sustainable land use change in a given context or if they make a significant contribution to filling crucial knowledge gaps to enhance the understanding of the issue under investigation.

Identifying the "right" KALU is one of the key tasks in SLM. It requires a thorough stakeholder analysis that is backed up by scientific methods. The heterogeneity of the actor landscape evolves from the way in which land governance is performed in post-modern societies. Land as a commonality is the subject of an inextricable network of different regulations and actors' accesses and interests. A multitude of institutions (usually at different scales; from international to local) may have a direct or indirect impact on land use on a certain plot of land. In addition, context conditions such as market forces or environmental conditions such as climate change have an indirect (if not direct) impact on land use practices. Also, as in most democratic systems, significant changes in land use regulations presuppose democratic negotiation processes that can substantially extend project time schedules. In sum, every change in this highly regulated action arena must consider the limiting borders, as well as the enacting and supportive forces. Apart from the fact that this should be part of an expectation management process conducted in every project at an early stage, I suggest three **strategic policy approaches** (or impact pathways) that provide orientation and support strategic project decisions for T&I.

Each approach constitutes an ideal path to influence land use decisions made by KALUs. They should not be considered as an exclusive strategy that leaves other approaches unnoticed. On the contrary, depending on the goal(s) of the research project, these paths might be used alternatively or even in combination, as they may generate synergies. Any given measure along these paths should be continuously evaluated according to its impact and (if necessary) readjusted accordingly.

1. The **actor-centric approach** (or persuasive transfer approach) involves measures that address KALU directly and integrate them into communication and learning processes. Institutionalised agricultural advisory services have followed this approach for decades, addressing farmers to communicate information, produce knowledge and enhance skills (Labarthe et al. 2013). By pursuing this path, research projects seek to improve individuals' knowledge base on which they base their decisions. Depending on the knowledge base of the targeted KALUs, the communicative goal may vary between raising awareness for a hitherto neglected issue (e.g. effects of land sealing), concrete policy options or even hands-on training via "how-to-do SLM" manuals. If considered as a two-way communication setting, research projects may also benefit from feedback. This transfer option, however, requires exact knowledge of the setting of target groups in the investigated field and presupposes stakeholder analysis methods. Since persuasion is a relatively "soft tool" for initiating change, activities in this direction should be persistent and repetitive in order to have an effect. Assessing the effectiveness of persuasive methods is equally challenging.

2. Another transfer approach can be described as an indirect form of the persuasion of KALU. In the **prototype approach**, transfer measures aim to establish technological, procedural or institutional innovations in the form of pilot projects. This approach picks up the diffusion of innovations theory by Rogers (1995) and the Multi-Level Perspective framework (Geels 2002) by designing innovations in niche contexts first that, eventually, diffuse into a broader context. The niche innovation is nurtured by specific actors at the micro scale that have been withdrawn from the influences of market and regulation.

   Besides being particularly feasible for projects that focus on techno-societal innovations, this transfer path is increasingly applied to social innovations, too. It is worthwhile cooperating with businesses in this case, since they have a genuine interest in driving innovations that invoke marketing potential. However, cooperation with economic partners also harbours a number of risks for research projects (e.g. confidentiality of results vs. scientific publications).

   Projects that apply the prototype approach may benefit from the cognitive phenomenon that artefacts (such as prototypes) suggest visible and tangible progress, and convey the feeling of being able to "get a grip" on the problems that are being investigated across research groups (Ukowitz 2012, p. 303f.). This effect seems to be magnified for haptic interactions that users may experience when testing a prototype or tool. Also, it inhibits positive effects on the visibility and legitimacy of research on the ground, and often corresponds to the stereotype of the scientist as an inventor (ibid.).

   The adoption potential of a new technique for tillage, for example, that has been designed and tested in a project context is greater if test sites show the effects to other farmers. These innovative products, processes or institutions alone, however, are not the result of SLM but, in turn, are only further persuasive vehicles to indirectly influence land use decision makers. A successful prototype approach should therefore implement suitable instruments from the actor-centric approach to increase its effectiveness. This might include field days for farmers or farming business representatives, on-site performances and corresponding Q&A sessions with an expert farmer who may also be a research partner for the new technique.

With this approach, target group sensitivity is therefore a *conditio sine qua non* for the required knowledge transfer. It necessitates a thorough analysis of the connectivity potential of such prototypes, its (unwanted) side-effects and trade-offs. If applied, design-thinking methods, boundary objects or adjacent creative methods may be suitable. Depending on the research question, the prototype approach can boost the transfer potential of projects.

3. Since land use is embedded in a highly regulated environment (market, laws, regulations, planning, etc.) that operates at all spatial levels, another transfer approach exists for influencing KALUs: **the context management approach**. Steering the development of a given system by stimulating its given framework conditions is adapted from systems theory thinking (Willke 2014). Among the described transfer paths, changing the regulative framework for land use is certainly the most demanding and least manageable option for research projects. Nonetheless, it holds the biggest potential of transformation as it sets the rules for post-project timetables.

The context management approach seeks to involve actors and institutions that set the framework conditions for land use or interact with intermediate institutions that prepare political decision-making, feed in recommendations to decision-making levels, advance agenda-setting processes or form opinions. Interestingly, the framework approach is still deemed to be the conventional pathway for scientific knowledge that "leaks" into society. In this concept, "excellent scientific work" either catches the attention of science-policy networks that perform at higher scales (such as IPCC or IPBES) or directly finds its way to decision-makers. In our introductory example of the ozone hole, knowledge transfer took that path as the news of a global danger "quickly" found its way into the highest levels of global politics. I must stress that this example represents an extreme exception.

However, the context management approach in SLM is slightly different as it adds another dimension by proactively seeking to change context conditions that, for instance, slow down the implementation of sustainable practices. Also, research projects actively seek to advance options that hold more potential for sustainability (e.g. by simplifying approval processes, and setting standards and incentives). Other target groups of T&I activities could also be politicians, administrations, spatial planning institutions, companies and citizens, as they influence land use indirectly (e.g. market conditions, public discourse). The degree of influence varies greatly between actor groups as well as the possibility of influencing existing regulations and assessing the impact of the transfer measure taken. It is certainly easier to launch a regional campaign for sustainable land use in public than to influence existing legal framework conditions, especially if these are set at supraregional level.

The conceptual model of strategic T&I in SLM described may be used by project architects and coordinators at different stages of the research cycle. It is of particular value as a "tool" for planning outreach activities during the starting phase of projects. However, it can also be used as a project management method that sparks debate between project partners on the desired impact and the realistic output of SLM projects, i.e. to develop a mutually shared "model of change" in real-world settings.

Nonetheless, the model invokes various limitations that should be briefly noted. First, it depicts an idealised form of a complex "transfer reality". Actors cannot be classified into rigid roles as they appear in the model, nor do they think and act consistently and rationally in the sense of sociological institutionalism. Transdisciplinary research means scrutinising existing structures and modes of operation, and dissolving boundaries between actors. In addition, addressing a broad spectrum of actors is a practical challenge that is virtually impossible to operationalise in this form.

Second, additional actors of knowledge transfer that influence research projects are left out, such as traditional forms of academic knowledge transfer in which academia constantly feeds collaborative research with current developments in research. In addition, the quality and salience of research results are fundamental if research wants to have an impact on society. Economic institutions that are left out play an important role in rather technology-based solutions, especially when it comes to marketability and, ultimately, to stabilising solutions after the end of the project cycle.

Third, an evaluation of token T&I measures is not easy to apply as time lags prevail, i.e. diffusion effects might be visible after a certain period. As stated above, changing the regulative and institutional framework that influences land use usually presupposes democratic processes at multiple hierarchy levels. Also, attribution between measure and effect may not be clearly identified. If, for instance, a whole set of different transfer activities led to a desired effect, then how do we know which of them was effective? Both aspects complicate the adaptive management of T&I processes during an ongoing project.

Each transfer approach pathway is not exclusive and isolated from each other. Instead, actors, measures and framework conditions mutually influence each other in the transfer process. A combination of different approaches therefore seems necessary. It is therefore strongly recommended to integrate different perspectives for the area of SLM as they hold the potential for synergies and catalytic effects (see Leach et al. 2010; Stringer and Dougill 2013). The choice of path is determined by efficiency criteria on the one hand and the actual content of the research project on the other.

13.6 Conclusion

In this chapter, I have argued that the conventional "loading dock approach" of science transfer is not applicable for research projects that proactively seek to drive societal change towards sustainability. Conventional transfer channels of scientific results are still viable, but are inappropriate in SLM settings if they are the only strategic approach to transfer. Bi-directional communication channels that involve science and societal actors are needed to enhance the societal relevance and applicability of outputs. Moreover, innovative ideas in SLM require an interpersonal level in which key actors of land use (KALU) can share, interpret and contextualise information.

In a mental model, three strategic impact pathways for T&I activities in SLM have been presented. Each represents an ideal, heuristic approach towards KALUs that had been identified as the main addressees of T&I activities. These actors differ widely depending on the research context, and can impersonalise individuals and institutions alike. Notable resources in project planning should be tied to activities that analyse the environment of stakeholders and the system under investigation (and its inherent ties and interrelations), in order to identify crucial actors that should be approached.

T&I are activities that should be considered continuously during a research process, and be an element of strategic project management from the very beginning of the project cycle. Undoubtedly, T&I activities tie up additional management resources and cause trade-offs with other activities in the research project, which must be considered. T&I activities need professional experts who are paid and recognised for their work. Conceptualising and conducting T&I activities as described in the previous section requires professional expertise that is rarely taught in academic curricula as yet. If taken seriously, T&I should not be done part-time and in addition to a "real" function within a research project.

Moreover, there is no guarantee that T&I activities (even if conducted to perfection) will lead to a desired impact in society because external factors that cannot be influenced make transdisciplinary research projects vulnerable. Thus, T&I should be seen as activities that increase the potential for research impact only. After all, it is not the product, the process or the model of a prototype that stands for successful transfer alone; it is the knowledge of an innovation and its functioning, and its strengths, weaknesses, risks and opportunities that influence or slow down the dissemination of innovations.

Acknowledgements I thank Thomas Weith, Thomas Aenis, Jana Zscheischler and Andrea Rau for their scientific advice and support. I also thank numerous (not further mentioned) colleagues for noteworthy inputs and fruitful discussions.

References

Ascher, W., Steelman, T. A., & Healy, R. G. (2010). *Knowledge and environmental policy. Reimagining the boundaries of science and politics.* Cambridge, Mass: MIT Press (American and comparative environmental policy).

Beckers, K. (2012). *Kommunikation und kommunizierbarkeit von Wissen.Prinzipien und Strategien kooperativer Wissenskonstruktion.* Berlin: Erich Schmidt Verlag [Philologische Studien und Quellen; 237].

Best, A., & Holmes, B. (2010). Systems thinking, knowledge and action: Towards better models and methods. *Evidence and Policy: A Journal of Research, Debate and Practice, 6*(2), 145–159.

Blümel, C. (2016). Der Beitrag der Innovationsforschung zur Wissenschaftspolitik. In D. Simon, A. Knie, S. Hornbostel, & K. Zimmermann (Eds.), *Handbuch Wissenschaftspolitik* (2nd ed., pp. 175–190) VS Verlag für Sozialwissenschaften: Wiesbaden.

Böcher, M., & Krott, M. (2013). Mit Wissen bewegen! Erfolgsfaktoren für Wissenstransfer in den Umweltwissenschaften. oekom Verlag: München.

Bush, V. (1945). *Science—The endless frontier.* Washington DC: The National Science Foundation.

Cash, D. W. (2006). Countering the loading-dock approach to linking science and decision making: Comparative analysis of El Nino/Southern Oscillation (ENSO) forecasting systems. *Science, Technology and Human Values, 31*(4), 465–494.

de Jong, S. P. L., Wardenaar, T., & Horlings, E. (2016). Exploring the promises of transdisciplinary research: A quantitative study of two climate research programmes. *Research Policy, 45*(7), 1397–1409.

Farman, J., Gardiner, B., & Shanklin, J. (1985). Large losses of total ozone in Antarctica reveal seasonal ClO_x/NO_x interaction. *Nature, 315,* 207–210. https://doi.org/10.1038/315207a0

Fazey, I., Evely, A., Reed, M., Stringer, L., Kruijsen, J., White, P., et al. (2013). Knowledge exchange: A review and research agenda for environmental management. *Environmental Conservation, 40,* 19–36. https://doi.org/10.1017/S037689291200029X

Fry, P., Seidl, I., Théato, C., Bachmann, F., & Kläy, A. (2003). Vom Wissenstransfer zum Wissensaustausch. Neue Impulse für den Boden- und biodiversitätsschutz in der Landwirtschaft. *GAIA—Ecological Perspectives for Science and Society, 12*(2), 148–150. Accessible online: https://sag ufv2.scnatweb.ch/downloads/gaia12_2.pdf.

Geels, F. W. (2002). Technological transitions as evolutionary reconfiguration processes: A multi-level perspective and a case study. *Research Policy, 31,* 257–1273.

Granovetter, M. (1973). The strength of weak ties. *American Journal of Sociology, 78,* 1360–1380.

Gräsel, C., Jäger, M., & Wilke, H. (2006). Konzeption einer übergreifenden Transferforschung und Einbeziehung des internationalen Forschungsstandes. Expertise II zum Transferforschungsprogramm. In R. Nickolaus & M. Abel (Eds.), *Innovation und Transfer. Expertisen zur Transferforschung* (pp. 445–566). Baltmannsweiler: Schneider Hohengehren.

Guston, D. H. (2001). Boundary organizations in environmental policy and science: An introduction. *Science, Technology, and Human Values, 26*(4), 399–408. https://doi.org/10.1177/016224390102 600401

Hägerstrand, T. (1952). The propagation of innovation waves. *Lund Studies in Human Geography, Series B, 4,* 3–19.

Ibert, O. (2003). Innovationsorientierte Planungverfahren und Strategien zur Organisation von Innovation. Opladen.

Kaiser, D. B., Gaasch, N., & Weith, T. (2017). Co-production of knowledge: A conceptual approach for integrative knowledge management in planning. *Transactions of the Association of European Schools of Planning, 1.* https://doi.org/10.24306/TrAESOP.2017.01.002.

Labarthe, P., Caggiano, M., Laurent, C., Faure, G., & Cerf, M. (2013). *Concepts and theories to describe the functioning and dynamics of agricultural advisory services: Deliverable WP.2–1 of the PRO AKIS project.* Available at: https://www.proakis.eu/publicationsandevents/pubs.

Leach, M., Scoones, I., & Stirling, A. (2010). Governing epidemics in an age of complexity: Narratives, politics and pathways to sustainability. *Global Environmental Change, 20*(3), 368–377.

Lux, A., Schäfer, M., Bergmann, M., Jahn, T., Marg, O., Nagy, E., et al. (2019). Societal effects of transdisciplinary sustainability research—How can they be strengthened during the research process? *Environmental Science and Policy, 101,* 183–191.

Newig, J., Jahn, S., Lang, D. J., Kahle, J., & Bergmann, M. (2019). Linking modes of research to their scientific and societal outcomes. Evidence from 81 sustainability-oriented research projects. *Environmental Science and Policy, 101.* https://doi.org/10.1016/j.envsci.2019.08.008.

Nohlen, D. (Ed.). (1998). *Lexikon der Politik* (Vol. 7). München: Politische Begriffe.

Partidario, M. R., & Sheate, W. R. (2013). Knowledge brokerage—Potential for increased capacities and shared power in impact assessment. *Environmental Impact Assessment Review, 39,* 26–36.

Pohl, C., & Hirsch-Hadorn, G. (2006). Principles for designing transdisciplinary research.

Pregernig, M. (2000). Putting science into practice: The diffusion of scientific knowledge exemplified by the Austrian 'Research initiative against forest decline.' *Forest Policy and Economics, 1*(2), 165–176.

Rogers, E. M. (1995). *Diffusion of innovations* (4th ed.), New York, NY [U. A.]: Free Press.

Schröder, T., Huck, J., & De Haan, G. (2011). Transfer sozialer Innovationen, Wiesbaden.

Stringer, L. C., & Dougill, A. J. (2013). Channelling science into policy: Enabling best practices from research on land degradation and sustainable land management in dryland Africa. *Journal of Environmental Management, 114,* 328–335.

Ukowitz, M. (2012). *Wenn Forschung Wissenschaft und Praxis zu Wort kommen lässt. Transdisziplinarität aus der Perspektive der Interventionsforschung.* Marburg: Metropolis-Verlag.

Weith, T. (2011). (2011): Evidenzorientierung als Beratungsperspektive. *IzR, 7*(8), 487–495.

Weith, T., Rogga, S., Zscheischler, J., & Gaasch, N. (2019). Beyond projects: Benefits of research accompanying research: Reflections from the research programme *Sustainable Land Management. GAIA—Ecological Perspectives for Science and Society, 28*(3), 294–304. München:oekom.

Willke, H. (2014). *Regieren. Politische Steuerung komplexer Gesellschaften.* Springer VS: Wiesbaden.

Zscheischler, J., Rogga, S., & Busse, M. (2017). The adoption and implementation of transdisciplinary research in the field of land-use-science—A comparative case study. *Sustainability, 9,* 1926.

Zscheischler, J., Rogga, S., & Lange, A. (2018). The success of transdisciplinary research for sustainable land use: Individual perceptions and assessments. *Sustain Science, 13,* 1061. https://doi.org/10.1007/s11625-018-0556-3.

Part III
Co-Evolution: New System Solutions and Governance

Chapter 14
Small-Scale System Solutions—Material Flow Management (MFM) in Settlements (Water, Energy, Food, Materials)

Peter Heck

Abstract The complexities of the present energy-climate era coupled with the ambitious targets of the 2030 Sustainable Development Goals (SDGs) demand approaches to resource management that transcend conventional strategies. Material Flow Management (MFM) can be a vital tool in sustainable resource management (SRM) in complex systems. It contributes, among other things, to the protection of land, the conversion of abandoned land and the upcycling of degraded land. Despite its relative novelty, its usefulness has already been demonstrated. This chapter presents the practical application of MFM in small-scale systems, which are characterised by decentralised material and energy flows. It attempts to highlight the utility of MFM in SRM. The chapter gives special attention to the augmentation of source and sink capacity, employing MFM to reduce impacts on ecosystems both upstream and downstream—i.e. on the use of resources as well as on the amount of emissions.

Keywords Material flow management · Regional added value · Potential analysis · Zero emissions campus

14.1 Introduction

14.1.1 Anthropogenic Systems, GDP Growth, and Material Flow Management

Contrary to the (wise) premise, we, "the wise"—or *Homo sapiens,* have always taken an anthropocentric[1] stand toward everything we have done so far. By any reasonable judgement, this will largely stay the norm for the next couple of decades as well.

---

[1] The view/belief that human beings are the most important entity in the universe. The analysis of Kopnina et al. (2018) may be useful for some insights and perspectives.

P. Heck (✉)
FB Umweltwirtschaft/-recht–FR Umweltwirtschaft, Hochschule Trier, Umwelt-Campus Birkenfeld, Campusallee, 55768 Hoppstädten-Weiersbach, Germany
e-mail: p.heck@umwelt-campus.de

© The Author(s) 2021
T. Weith et al. (eds.), *Sustainable Land Management in a European Context,*
Human-Environment Interactions 8, https://doi.org/10.1007/978-3-030-50841-8_14

The taming of nature and the extraction/consumption of ever-increasing quantities of resources have been fundamental aspects of all progressive civilisations thus far. It is also fundamental to the current model of economic growth and development that we follow to the hilt. As a consequence, severe environmental degradation and the depletion of resources have occurred in tandem. Adding insult to injury, globalisation and population growth coupled with the fallacious chase for GDP growth have aggravated this situation even further.

However, since the mid-twentieth century, the *manmade* environmental calamities such as the Great Smog of '52, the threat of DDT in the '60s, the ozone depletion of the '70s, and the largest ever oil spill in human history—the Persian Gulf oil spill in '91—drew humanity's attention to the underlying issues, prompting a sluggish, nevertheless noticeable shift in this anthropocentric order of business toward an ecocentric direction.

The extensive consumption of fossil resources along with the application of synthetic chemistry/biology—which are also liberally attributed to the aforesaid calamities—has triggered a frenzy of industrial and economic development (see Fig. 14.1: GDP as a proxy indicator), which as a result has increased humanity's environmental footprint like never before.

Intensive pollution—land, air and water—was widely recognised as a symptom of the underlying societal metabolic disorder; this gave impetus to the global green movement in the twentieth century. Adding momentum, the contemporary debate on global climate change, which is attributed to anthropogenic global warming, has

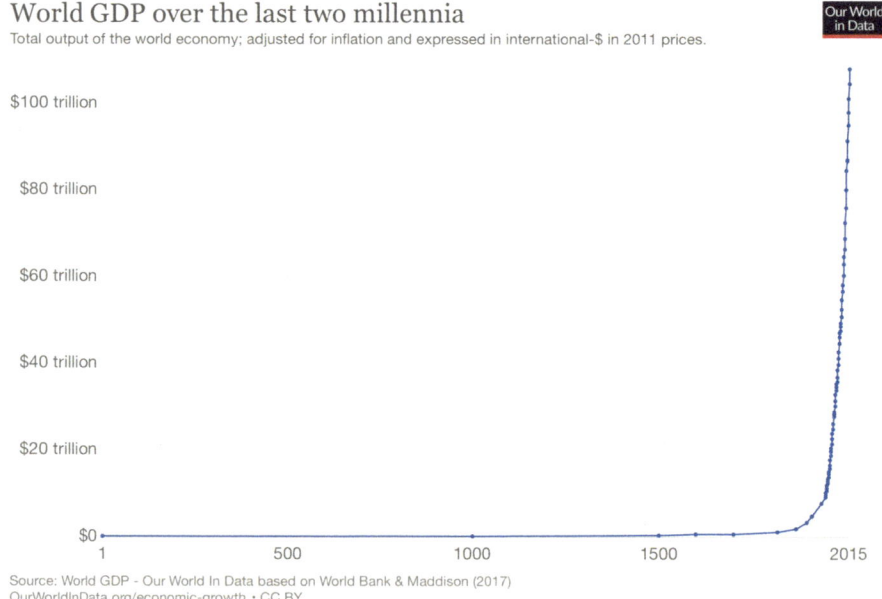

Fig. 14.1 World GDP growth. *Source* Roser (2019)

accelerated the pace of application of *green [environmental] solutions* globally in recent decades.

To create any significant impact, such solutions need applications on different levels and scales. Clearly, anthropogenic systems are complex and come in all shapes and sizes (e.g. a business, a municipality, a political system, the European Union, etc.). Systems are characterised by a system boundary and the flow/flux of material and energy across and within the system boundary. Inputs such as raw materials and energy, and outputs such as products, services and waste/emissions are inherent to anthropogenic systems. These various inputs and outputs further distinguish various systems. In dealing with complex anthropogenic systems, special attention should also be paid to the output of money as a valuable resource, as many of the problems originate from an incorrect allocation of money and the concurrent loss of economic power and opportunities related to financial waste from systems (see Fig. 14.3).

Despite the lack of consensus on the definition of "small-scale systems", in this text, we focus on small-scale political and administrative systems such as villages, municipalities and business entities such as farms, factories, SMEs, etc. Small-scale systems play an important role in rural areas of the world, in particular by provisioning ecosystem services, clean water and healthy food. They also help to conserve biodiversity and facilitate multifunctional land use, etc.

14.1.2 The Throughput Society and the SDGs

Currently, the predominant practice of resource *management* involves extracting, making, using, and wasting/throwing away. This practice is denoted as the throughput system (also known as the throughput metabolism). Accordingly, we can describe present human society as "the throughput society" and the current global economy as "the linear economy". Characteristics of these systems include the massive input/use of virgin resources, low levels of resource productivity/product efficiency, and the generation of gargantuan amounts of emissions or unwanted by-products that are usually termed *waste*. Thanks in large part to this throughput system of resource management, "global primary material use, and thus global primary material extraction, is projected to double in the coming decades… from 79 Gt in 2011 to 167 Gt in 2060" (see Fig. 14.2).[2] From a land use perspective, this foresees a massive increase in the use of valuable land resources for resource extraction, agriculture and urban development, collectively resulting in permanent degradation of ecosystems due to, among other things, the depositing of *waste materials*. From an anthropocentric point of view, ecosystems are *essential* for provisioning (e.g. food and water) and regulatory (e.g. flood control, climate) services. To use them as waste deposits or sinks leads to the loss of (sometimes irreversible) service capacity.

The constant bombardment of news about environmental calamity, resource scarcity, social inequality, economic downturn, etc.—inevitable consequences of

---

[2]Note: indicates the net effect of the three trends.

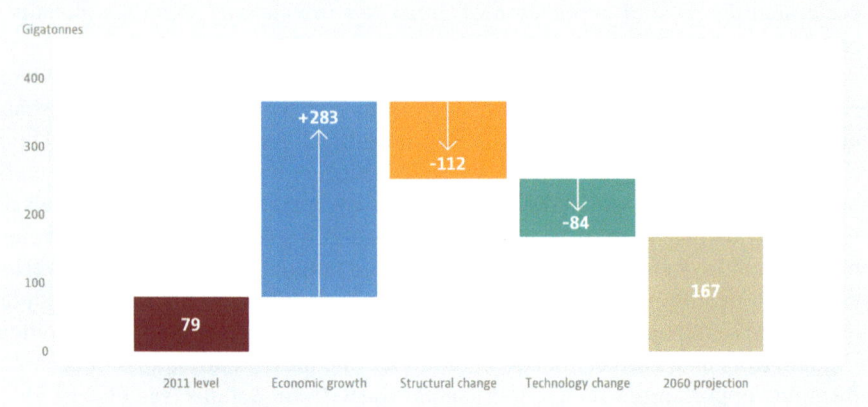

Fig. 14.2 Forecast of global material use. *Source* OECD (2019)

the throughput system—remind us that humanity is facing an existential threat, for which achieving sustainability has been projected as the panacea. It was for this reason that UN policymakers set out to achieve environmental, social and economic sustainability (popularly termed "triple-bottom-line sustainability") as humanity's goal for the current century, employing the Sustainable Development Goals (SDGs) in the short term for the endeavour.

The underlying matrix that forms the objectives of the SDGs also includes the following: achieving social and intergenerational equity, extracting and consuming resources in accordance with the planetary boundaries, and, at the same time, achieving economic growth and prosperity while minimising negative environmental consequences.

Though the exact origin of greening for sustainability is somewhat nebulous, its effects have become increasingly common over time, presenting a broad array of solutions for the aforementioned socioeconomic and environmental concerns. These solutions include green products (e.g. green chemicals) and services (e.g. green IT, green design and green certification); green infrastructure (e.g. green buildings); green energy (e.g. carbon–neutral/renewable energy); green processes (e.g. green manufacturing, green chemistry); green policies (e.g. green public procurement), etc. The goal is to establish green jobs and green cities with the ultimate aim of introducing green economies, where sustainability is the fundamental value.

14.1.3 MFM and Associated Tools

The circular economy, material flow management and zero emissions are different aspects of a new nexus in the world. The circular economy is the new paradigm, material flow management is the management tool for implementation, and zero emissions

Fig. 14.3 The transformative approach of MFM. *Source* IfaS (2019)

are the ultimate target. "Nexus" in this context emphasises the new holistic view of managing systems as such, instead of optimising individual components in linear flows. Closed-loop economies need to pay attention to embedded energy, virtual water, carbon footprints, levelised costs of service, etc. This segmented view of our society hinders economic efficiency, resource efficiency and emission reduction. As often observed in such an approach, the suboptimal allocation of financial resources also leads to negative incentives for unsustainable investments.

14.1.3.1 Material Flow Management

Material flow management (MFM) strives to change the throughput metabolism and helps users to develop technical and economic alternatives designed to improve the system's conditions and reduce the outgoing material and energy flows—more commonly termed *emissions* or *waste*. MFM could be applied to any typical consumption and production system insofar as its primary goal is to optimise material and energy flows according to given objectives. Figure 14.3 illustrates the material and energy flow of a typical throughput system that, by implementing MFM, circulates resource flows while reducing virgin and non-renewable resource inputs, minimising the loss of financial resources and reducing environmental impacts due to emissions. Ideally, the resulting new system could be a zero-emissions (ZE) system, depending on the targeted objectives.

Typically, material flow analysis (MFA) precedes MFM, during which a thorough system analysis is performed to qualify and quantify the resource flows through and within the defined system boundary temporally and spatially.[3] The MFA not only collects and assesses the pertinent data, but also makes this information visible and transparent. This, in turn, allows policymakers and scientists to simplify and elaborate on the problems and their subsequent management options. Figure 14.4

---

[3] MFA and MFM proper are the two methodological steps of material flow management. Material flow management is a yet unpublished but often used management strategy developed at ECB by Professor Dr. Peter Heck.

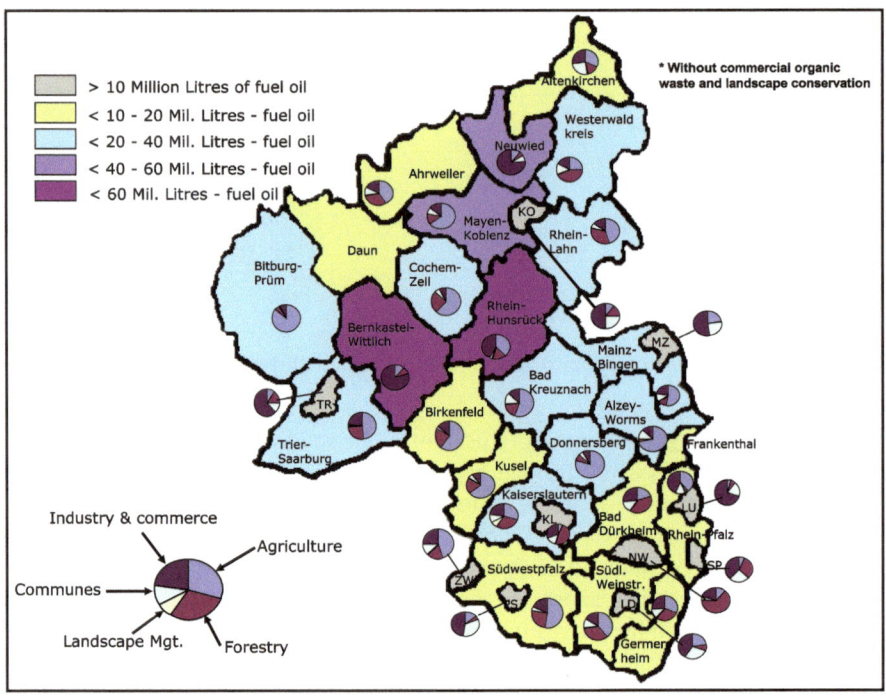

Fig. 14.4 Available biomass potential in the counties of Rhineland Palatinate (Heck and Hoffmann 2001). *Note* The biomass estimates include various sources, such as forestry, landscaping, agriculture and organic household waste. Various conversion technologies such as incineration and digestion have been used to calculate the oil equivalent. Only biomass that is technically and legally available as well as economically viable is shown

shows an analysis of biomass potential in the state of Rhineland-Palatinate. The MFA clearly reveals the enormous potential of biomass, expressed as the availability of oil equivalents per year. In this way, MFA leads to more transparency of systems with regard to their potential. MFA also illustrates the current states of systems (or the status quo), as illustrated in Fig. 14.5.

MFM usually details a comprehensive plan for the specific management and financing of individual projects that optimises specific resource flows; together these projects lead to system change. As mentioned earlier, one ideal system optimisation target could be a ZE system, in which emissions flows are utilised within the system's boundaries or connected to adjacent subsystems as valuable raw material inputs (such as in the case of industrial symbiosis), creating closed loops of material and energy flows—i.e. a *circular* system. This ultimate system state is usually referred to as the circular economy (CE) model (as opposed to the "linear" model of the economy mentioned above), which is environmentally, socially and economically sustainable. Typically, the holistic sustainability results of such an optimised system can be measured in terms of regional added value (RAV). RAV presents/quantifies

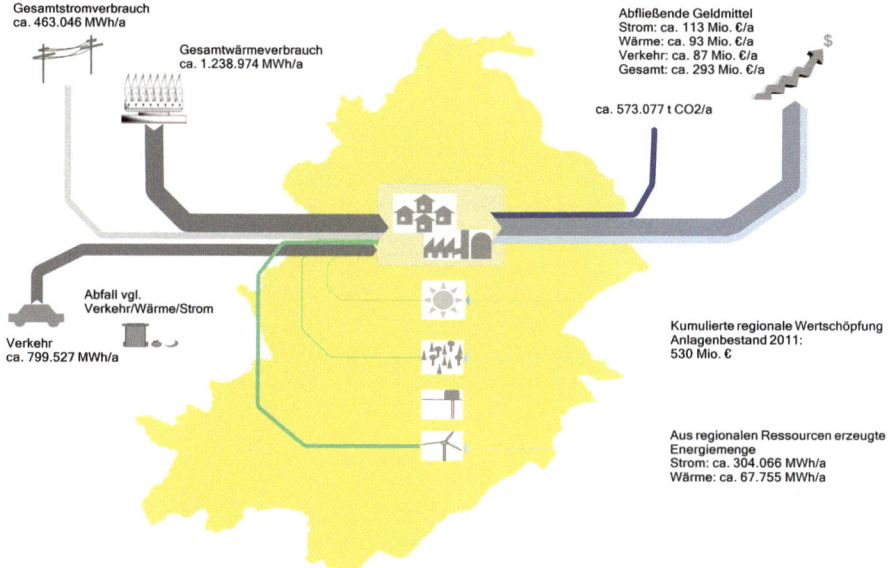

Gesamtstromverbrauch
ca. 463.046 MWh/a

Gesamtwärmeverbrauch
ca. 1.238.974 MWh/a

Abfließende Geldmittel
Strom: ca. 113 Mio. €/a
Wärme: ca. 93 Mio. €/a
Verkehr: ca. 87 Mio. €/a
Gesamt: ca. 293 Mio. €/a

ca. 573.077 t CO2/a

Abfall vgl.
Verkehr/Wärme/Strom

Verkehr
ca. 799.527 MWh/a

Kumulierte regionale Wertschöpfung
Anlagenbestand 2011:
530 Mio. €

Aus regionalen Ressourcen erzeugte
Energiemenge
Strom: ca. 304.066 MWh/a
Wärme: ca. 67.755 MWh/a

Fig. 14.5 MFA for the Rhein Hunsrück District (IfaS–Heck et al. 2011). *Note* The Sankey diagram shows financial losses (in €/year) and the CO_2 emissions incurred (in t CO_{2e}/year) on the upper left side. The main energy sources are still fossil fuels, as can be seen by the thin green lines for renewable energies and the rather thick black/grey lines for conventional electricity and heat sources

both monetary and non-monetary benefits derived from MFM. Non-monetary benefits include, among others, lower pollution levels, increased biodiversity, improved aesthetics, increased innovation, an enhanced public image, etc. Monetary benefits include increased labour and employment opportunities, increased savings, lower costs, increased revenues from new business ventures, new sales options, etc. As can be seen, the key objective of material flow management (MFM) is to optimise systems in order to achieve more systemic added value, while achieving triple bottom line sustainability.

The starting point of current models of targeted sustainable economies—specifically the circular economy (CE), the bio-economy (BE) and the green economy (GE) as depicted in Fig. 14.6–is the use of MFM and ZE technologies and strategies. Despite the size of the anthropogenic system targeted or the sustainable economic model to be followed, these tools are intrinsic; therefore, here we investigate the concept of ZE and its implications on CE.

14.1.3.2   Zero-Emissions

As remarked on earlier, the throughput society of today extracts ever-increasing amounts of resources from the earth's ecosystems and turns the bulk of it into different

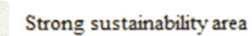

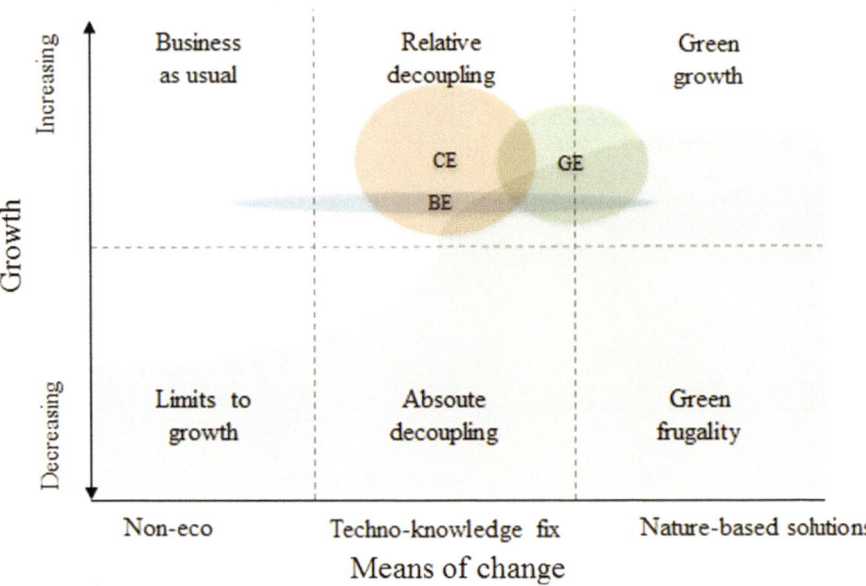

Fig. 14.6 Relative positions and associations of the GE, the BE and the CE. *Note* Fig. 14.6 is intended to highlight the close association and relative position of the three concepts. The size of the bubbles in the figure does not pertain to the current work. Franceschini and Pansera (as cited in D'Amato et al. 2017)

kinds of waste products such as solid waste, wastewater, waste gas, waste heat and greenhouse gases. These create enormous environmental pressures, often with lasting negative impacts. As resources become depleted and sinks become increasingly overloaded, the reduction and/or the avoidance of emissions will be a vital measure to protect resources and sinks, and thus prevent land and ecosystem degradation.

In that context, the approach of zero emissions (ZE) calls for the systemic, overarching optimisation of processes, incorporating elements of sufficiency, efficiency and substitution. Zero emissions do not mean avoiding all emissions as such. Instead, it means avoiding the types of emissions that lead to negative impacts on neighbouring (eco) systems, such as rivers, wetlands, agricultural land, etc. It also stipulates changing system metabolism in such a way that by-products are either upcycled or improved as a result of secondary use options, or drastically reduced in terms of volume. Another key element of the zero emissions strategy is a focus on economics. Avoiding emissions makes economic sense, even without considering externalities. In a world with dwindling resources and shrinking sinks, the efficient use of both could save money and create competitive new business opportunities with better bottom lines. For small-scale, financially strained systems such as rural villages and municipalities, SMEs, etc., a ZE strategy can assist in boosting the local economy and/or cash flow, while reducing the environmental burdens arising from consumption and production systems.

14.1.3.3 The Circular Economy

The CE is an alternative route to holistic sustainability, though it is still in an early phase of adoption. Developed to emulate the energy and material flow management model in biological systems, its supporters position the CE as an alternative to the current take-make-waste extractive industrial model. CE aims to redefine growth, focusing on positive society-wide benefits. It entails gradually decoupling economic activity from the consumption of finite resources and designing waste out of the system. Underpinned by a transition to renewable energy sources, the circular model builds economic, natural, and social capital (Ellen MacArthur Foundation 2015).

The value of the CE stems from its explicit focus on the economy. Compared to sustainable development (which is widely seen as an environmental initiative, even though by definition, it is not), the dominance of economic thinking within CE concepts is clearly visible.

As hinted at earlier, according to Elia et al. (2017) and the European Environmental Agency (EEA 2016), the CE is characterised by its ability to reduce the input and use of natural resources; reduce emission levels; reduce valuable material losses; increase the share of renewable and recyclable resources; and increase the durability of products. It is based on three simple principles: design waste and pollution out of the system; keep materials and products in use as long as possible and as economically as possible; and regenerate natural systems.

Despite the CE's relative novelty, its unambiguous and application-oriented nature is a *positive* in fostering action toward sustainability at local, national and international levels. The CE is perhaps still a road less travelled. But, analogous to the German Autobahn, the CE is a smooth, straight, obstacle-free, high-speed freeway to sustainability. As exemplified in many domains in the European Union—the predominant promoter of the CE at present—the CE is not just another fancy term for waste management. The CE would help to reduce virgin resource extraction/input for economic processes, while also reducing the associated environmental impacts. As opposed to other economic models, the utility of the CE has been tangibly proven in applications in the EU. Accordingly, one recent estimation has suggested that CE practices such as chemical leasing, nutrient recovery in agriculture, materials substitution in the construction sector, and shared ownership models in transport systems could reduce up to 7.5 billion tonnes of CO_{2e} globally. This would bridge half of the existing emissions gap to reach the 1.5 °C target as outlined under the Paris Agreement (Schroeder et al. 2019).

Optimising material flows according to the principles of the CE is important for easing the pressure on land. In addition to this indirect positive effect on land use, the CE enables strategies for new land use systems based on the cascade approach and system thinking. For example, the "More Value from a Hectare" project, conceptualised and applied by the Institute for Applied Material Flow Management (IfaS) at Trier University of Applied Sciences for a new rural bio-economy strategy, is designed to enhance the resilience of agricultural systems, with a special focus on land and soil, while provisioning more services—or value—from each hectare of land utilized (Böhmer et al. 2019).

The CE would also create innovative business models. That means, besides generating profits, the CE would create employment opportunities—in other words, it contributes to social sustainability targets (see Schroeder et al. (2019) for some insights). Concerning economic aspects, according to the Ellen MacArthur Foundation (2015), a shift to a CE would reduce net expenditures on resources by €600 billion per year, improve resource productivity by 3%, and generate €1.8 trillion per year of net benefits in the EU by 2030.

In light of its origins and the compatibility of its transformative tools (i.e. MFM, ZE, etc.), the CE model's applicability seems universal. Clearly, the CE provides a very practical option to treat the societal metabolic disorder modern civilisation suffers from. Its versatility in solving developmental and environmental challenges simultaneously is also worth considering when promoting the CE as an effective tool in achieving the UN's SDGs.[4] Given the anticipated severity of impending resource and environmental crises (see Fig. 14.7 for some insights), the CE seems to be one of the best alternative paths to follow.

According to the OECD (2019), the material intensity of economies—in particular in OECD countries—is set to decline (by 2060); furthermore, growth in the recycling sector (i.e. use of secondary materials) will surpass that of the mining sector as recycling becomes more price competitive than mining. This is in part attributed to the strong presence and growth of the CE. However, citing a 2015 report by the European Academies' Advisory Council, Schroeder et al. (2019) have pointed out that transforming the current linear economic model to a CE model has been stymied by "a skills gap in the workforce and lack of CE programmes at all levels of education."

Nevertheless, the efforts of Europe's research and higher education institutions such as the Institute for Applied Material Flow Management (IfaS) of the Trier University of Applied Sciences[5] in Germany have been highly regarded locally and internationally by representatives of industry, academia and the public sector alike. For nearly two decades, IfaS has deployed its expertise on the CE on practical projects on five continents, offered graduate and postgraduate level education on the CE through dedicated degree programmes,[6] and continually disseminated applied knowledge regarding the CE through its signature events platform: the International Circular Economy Week and Conference.[7]

---

[4] See Schroeder et al. (2019) for some in-depth insights on this aspect.

[5] Find out more at https://www.umwelt-campus.de/ucb/index.php?id=home&L=1 and https://www.stoffstrom.org/en/.

[6] https://www.imat-master.com.

[7] https://icew.de.

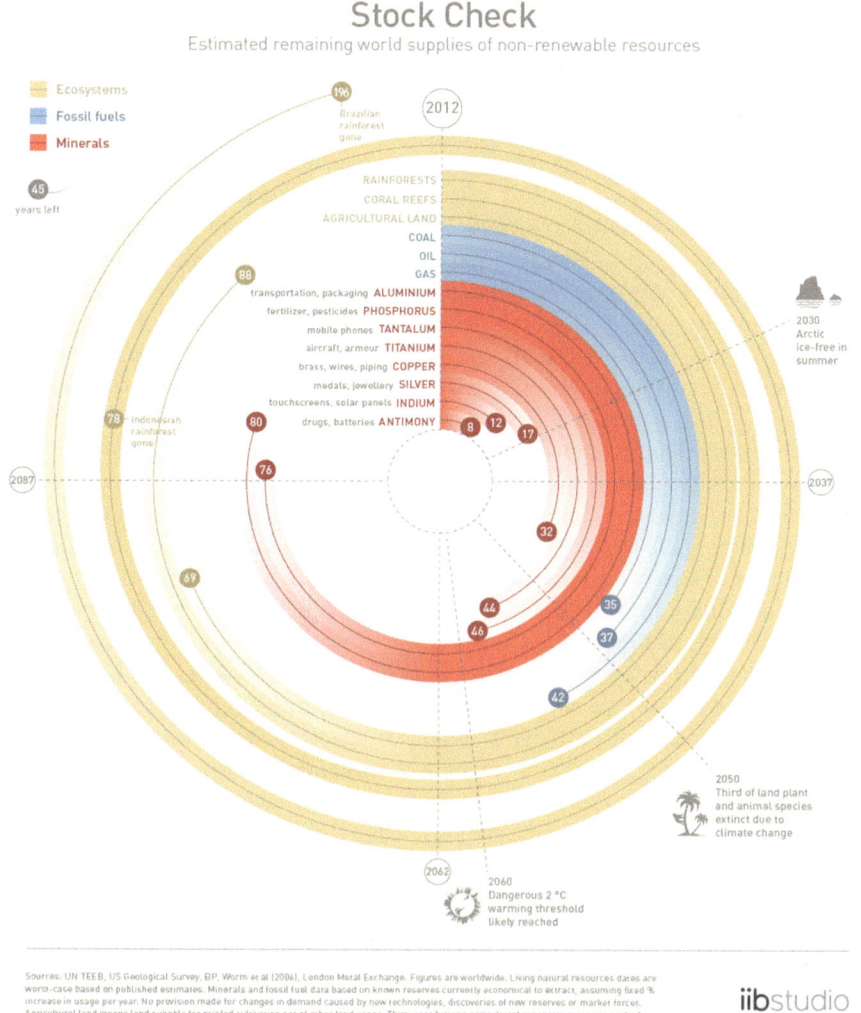

Fig. 14.7 Remaining non-renewable stock of resources (Source of image BBC Future: https://www.bbc.com/future/story/20120618-global-resources-stock-check)

The home base and foundation of this innovative education programme is the Environmental Campus Birkenfeld (ECB). ECB itself is a small-scale, decentralised and sustainable system of energy and material management; it has been purposefully designed as an object of study and a living laboratory for the specialised streams of scientific research pertaining to these areas. The following section explores the unique features of ECB in MFM.

14.2 Europe's First Zero Emission Campus: The Environmental Campus Birkenfeld (ECB)

This section gives an overview of the pertinent features and characteristics of the Environmental Campus in Birkenfeld (ECB) as a case study. Located in the Hunsrück Hochwald mountain region of the Rhineland-Palatinate, ECB itself represents a small-scale, decentralised system of material and energy management. ECB's energy and material management system also qualifies it as one of the largest of Germany's bio-energy villages (BEV), where at present nearly 2,600 people—including students, academics, researchers, administrative staff, private sector employees, etc.—enjoy CO_2-neutral electricity, heating and cooling. ECB can boast about its status as the first zero emission campus in Europe. Furthermore, it has been officially recognised as the Greenest University Campus in Germany since 2016. ECB is a unique higher education facility, where zero emissions system design is not only taught in theory, but is also implemented in practice by converting a former US reserve military hospital into an actual zero emission university campus. Hence, the zero-emission campus facility, with its innovative technology infrastructure, is not only home to students and environment-related study programmes, but is also an object of study itself. Especially in the context of land use, the repurposing of abandoned military brownfields into a centre of higher education and sustainability creates a perfect example of the CE.

The conversion process took place from 1994 to 1996 (Fig. 14.8b), following ecological construction principles and applying cutting-edge environmental technologies in the areas of sustainable repurposing of existing buildings as well as energy-efficient and material-efficient new construction. Energy aspects of the buildings such as heating, cooling and electricity supply have been entirely based on renewable energies. Moreover, the campus features biotopes and rainwater recycling infrastructure that contribute to the sustainable water management system of ECB.

In the vicinity of the campus, an eco-industrial park was also constructed to optimise regional material and energy flows connecting the campus via district heating and a low-voltage transmission grid. In 1997, a wood-chip power station was commissioned with an installed thermal capacity of 28 MW utilising about 65,000 tons of low-level and highly contaminated waste wood from forestry, agriculture, landscaping and industry annually; the power station can produce up to 8 MW of heat, 37.5 t/hour of steam and up to 8.3 MW of electricity for the environmental campus, neighbouring industrial facilities and the national electricity grid. In addition, the cogeneration units at the wood-chip power station utilise the biogas output of the nearby anaerobic digestion plant, which annually treats about 40,000 tons of municipal organic solid waste collected from the administrative districts of Birkenfeld and Bad Kreuznach. As a result, these local energy sources utilising regional biomass residues end up supplying a significant share of the campus' total energy demand. The remaining energy demand is covered by renewable energy installations on the campus itself. Various photovoltaic (PV) systems installed on rooftops and on the building

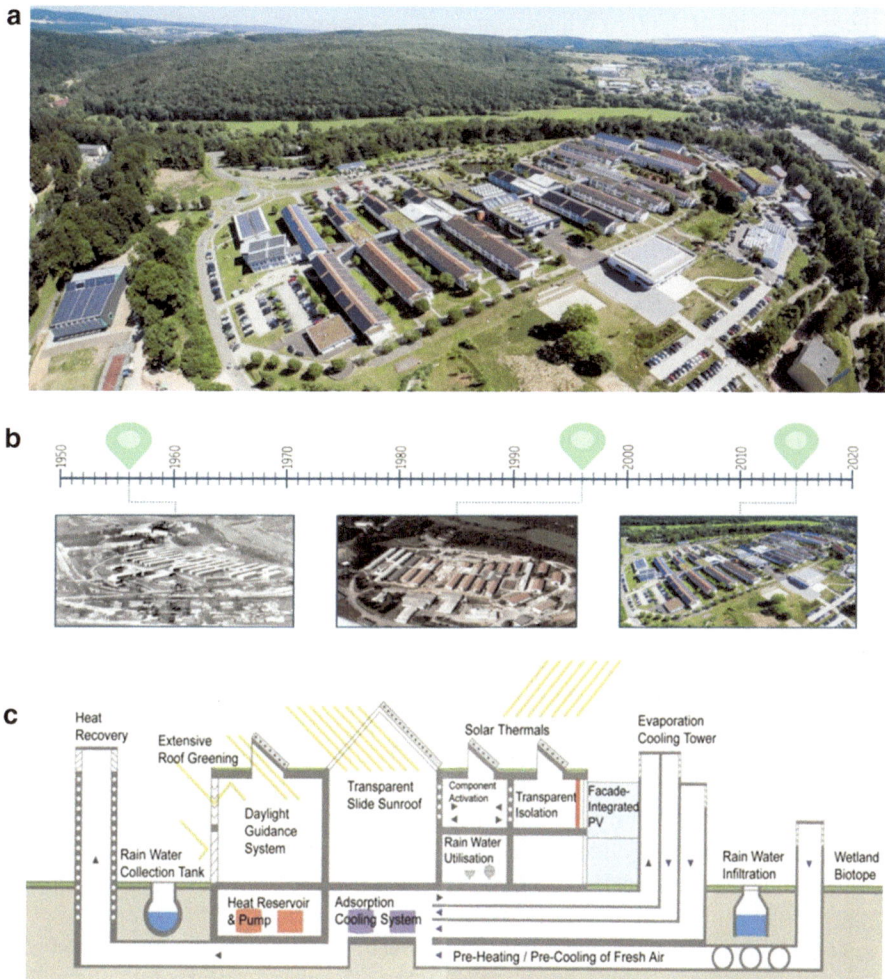

Fig. 14.8 a Bird's-eye view of ECB 2018. **b** bird's-eye views of ECB in 1950, 1996 and 2016; and **c** technical flow chart illustrating some of the energy innovations of ECB. Source: IfaS (2018) ECB and (2019). Green Campus Concept. https://www.umwelt-campus.de/en/campus/life-on-cam pus/green-campus-concept

façade, with an installed capacity of 510 kWp, cover another 40% of the total electricity demand. The performance of different PV module types and mounting systems are continuously monitored, displayed and employed as a subject for teaching and research. The use of solar energy is complemented by various solar thermal installations with a total capacity of 135 kW and a collector area of 270 m^2 that augment the heating system during the cold months. This solar thermal system also supports the solar adsorption cooling system, which has an installed cooling capacity of 175 kW. Thus, the total annual electricity demand of 1000 MWh and the total heating/cooling energy demand of 3700 MWh are completely supplied by renewable energies. With a new 99 kWp solar carport, a 100 kWh battery storage system, and five charging points for ten IfaS-owned electric cars, the campus is also progressively tackling the transport issue, which is still largely based on fossil energy (ECB 2019).

In the repurposed (and refurbished) buildings, a computer-controlled building automation system controls the natural lighting—maximised by daylight guidance systems and skylights—and the artificial lighting as well as the ventilation, heating and cooling of the buildings by employing motion detectors, luminosity sensors and CO_2 sensors. New buildings, such as the ones for student housing and the communications centre, have been built according to passive-house and/or energy-plus standards. The combination of the highest insulation and airtightness standards together with a roof-mounted PV system ensures that these buildings produce more energy than they consume. Heat exchangers for the recovery of waste heat in the used air coupled with geothermal preheating and cooling systems complete ECB's innovative HVAC system.[8]

The overall rational use of energy has been extended to other emerging areas such as street and building lighting. While various LED-based street illumination prototypes had already been in use and researched for many years, recently an entire street was retrofitted with state-of-the-art, highly efficient LED street lights, including a demonstration and performance control unit.

Aside from showcasing CE land use by converting an abandoned military complex into a high-tech university site, ECB also demonstrates aspects of the CE through sustainable landscape management strategies by developing biotopes based on separately collected rainwater. The ZE concept of ECB is extended to biotopes and biodiversity management, creating and maintaining protected areas and greened rooftops. Rainwater is collected on retention surfaces and redirected into natural water bodies or allowed to be channelled towards the aquifer. Part of the rainwater is collected, after mechanical purification, in two tanks and used for various greywater applications such as toilet flushing, irrigation and even as a coolant for an adsorption cooling system. The present system already avoids the discharge of rainwater into the sewer system. Nevertheless, modernisation of the wastewater infrastructure is currently underway, including water use minimisation strategies by installing vacuum toilets. The reduction of wastewater by separation of urine and brown water in one student dormitory is ongoing but not yet finished. In case of successful implementation, this system will be implemented to all dormitories on campus. With the complete installation of waterless urinals, the first step towards efficient water and nutrient management has already been taken.

---

[8]https://wissenszentrum-energie.ludwigsburg.de/,Lde/start/Bauen_Sanieren/gebaeudeenergiesta ndards.html, https://www.umwelt-campus.de/fileadmin/Umwelt-Campus/Oeffentlichkeit-sarbeit/ Tech-Broschuere/2016_05_04_Broschuere-GrueneTechnologien-am-Campus_klein.pdf.

14.3   Sustainable (Bio)energy Villages and Communities

Following the example of ECB and partially motivated by its success, many communities in Germany have begun implementing similar strategies based on the idea of the CE and MFM. Communities all over the world face a multitude of critical socioeconomic and environmental issues in this energy-climate era due to their fossil-based primary energy use and their unsustainable land use patterns. Tackling these issues at the smallest political entity—the village—was envisioned and pioneered in Germany, where material and energy flow optimisation has taken the lead in creating self-sufficient and sustainable villages.[9] The bulk of the energy needs of these "sustainable villages" come from locally available, renewable energies like biomass, wind and solar power.

Due to the limited availability of biomass, more and more villages are relying on solar and wind resources for their energy supply (accordingly, the prefix *bio* is parenthesised in the terminology). As noted above, this original concept has since evolved to an advanced state where (bio) energy villages (BEV) tend to fulfil all their own energy needs, such as electricity, heating and cooling, and transportation from regionally available renewable energy (REN) sources such as wind, solar power, hydropower, and geothermal sources, as well as biomass; this makes them self-sufficient in terms of energy as well as independent from the national grids. Moreover, this initiative puts great emphasis on developing the land used for harnessing new sources of energy in a way that keeps resilience and biodiversity in mind.

Sustainable energy development is crucial to developing a BEV where regional added value, stemming from regional material flow management (rMFM), plays a pivotal role in achieving more general sustainability goals. The energy autonomy characteristic of BEVs is due not only to their high utilisation of regional renewable energy sources, but also to their high energy conversion and use efficiencies. In addition, features such as integrated water management, sustainable agriculture through sustainable soil management, and the use of bio-fertiliser and soil amendments—such as biochar—are integral parts of its design fabric. The (re)circulation of material and energy flows in a BEV unleashes enormous regional potential, referred to as regional added value (RAV), which would be otherwise unutilised or underutilised. Key elements of RAV include (but are not limited to) the reduction of greenhouse gas emissions (GHG); the creation of new products and services; the creation of employment opportunities; the reduction of the emigration of vital human resources; the reduction of environmental degradation and the loss of biodiversity; and the preservation of cultural heritage and indigenous knowledge by integrating them into the system.

In summary, BEVs based on zero emission strategies are much more than simply a new approach in environmental and climate protection. They provide the basis for

---

[9]There are about 150 established (bio)energy villages and another 43 under development in Germany under the FNR initiative. (See further details at https://bioenergiedorf.fnr.de/bioenergiedoerfer/liste/). Furthermore, the IfaS alone is currently in the process of developing another 94 new (bio)energy villages.

a sustainable economic policy based on the local potential that promotes innovation-driven modernisation of developed and emerging economies alike.

The utility of BEVs is well understood, but planning, developing and implementing BEVs is a complex process. Starting with the MFM process, a substantial amount of time and money is required to set up a BEV. Moreover, the required level of human capital, which can create opportunities out of problems, is large. However, the benefits of BEVs far outweigh the costs and complexities of implementation. Let us illustrate this point taking a common issue in any community or village: solid waste. Organic waste is a bothersome and expensive problem for communities in a conventional waste management context. However, in material flow management, organic waste is a vital resource and input for energy and fertiliser production.[10] Conventionally, this unused material flow is dealt with in compliance with environmental management and pollution prevention requirements. This siphons off scant financial and labour resources without generating any added value. As depicted in Fig. 14.3, *The Throughput System*, this would lead to a net loss of financial resources and further perpetuate an unsustainable system. As opposed to the aforementioned status quo system, should the principles of MFM and the CE be applied to this, the waste stream would no longer exist, and instead a resource stream for energy and fertiliser would be established. This, in turn, would create income, environmental benefits and employment opportunities as well. Such applications are the building blocks of BEVs.

This example can be transposed to most other socio-industrial metabolic systems as well, such as food, transportation, land use, etc., and can cover resource streams such as water, wastewater and energy. It is clear, therefore, that BEVs are sustainable villages with regard to energy consumption, energy provision, and the participation of local inhabitants in the energy management system. In the case of biomass use for energy production, land use must be decoupled from the pressure on biodiversity, soil erosion and nutrient oversupply. In this energy-climate era, it is necessary to consider carbon and water storage in appropriate respective sinks/reservoirs to avoid undesirable externalities. This is already happening in Germany.

An important question worth asking at this point is if it is possible to achieve sustainable, decentralised community development. As mentioned earlier, it is a good idea to perform a thorough systems analysis (usually an MFA) at the outset to determine the required level of human and financial capital. This helps clarify the reasons for the poor allocation of resources—energy, water, money, human capital, etc.—and as a consequence helps to prepare, design and execute technical and administrative solutions. As experienced in many of IfaS' applied projects, the resources needed to organise this change or shift can be retrieved within the system itself. Otherwise, market or public/government programmes may provide the necessary resources. The

---

[10]For example, 1 ton of organic material is equivalent to 100 m^3 of biogas with at least 50% CH$_4$ content, which means approximately 50 L of oil. The fermented residues contain in addition approximately 600 kgs of organic fertiliser.

latter, of course, is the most convenient, yet not always the most efficient, way to approach the critical issue of transaction costs.[11]

The following section presents two case studies on particular aspects of sustainable resource management: transferring know-how and knowledge on sustainable development to German communities. These government-funded projects, carried out by IfaS, follow two tracks. First, *The Sustainability Roadshow* is a communication programme specifically developed to disseminate research findings and communicate knowledge and know-how on decentralised sustainable development. Second, the *Dorfkern Limited* project is an example of a market mechanism to support sustainable resource management (SRM) and sustainable development (SD), demonstrating a public–private partnership (PPP) approach to securing necessary funds and allocating the necessary human capital to the sustainable development of small villages.

14.4 Programmes and Concepts

14.4.1 The Sustainability Roadshow

Since 1999, the German Ministry of Research and Education (BMBF) has offered research grants for sustainability research under the *Forschung für Nachhaltigkeit* (FONA) programme. After spending billions of euros and commissioning over a hundred projects,[12] questions were raised on the impact of the projects in the real world.

IfaS was contracted in 2018 to design and implement a research project on the transferability of FONA products to communities in Germany. The project was divided into four areas of activity:

1. The analysis and evaluation of FONA products with regard to their transferability;
2. The selection of feasible products for practical transfer;
3. The presentation of these products to German communities in six major Roadshow conferences;
4. The selection of 27 model communities from among Roadshow participants.

The analysis of the research projects from the FONA database immediately showed the difficulty of maintaining sustainability and transferring the concept to communities. Only a few products were left for transfer and even fewer were developed for practical application. The definition of sustainability and sustainable projects needed to be clarified, as some projects and outcomes were of questionable value according to the 17 SDG indicators. Using new software tools to optimise and accelerate the search for available commercial real estate in communities might help the local economy, but otherwise interfere with other aspects of sustainable land use.

---

[11] Based on our own experiences.
[12] €3.3 billion during the period from 2005 to 2014, with another €2 billion from 2015 to 2020.

The selection of feasible products was done in close cooperation with the participants of the projects, specifically researchers, academics/professors, consultants, and the ministry itself. Out of 87 projects, 62 were selected in the first phase. In the second phase, interviews with project directors and researchers led to a final 26 projects with a total of 32 specific products.[13]

We classified the projects into four categories: water/wastewater, energy, land use, and general issues. The "general issues" category was designed to introduce sustainability, zero emissions approaches and material flow management to the audience, and to offer a glimpse into the future, such as by introducing examples of best practices.

We then assembled these into presentations. One of these general presentations described successful project examples, model communities, added value assessments, and policy support. Another one dealt with specific financing options.

The presentation to German communities was designed to cover most of Germany. Six different cities—Nuremberg, Hannover, Stuttgart, Leipzig, Schwerin and Emsdetten—each hosted a one-day or two-day conference. The conference venues were selected to allow as many mayors and public experts as possible to attend.

372 participants participated in four meetings; 178 were representatives from small villages and towns. The conferences were organised as one-day and two-day events. Two-day conferences offered attendees the opportunity to meet in the evening, reflect on the day's information, and start networking for implementation. This was crucial, because the projects energised and intellectually challenged the audience. Accordingly, there were many requests for more information, especially for implementation support. Out of these 178 participating communities, 27[14] model villages and towns were selected for intensive transfer of know-how and coaching within the year.

The next steps of the process will involve a detailed data analysis of each community, followed by stakeholder interviews. This research will then be used to formulate ideas for the transfer of existing concepts or products, and transfer them to communities. At the end of the process, which will involve town hall meetings to facilitate ownership among local residents as well, the respective town councils will decide on implementing a proposed product or strategy.

14.4.2 "Dorfkern" Companies

A "*Dorfkern*" company is a new strategy developed by IfaS to cope with the high transaction costs and complex efforts to achieve a sustainable village, a resilient village or a (bio) energy village (Heck and Blim 2019).

---

[13]Some projects had multiple products or solutions that could be transferred.

[14]One additional model village was selected due to the amount of excellent applications.

In this approach, several small villages set up a not-for-profit company, which mobilises the resources to do a proper MFA, and develop detailed projects in the fields of energy, water, land use and infrastructure. The non-profit company then uses the MFA as a guidebook to develop projects and prepare the implementation legally and economically. In addition, the *Dorfkern* is designed to support the aspects of technology expertise and financing. *Dorfkern* is mainly owned by the consortium of villages; however, there are provisions for inviting other shareholders.

Dorfkern is a clear solution to addressing the issues that a small village or a community cannot address alone. Individual villages or communities have options, but alone they are too small to recognise issues and implement solutions. *Dorfkern* further rectifies the problems of the lack of human capacity, know-how, and legal and financial resources as a large, collective entity, effectively.

A *Dorfkern* company's success comes from the capitalisation of unused or non-monetised potential in villages and communities. These opportunities include the installation of solar panels on roofs and other free space, LED street lighting, the installation of heat pumps, increased building insulation, implementation of district heating, increased capacity for electric transport, the changing of land use with new marketing ideas such as solidarity farming, harvesting biochar from farm hedges, etc. The MFA helps stakeholders find opportunities and turn them into assets, which they can subsequently offer on the market. As the first stakeholder discussions have shown, potential investors have been interested in buying into these opportunities and making a good return on investment.[15]

14.5   Conclusions

Small-scale systems are direct victims of the worldwide plundering of energy and resources, and also have been severely affected by the throughput system of economy. Accordingly, resources and sinks have been overused, and neither efficiency nor effectiveness has been given sufficient attention. As a result, small-scale systems have lost opportunities and competitiveness, and in turn, have become a burden. In this energy-climate era, these issues are becoming even more severe, impacting all aspects of sustainability. In rural areas—where these small-scale systems are dominant—changes in precipitation regimes, frequent drought, increased soil erosion and biodiversity loss have pushed stakeholders to their limits. Furthermore, and characteristically for small-scale systems, the know-how, money and management skills to organise *change* are scarce.

A shift from a throughput society toward a more sustainable metabolic system such as the CE would pose major challenges not only for small-scale systems but for all systems. However, using an MFM methodology would help turn these challenges into targeted opportunities for the decentralised management of resources and sustainable development. As this chapter has pointed out, small-scale systems such as small

---

[15]Project talks with Naturwind, Energie Leipzig about financing Dorfkern GmbH, 2019.

villages and communities have multiple opportunities similar to those of large-scale systems. However, strategic planning and management solutions are necessary at the level of project deployment and resource mobilisation, as exemplified in the case of ECB and *Dorfkern* companies.

In light of that challenge, we presented two randomly chosen management approaches to turning small-scale systems into resilient, sustainable systems—i.e. BEVs or communities. The Sustainability Roadshow has transferred thus far unused or little-used research results to small communities and towns. Best-practice and next-practice solutions developed using BMBF research grants have been interwoven with the challenges and problems of small systems. Regional conferences and the selection of model communities push research directly into the heart of small systems. Stakeholders become motivated and informed, and receive implementation coaching. This is an ongoing project, which so far has demonstrated very positive results.

A second, more market-driven example, the *Dorfkern* company, demonstrates how to change small-scale systems based on a throughput society into sustainable systems by employing a business approach. Achieving SRM and the SDGs via a market-based small-scale system by bundling resources and capabilities is an innovative approach that could be deployed in almost any (rural) geographic location in the world.

In summary, there is no lack of potential and solutions, but rather a general lack of management and strategy, as well as courage. Referring to courage, a quotation from Nelson Mandela seems appropriate to end this work: "It always seems impossible until it's done!".

References

Böhmer J., Becker J., Bentkamp C., Wagener F., Rupp J., Heinbach K., Bluhm H., Heck P., & Hirschl B. (2019). *Ländliche Bioökonomie—Stärkung des ländlichen Raums durch eigene dezentrale bioökonomische ansätze* (p. 43). Neubrücke: Hochschule Trier, Institut für angewandtes Stoffstrommanagement. https://laendliche-biooekonomie.de/wp-content/uploads/2019/03/LBio_Download.pdf.

D'Amato, D., Droste, N., Allen, B., Kettunen, M., Lähtinen, K., Korhonen, J., & Toppinen, A. (2017). Green, circular, bio economy: A comparative analysis of sustainability avenues. *Journal of Cleaner Production, 168,* 716–734. https://doi.org/10.1016/j.jclepro.2017.09.053

ECB. (2019). *Green campus concept.* https://www.umwelt-campus.de/en/campus/life-on-campus/green-campus-concept/.

EEA. (2016). Circular economy in Europe—Developing the knowledge base: Report II. *European Environment Agency.* https://doi.org/10.2800/51444

Elia, V., Gonin, M. G., & Tornese, F. (2017). Measuring circular economy strategies through index methods: A critical analysis. *Journal Clean Producao, 142,* 2741–2751. https://doi.org/10.1016/j.jclepro.2016.10.196

Ellen MacArthur Foundation. (2015). *Delivering the circular economy: A toolkit for policymakers.* Cowes, UK: Ellen MacArthur Foundation. https://www.ellenmacarthurfoundation.org/circular-economy/concept.

Heck, P., & Hoffmann, D. (2001). *Biomasse Masterplan Rheinland-Pfalz.* Birkenfeld.

Heck, P., & Blim, M. (July 2019) Dorfkern Strategiepapier für das Energieministerium Mecklenburg-Vorpommern, unpublished.

Heck, P., Anton, T., Thome, P., Pietz, C., Latzko, S., Schaubt, M., et al. (2011). *Integriertes Klimaschutzkonzept für den Landkreis Rhein-Hunsrück.* Kreisverwaltung Rhein-Hunsrück-Kreis, Birkenfeld: Simmern.

IfaS. (2018) and ECB (2019). Green campus concept. Retrieved 22 Jan 2020 from https://www.umwelt-campus.de/en/campus/life-on-campus/green-campus-concept.

IfaS. (2019). *Circularizing flows: Material flow management concept.* Unpublished work.

Kopnina, H., Washington, H., Taylor, B., et al. (2018). *Journal of Agricultural and Environmental Ethics, 31,* 109. https://doi.org/10.1007/s10806-018-9711-1

OECD. (2019). *Global material resources outlook to 2060: Economic drivers and environmental consequences,* Paris: OECD Publishing. https://doi.org/10.1787/9789264307452-en.

Roser, M. (2019). *Economic growth.* Published online at OurWorldInData.org. Retrieved from 22 Jan 2020. https://ourworldindata.org/economic-growth.

Schroeder, P., Anggraeni, K., Weber, U. (2019). The relevance of circular economy practices to the sustainable development goals. *Journal Industrial Ecology, 23-1,* 77-95. https://doi.org/10.1111/jiec.12732.

Chapter 15
Multifunctional Urban Landscapes: The Potential Role of Urban Agriculture as an Element of Sustainable Land Management

Kathrin Specht, Julian Schimichowski, and Runrid Fox-Kämper

Abstract Urban agriculture (UA) has long been the subject of civic and scientific debate, since it provides cities with a diverse range of functions and services. UA is thought to have a positive effect on sustainable urban development in environmental, economic and social areas. Although most of the effects attributed to UA are positive, there are also critical aspects and concerns. For example, doubts have been raised about the quality of the products grown, considering the prevalence of air pollution and contaminated soils; and there are doubts about the contribution that urban agriculture makes to feeding urban populations, owing to the small quantities produced. Moreover, there are conflicts surrounding land use, especially in big cities; and in some cases, agricultural activities are undertaken in the city without the necessary building approvals. The concept of co-production and sharing as a (business) model is increasingly being applied with reference to urban gardens. This is particularly the case with volunteer-led community gardens, which are tremendously open to gardening enthusiasts and are renowned for the sharing of resources and space. Rather than seeking to make a profit, many of these initiatives operate under the principles of a non-profit or sharing economy. This chapter explores how UA can contribute to sustainable land management and co-production. To this end, background information is given on the (re-)emerging phenomenon of urban food production and on what motivates those involved to implement collaborative practices. The functions and services provided by UA as an element of sustainable land management are then explored using the three pillars of sustainability.

K. Specht (✉) · R. Fox-Kämper
Research Institute for Regional and Urban Development (ILS), Brüderweg 22–24, 44135 Dortmund, Germany
e-mail: kathrin.specht@ils-forschung.de

R. Fox-Kämper
e-mail: runrid.fox-kaemper@ils-forschung.de

J. Schimichowski
Master Graduate Spatial Planning - Technische Universität Dortmund, August-Schmidt-Straße 1, 44227 Dortmund, Germany
e-mail: jschimichowski@gmail.com

© The Author(s) 2021
T. Weith et al. (eds.), *Sustainable Land Management in a European Context*,
Human-Environment Interactions 8, https://doi.org/10.1007/978-3-030-50841-8_15

Keywords Alternative Food Networks (AFN) · Co-Production · Sustainability · Consumer-producer relationships · Urban gardening

15.1 Background on Urban Agriculture

There has been an enormous increase in UA activities throughout the world in the recent past, as evidenced by the growing interest in allotment gardening and urban gardening (BMUB 2015). And yet the production of food in the city is not a new phenomenon – it has been common practice ever since cities evolved. According to Colding and Barthel (2013), the emergence of UA is closely linked to crisis-prone developments in the city. The term "crisis" refers primarily to times of economic upheaval, although it can also be extended to social transformations (Fox-Kämper 2016). Thus allotments emerged around the mid-nineteenth century in response to the poor housing and supply conditions experienced by the urban population of European cities that swelled as a result of industrialisation, whereas the growing popularity of community gardens since the mid-1970s is partly a response to the growing polarisation, fragmentation and segregation of the urban population (Calvet-Mir et al. 2016). In view of globalised food markets, the development of current types of urban agriculture also reflects an increasing desire for traceability and knowledge about the origin of food and the interplay involved in its production.

Especially in view of shrinking cities and regions – for example owing to post-reunification phenomena in Germany or structural change – people are currently in search of approaches that can bring about stabilisation and improvement. New ideas and approaches are needed to embrace today's urban reality, which is marked by the coexistence of shrinking and growing regions. The concept of "Continuous Productive Urban Landscapes" (CPUL), proposed by architects and urban designers Katrin Bohn and André Viljoen, seeks to integrate agriculture into urban planning, and to make it an essential element of sustainable infrastructure (Schulz et al. 2013; Viljoen et al. 2005). This example illustrates today's continued relevance of UA for urban planning. In contrast to the past, however, UA is only rarely pursued to enable the urban population to become self-sufficient (Schulz et al. 2013). As indicated above, UA was popular whenever populations suffered deprivation and hardship. It is therefore all the more surprising that UA has become such a widespread phenomenon in prosperous cities in developed countries of the Global North, resulting in the evolvement of a "new urban agriculture" (Karge 2015). Appel et al. (2011) identify the emerging green movement in the 1970s and the effects of the Chernobyl disaster in the mid-1980s as starting points for the renaissance of UA (Appel et al. 2011). This was a time when alternative lifestyles and economic systems were being debated and put to the test, particularly in Germany, because the Western economic model had led to mistrust and uncertainty among the population, owing to an increase in fragility. In the light of concerns about the finite nature of fossil resources, growing numbers of people asked themselves how they, as individuals, could lead more sustainable lives (Lohrberg 2011). Growing food can therefore be regarded as an alternative to globalised food markets and a consumerist lifestyle. Then there are social motivations – many urban gardens are tended communally, and serve as places of integration and openness towards the urban environment (Lohrberg 2011).

Internationally, "urban agriculture" is considered to be an umbrella term for all types of food production in the city. Whereas central importance was attached to food production in early definitions of the term (UNDP 1996), more recent definitions recognise the diversity of the different types of UA, which provide the city with other services apart from food:

> Urban agriculture spans all actors, communities, activities, places and economies that focus on biological production in a spatial context, which – according to local standards – is categorized as 'urban'. Urban agriculture takes place in intra- and periurban areas [...] urban agriculture is structurally embedded in the urban fabric; it is integrated into the social and cultural life, the economics, and the metabolism of the city. (Vejre et al. 2016, 21).

The following table lists examples of different types of UA. Their boundaries are blurred or may overlap in some cases, and are not always clear-cut (Table 15.1).

15.2 Motivations for UA: New Consumer-Producer Relationships and Co-production

The notion of sharing as a (business) principle is increasingly applied with reference to urban gardens. This is particularly the case with volunteer-led community gardens, which are characterised by tremendous openness towards gardening enthusiasts and by the sharing of resources and space (Zoll et al. 2018; Opitz et al. 2017). Rather than seeking to make a profit, many of these initiatives operate under the principles of a non-profit or sharing economy (Piorr et al. 2018; De Cunto et al. 2017). The principle of sharing is evident in collaboration among the volunteers and in the sharing of resources (land, money, labour and means of production) as well as crops and knowledge.

Figure 15.1 shows the focus of different types of UA. While some types of UA (such as home gardens) are merely practised in a private setting, with no engagement involved, other types of UA are largely shaped by elements of collaboration and co-production. For example, interaction between members is a key element of community gardens, which are geared towards social networking, whereas interaction between producers and consumers is the core of the business model when it comes to community-supported agriculture, for instance.

The main reason for the particular relevance of collaboration and co-production in UA is that the marketing of food produced in UA differs considerably from that in commercial agriculture. Food grown in urban settings is usually marketed locally, i.e. it is consumed in the city where it was grown. Whereas the food produced by established large farms is usually traded on the world market, UA is characterised by direct marketing (RUAF Foundation). While urban gardens tend to be small-scale

Table 15.1 Different types of urban agriculture, and their main characteristics

Type of urban agriculture	Main characteristics
Home gardens	• Affiliated with individual households • Not institutionally embedded
Allotments (also referred to as family gardens)	• Plots for individual use, e.g. for the non-commercial production of food, usually in specific complexes • Institutionally embedded in acts and planning law
Self-harvest gardens	• Plots that have been pre-tilled by farmers, and are individually tended by urban residents who grow their own vegetables • Leases, usually for a season
Community gardens	• Volunteer-led gardens for communal use • Often as interim use concepts on brownfield or infill sites; no institutional backing
Intercultural gardens	• A special type of community garden with the aim of fostering integration
Therapeutic gardens	• Often affiliated with psychotherapy institutions • Doing gardening and experiencing the garden may have a healing effect on patients
School gardens	• Usually affiliated to schools • The aim is to teach children about how to grow, process and prepare food, and about the environment
Social farming	• Managed farms operated by institutions responsible for social welfare establishments • By handling farmed animals or plants, individuals are given the opportunity to relax and perform work-related activities
School farms	• Further development of the traditional use of farms • The aim is to provide children with an educational and recreational facility in a near-natural environment
Community-supported agriculture	• A community of producers and consumers comprising farms and private households • Seasonal contractual relationship between producers and consumers
Local direct marketing	• Direct marketing of agricultural products – generally in farm shops

projects that may meet a few households' demand for food, market oriented urban agriculture can also be practised on large-scale farms (Opitz et al. 2016).

In the process, UA does not only produce material goods. On the urban periphery in particular, farms have adopted strategies to adapt their range of services, which are usually aimed at local or regional markets. The strategies for UA mentioned by Piorr et al. (2018) include specialisation, differentiation and diversification. Specialisation in a few specific products enables production costs to be cut and the functional proximity to urban areas to be exploited. In this way, for example, highly perishable products that are only feasible for short transport distances can be marketed with a

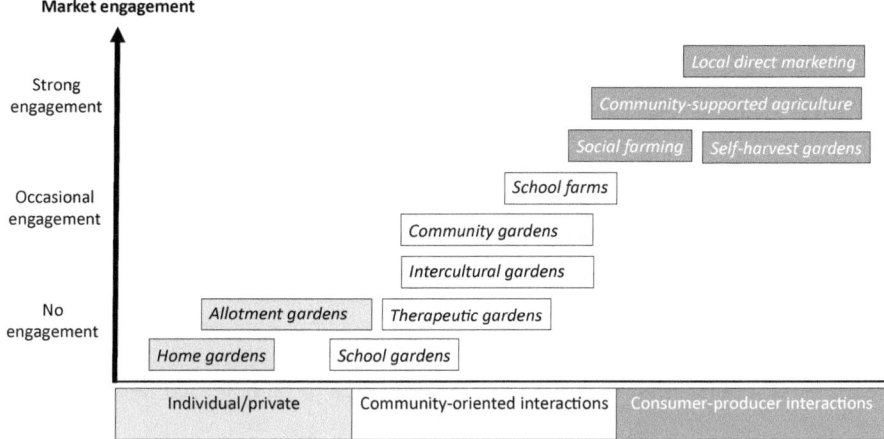

Fig. 15.1 Collaboration and level of market engagement in different types of UA (source: authors)

competitive advantage over rural agriculture. Differentiation can involve growing old varieties or especially high-quality products, for instance. Diversification enables UA projects to expand their portfolio by offering services such as horse management or educational activities in addition to food (Piorr et al. 2018; van der Schans et al. 2016). All operational strategies place emphasis on strong regional networking structures.

In the historical context, UA was mostly linked to the motivation of subsistence. The primary actors in UA today have different reasons for growing food in an urban setting. The three typical types of motivation named by Berges et al. (2014) are (1) an emphasis on subsistence, (2) a socio-cultural emphasis and (3) an emphasis on commerce (Berges et al. 2014). Those who place an emphasis on subsistence pursue the goal of producing their own food by their own efforts, along with the added monetary, health and recreational benefits. This also ties in with the motivation to supply oneself with fresh produce, and to create and pass on knowledge about food production (van der Jagt et al. 2017). Environmental considerations are another important reason for getting involved in agricultural activity, enabling possibilities for more sustainable urban food production to be tested in cities that are apparently becoming increasingly disconnected from nature (van der Jagt et al. 2017; Smit et al. 2001; Vejre et al. 2016). The key objectives of activities with a socio-cultural emphasis are community-building, cultural exchange, education and social inclusion (Berges et al. 2014, 12). Such activities are often politically motivated. An anti-capitalist stance and a rejection of the globalised food system are reflected in some UA initiatives (McClintock and Simpson 2017). Other UA activities seek to appropriate urban space, and to demand the "right to the city" (Horst et al. 2017, 283; Thibert 2012; Holm 2011; Purcell and Tyman 2015). The main objective behind the commercial emphasis is to generate income and profit, create jobs, and develop new markets.

With reference to UA, Appel et al.(2011) mention, in particular, the possibility to create a new type of open space "that may be the starting point for collective grass-roots processes, practised autonomy and networking, and that may play an important role in promoting (local) politics" (Appel et al. 2011, 152, translated by authors). This enables UA stakeholders to actively shape a part of the city, based on their own ideas (Appel et al. 2011).

Those engaged in UA can broadly be classified into two groups. The first group comprises people who carry out farming as their usual main occupation, whether part-time or full-time. They may have deliberately chosen the proximity to an urban area, or may have experienced urban growth encroach on their farm over time. The second group comprises individuals who practise UA as a hobby in their spare time (Vejre et al. 2016). These two UA subgroups differ. Above all, urban gardens are usually tended as a hobby, whereas urban agriculture tends to be practised by professional actors.

15.3 Functions and Services Provided by UA Concerning Sustainable Land Management

The functions and services of UA mentioned in the scientific debate are extremely varied. As a multifunctional service, UA has, in principle, the potential to have a positive impact on social, ecological and economic concerns in cities (Schulz et al. 2013). Based on Pearson et al. (2010), the services are presented below using the three pillars of sustainability. Where possible, reference is made to sustainable land management.

15.3.1 Social Functions and Services

Various studies have documented that UA may contribute to strengthening the social capital of the city (Santo et al. 2016). In particular, the notion of multifunctionality of the city should be stressed in this respect. There are social demands, for example, with regard to additional services provided in the areas of leisure, recreation and education. Being green spaces in a city, urban gardens may make an important contribution to the city landscape and to increasing the quality of life in urban spaces.

This is achieved by creating new networks between actors from different social backgrounds and milieus, which is particularly the case with community gardens. In this context, services conducive to the integration of disadvantaged social groups are attributed to UA projects (Santo et al. 2016; RUAF Foundation). These social services are also achieved by creating new places of interaction and encounter, which, particularly in neighbourhoods with a lack of green space, improve residential quality and may result in a higher level of identification with the neighbourhood (Appel et al.

2011). Added value is generated by creating leisure activities to enhance physical and mental health, promoting general health and well-being (RUAF Foundation, Piorr et al. 2018). As an element of sustainable land management, therefore, urban gardens make an important contribution to neighbourhood development. The result is a "new sense of community, prospects for shaping one's own life; and gardening encourages participants to become more deeply engaged in the neighbourhood" (BBSR 2015, 5; translated by authors). UA also makes a contribution to practical environmental education by enabling individuals to experience first-hand how food is produced (Gahm 2017; Appel et al. 2011). In that regard, UA induces a change in consumer behaviour, one that is more sustainable, because it enables passive consumers to become co-producers of food (Santo et al. 2016).

15.3.2 Ecological Functions and Services

UA increases the heterogeneity of urban land uses, contributing to the urban ecosystem (Sanyé-Mengual et al. 2018; Artmann and Sartison 2018). The additional vegetation grown increases biodiversity in urban spaces. In addition, inner-city urban gardens may help to balance the temperature and reduce heat islands (Santo et al. 2016). The fresh air produced also helps improve air quality (BMUB 2015). Moreover, UA generates new infiltration areas. After all, sealed brownfield sites are often converted into land for UA, which has a positive impact on the groundwater and on flood protection (Santo et al. 2016; Piorr et al. 2018). UA may also help to reduce waste by recycling organic waste and waste water as part of the urban material cycle or urban metabolism (Roggema 2016; de Zeeuw 2011). In general, the ecological functions described may make cities more resilient to climate change and, at the same time, contribute to mitigate climate change (Piorr et al. 2018; Santo et al. 2016). This is also achieved because local food production reduces transport distances – referred to as food miles – as well as the energy consumption involved, e.g. for cooling food, reducing the overall level of pollutant emissions (Piorr et al. 2018; Pearson et al. 2010).

15.3.3 Economic Functions and Services

First of all, UA performs economic services with respect to urban farmers and gardeners because food is grown, easing the financial burden on households. Farmers and gardeners either sell produce to this end, or consume it themselves (RUAF Foundation). However, the emphasis on subsistence is less important in post-industrial nations of the Global North. In these countries, economic advantages can be generated instead by developing business strategies tailored specifically to the framework conditions of urban spaces (Piorr et al. 2018). As such, UA may generate jobs and create employment opportunities in the local value chains (Santo et al. 2016). Furthermore, the concept of "social entrepreneurship" is common in UA, and alternative economic models that focus on greater cooperation between consumers and producers often come to fruition. UA therefore promotes a diversified urban economy and added value by way of urban production (Brandt et al. 2018; Pearson et al. 2010). This also helps cities to become more resilient in terms of food supply and food security (BMUB 2015). As with other green infrastructure, UA may help to upgrade urban neighbourhoods, stemming a decline in value of urban infrastructure and real estate (Voicu and Been 2008), particularly in regions affected by shrinkage. Municipalities also benefit from UA, such as when green spaces are tended communally by garden initiatives, reducing the financial burden on cities (Rosol 2006; Stierand 2012).

15.3.4 Limitations

Aside from these positive attributions, there are also voices in the academic debate that challenge the functions and services provided by UA, or at least limit them. For example, it is said that, although locally initiated urban gardens improve neighbourhoods, they may ultimately encourage gentrification processes and damage the neighbourhood's social fabric (Cohen and Reynolds 2014; Specht et al. 2017; Horst et al. 2017). There is also criticism that the oft-touted tremendous openness may not signify openness for everyone. After all, the use of former public open spaces for urban gardens excludes non-gardeners from using these spaces, raising doubts about whether they are indeed of interest to the general public (Sondermann 2014). With reference to the integrating effect of UA, mostly associated with intercultural gardens, it is noted that colonial ways of thinking are re-inscribed in that the better-off bring good food to disadvantaged social groups (McClintock et al. 2016).

Another key limiting factor that must be mentioned is the availability of land, which is a necessary condition for any kind of UA. Undeveloped areas are highly sought after, particularly in growing cities with a high influx of newcomers, resulting in strong competition for land use and high land prices. In this context, inner-city urban gardens particularly compete with investors usually keen to implement real estate projects, e.g. for residential purposes (Schmidt 2016; Berges et al. 2014; BMUB 2015). Since urban gardens are mostly projects that are initiated and run by residents, they are financially inferior to commercial real estate development and usually unable to compete successfully for urban open spaces (Berges et al. 2014). As a result, land reserves are increasingly being covered with buildings, and urban greenery and urban gardens are in conflict with other uses (BMUB 2015). This situation is exacerbated by the principle of "inner development before outer development" defined in the many national building codes - e.g for Germany in the Federal Building Code (BauGB) (Siedentop 2010; Koch 2017). The aim of this requirement is to reduce land consumption, which leads to the re-densification of central areas and fewer building land designations as urban expansion. Concerning the availability of land for urban gardens, this means that more infill sites are developed, more brownfields are covered with buildings, and more land is recycled, putting even greater pressure on urban open spaces (Schmidt 2016; BMUB 2015; Siedentop 2010). The bigger the city and the greater influx it experiences, the more this situation is exacerbated. This is especially true for cities that experience a high influx of newcomers and are particularly appealing to young people (Simons and Weiden 2016). Growth dynamics, and therefore the pressure on open spaces, are particularly high in such cities. Then again, limiting settlement development on the outskirts of the city leads to the securing of land for UA initiatives located in those areas, because it reduces the excess planning of agricultural land on the outskirts. UA is thus largely influenced by the availability of land (Berges et al. 2014).

The situation is opposite in shrinking cities, where there is ample land for UA (Dams 2012). In this respect, for example, the structural availability of brownfield sites in former industrial regions may potentially promote UA, whereas cities in prosperous metropolitan regions exhibit higher land prices and commercialisation pressures, making it more difficult for UA initiatives to develop and persist. Although this finding primarily relates to urban gardens located in urban areas, it can also be applied to peri-urban areas. UA in peri-urban areas runs the risk of farmland being converted into building land. After all from the urban planner's perspective, peri-urban agricultural areas were long regarded as land reserves for urban development concepts and planning processes (LWK NRW 2012).

The quality of land and its suitability for growing food is also of crucial importance. The availability of other resources such as water, organic waste and soil are the determining factors in this case. Then it could be the case, especially with inner-city areas, that soil has been contaminated by previous industrial use, for example, preventing food production or necessitating complex soil remediation measures (Specht and Siebert 2017; Howe et al. 2005).

There are also doubts concerning the independent economic viability of UA projects because they are often a recreational activity and therefore do not have to be economically sustainable (Schulz et al. 2013). The economic capacity of UA and its actual contribution to the food system are also cast into doubt. Compared to conventional farming, the contribution of UA to providing food for the urban population is considered minimal (McEldowney 2017).

In spite of the limitations described, it has been shown that UA may potentially be beneficial to a variety of broader development goals such as sustainability, quality of life, food justice and urban resilience (Horst et al. 2017).

15.4 Conclusion

Urban agriculture (UA) has long been the subject of civic and scientific debate. In connection with sustainable land management, UA is especially considered in the context of urban development with the simultaneity of suburbanisation and reurbanisation processes (Schulz et al. 2013; Piorr et al. 2018). The functions and services of UA are manifold. UA has a positive impact on sustainable urban development, e.g. by reducing food miles; economic benefits are created by increasing regional added value; social benefits are gained by way of integration and community-building; health effects are produced due to exercise, nutrition and recreation; and environmental advantages are achieved by improving the urban climate (Berges et al. 2014; BMUB 2015). These are just a few of the positive impacts of UA discussed in the scientific literature. The above-mentioned benefits do not apply in equal measure to all types of UA. Moreover, the actors involved differ, as do the motivations for undertaking UA activities, ranging from subsistence and socio-cultural motivations to commercial intentions (Berges et al. 2014). Although most of the effects attributed to UA are positive, there are also critical aspects and concerns. For example, doubts have been raised about the quality of the products grown, considering the prevalence of air pollution and contaminated soils; and there are doubts about the contribution that urban agriculture makes to feeding urban populations, owing to the small quantities produced. Moreover, there are conflicts surrounding land use, especially in big cities; and in some cases, agricultural activities are undertaken in the city without the necessary landuse and building approvals. Urban green spaces are under tremendous pressure: population growth and economic developments lead to a growing need for settlement areas, particularly in conurbations.

It is recognised that collaboration and sharing can be regarded as key elements of UA. The strong role that UA may play with regard to urban networking and interaction between a wide variety of actors can primarily be explained by the special forms of organisation and marketing that are geared towards the urban setting.

In order to be able to make a real contribution to sustainable land management, it is important for UA to overcome the boundaries and obstacles relating to agricultural production in cities. This includes seeing existing competitive usages as an opportunity by further developing unused areas for multifunctional uses. For example, many unused roofs are suitable for creating rooftop gardens. Also allotments, which have come under pressure in many places, could be made available for use as green spaces by a wider public, by testing new approaches and demonstrating greater openness towards non-gardeners, ensuring their continued existence as green oases in the heart of the city.

Particularly the notion of multifunctionality, which also includes social and ecological benefits that are hard to quantify (Schulz et al. 2013), may become an important guiding principle for sustainable land management in the future, in which UA may be acknowledged as a "multifunctional service".

References

Appel, I., Grebe, C., & Spitthöver, M. (2011). *Aktuelle Garteninitiativen. Kleingärten und Gärten in deutschen Großstädten.* Kassel: kassel university press.

Artmann, M., & Sartison, K. (2018). The role of urban agriculture as a nature-based solution: A review for developing a systemic assessment framework. *Sustainability, 10*(6), 1937. https://doi.org/10.3390/su10061937

BBSR (Ed.). (2015). Gemeinschaftsgärten im Quartier. BBSR-Online-Publikation. No. 12/2015. Bonn.

Berges, R., Opitz, I., Piorr, A., Krikser, T., Lange, A., Bruszewska, K., et al. (2014). *Urbane Landwirtschaft. Innovationsfelder für die nachhaltige Stadt?* Müncheberg: Leibniz-Zentrum für Agrarlandschaftsforschung (ZALF) e. V.

BMUB. (2015). *Grün in der Stadt − Für eine lebenswerte Zukunft.* Berlin: Grünbuch Stadtgrün.

Brandt, M., Gärtner, S., & Meyer, K. (2018). Urbane Produktion, Planungsrecht und dezentrale Finanzsysteme. *IAT Forschung aktuell 10/2018. Institut Arbeit und Technik (IAT).* Gelsenkirchen.

Calvet-Mir, L., March, H., Nordt, H., Pourias, J., & Čakovská, B. (2016). Motivations behind urban gardening – 'Here I feel alive'. In S. Bell, R. Fox-Kämper, N. Keshavarz, M. Benson, S. Caputo, S. Noori, & A. Voigt (Eds.), *Urban allotment gardens in Europe.* New York: Earthscan from Routledge. https://doi.org/10.4324/9781315686608.

Cohen, N., & Reynolds, K. (2014). Resource needs for a socially just and sustainable urban agriculture system: Lessons from New York City. *Renewable Agriculture and Food Systems.* https://doi.org/10.1017/S1742170514000210.

Colding, J., & Barthel, S. (2013). The potential of 'urban green commons' in the resilience building of cities. *Ecological Economics, 86,* 156–166. https://doi.org/10.1016/j.ecolecon.2012.10.016

Dams, C. (2012). Gärten gehören zur Stadt! Zur städtebaulichen Relevanz der urbanen Landwirtschaft. In *Urban Gardening: über die Rückkehr der Gärten in die Stadt, edited by Christa Müller* (5th ed., pp. 160–172). Oekom.

De Cunto, A., Tegoni, C., Sonnino, R., Michel, C., & Lajili-djalaï, F. (2017). *Food in cities: Study on innovation for a sustainable and healthy production, delivery, and consumption of food in cities.* Luxembourg.

de Zeeuw, H. (2011). Cities, climate change and urban agriculture. *Urban Agriculture Magazine, 25,* 39–42.

Fox-Kämper, R. (2016). Concluding remarks. In S. Bell, R. Fox-Kämper, N. Keshavarz, M. Benson, S. Caputo, S. Noori, & A. Voigt (Eds.),*Urban allotment gardens in Europe.* New York: Earthscan from Routledge. https://doi.org/10.4324/9781315686608.

Gahm, S. (2017). Urbane Landwirtschaft als Form der praktischen Umweltbildung für und mit sozial Benachteiligten. In *Bachelorarbeit an der Fakultät Angewandte Sozialwissenschaften an der Hochschule für angewandte Wissenschaften Würzburg-Schweinfurt.*

Holm, A. (2011). Das Recht auf die Stadt. *Blätter für deutsche und internationale Politik, 8,* 89–97.

Horst, M., McClintock, N., & Hoey, L. (2017). The intersection of planning, urban agriculture, and food justice: A review of the literature. *Journal of the American Planning Association, 83*(3), 277–295. https://doi.org/10.1080/01944363.2017.1322914

Howe, U., Viljoen, A., & Bohn, K. (2005). New cities with more life: Benefits and obstacles. In A. Viljoen, K. Bohn, & J. Howe (Eds.), *Continuous productive urban landscapes: Designing urban agriculture for sustainable cities* (pp. 56–60). London: Elsiever Architectural Press.

Karge, T. (2015). Neue Urbane Landwirtschaft - Eine theoretische Verortung und Akteursanalyse der Initiative Himmelbeet im Berliner Wedding. *Arbeitshefte des Instituts für Stadt- und Regionalplanung der Technischen Universität Berlin, 79.*

Koch, M. (2017). Möglichkeiten und Grenzen der Freiraumsicherung in urbanen Wachstumsräumen. In S. Kost & C. Kölking (Eds.), *Transitorische Stadtlandschaften: Welche Landwirtschaft braucht die Stadt?* (pp. 27–40). Wiesbaden: Springer.

Lohrberg, F. (2011). Masterplan Agrikultur. Städte müssen Dialog mit Landwirten suchen. *Stadt + Grün, 09,* 43–48.

LWK NRW (Ed.). (2012). Landwirtschaftlicher Fachbeitrag zum Regionalplan Metropolregion Ruhr. Daten, Fakten und Entwicklungsperspektiven der Landwirtschaft im urbanen und suburbanen Raum. Unna

McClintock, C., Miewald, E., & McCann, N. (2016). The politics of urban agriculture: Sustainability, governance, and contestation. In A. E. G. Jonas, B. Miller, K. Ward, & D. Wilson (Eds.),*The Routledge handbook on spaces of urban politics* (pp. 1–15). London: Routledge.

McClintock, N., & Simpson, M. (2017). Stacking functions: Identifying motivational frames guiding urban agriculture organizations and businesses in the United States and Canada. *Agriculture and Human Values, 35*(1), 19–39. https://doi.org/10.1007/s10460-017-9784-x

McEldowney, J. (2017). *Urban agriculture in Europe. Patterns, challenges and policies. European Parliamentary Research Service.* https://doi.org/10.2861/790177.

Opitz, I., Berges, R., Piorr, A., & Krikser, T. (2016). Contributing to food security in urban areas: Differences between urban agriculture and peri-urban agriculture in the Global North. *Agriculture and Human Values, 33*(2), 341–358. https://doi.org/10.1007/s10460-015-9610-2

Opitz, I., Zoll, F., Doernberg, A., Specht, K., Siebert, R., Piorr, A., et al. (2017). *Future|Food|Commons – Alternative food networks at the urban-rural interface.* Müncheberg: Agrathaer, ZALF.

Pearson, L. J., Pearson, L., & Pearson, C. J. (2010). Sustainable urban agriculture: Stocktake and opportunities. *International Journal of Agricultural Sustainability, 8*(1 & 2), 7–19. https://doi.org/10.3763/ijas.2009.0468

Piorr, A., Zasada, I., Doernberg, A., Zoll, F., & Ramme, W. (2018). Research for AGRI Committee – Urban and Peri-urban agriculture in the EU. Brussels: Policy Department for Structural and Cohesion Policies.

Purcell, M., & Tyman, S. K. (2015). Cultivating food as a right to the city. *Local Environment, 20*(10), 1132–1147. https://doi.org/10.1080/13549839.2014.903236

Roggema, R. (Ed.). (2016). *Sustainable urban agriculture and food planning.* New York: Routledge.

Rosol, M. (2006). Gemeinschaftsgärten in Berlin. Eine qualitative Untersuchung zu Potenzialen und Risiken bürgerschaftlichen Engagements im Grünflächenbereich vor dem Hintergrund des Wandels von Staat und Planung. In *Dissertation an der Mathematisch-Naturwissenschaftlichen Fakultät der Humboldt-Universität zu Berlin.* https://doi.org/10.1109/IGARSS.2015.7326091.

RUAF Foundation (Ed.). no date. Urban agriculture: What and why?" Accessed on February 2, 2019. https://www.ruaf.org/urban-agriculture-what-and-why.

Santo, R., Palmer, A., & Kim, B. (2016). *Vacant Lots to Vibrant Plots. A review of the benefits and limitations of Urban Agriculture.* Johns Hopkins Center for a Livable Future.

Sanyé-Mengual, E., Specht, K., Krikser, T., Vanni, C., Pennisi, G., Orsini, F., & Prosdocimi Gianquinto, G. (2018). Social acceptance and perceived ecosystem services of urban agriculture in Southern Europe: The case of Bologna, Italy. *PLOS ONE, 13*(9), e0200993. https://doi.org/10.1371/journal.pone.0200993.

Schmidt, D. (2016). Die Rolle der urbanen Landwirtschaft in der Stadtentwicklung: Übersicht und Umgang mit neuen Formen anhand von Fallbeispielen. In *Masterarbeit an der Fakultät Umweltwissenschaften der Technischen Univesität Dresden.*

Schulz, K., Weith, T., Bokelmann, W., & Petzke, N. (2013). *Urbane Landwirtschaft und "Green Production" als Teil eines nachhaltigen Landmanagements. Diskussionspapier No. 6. Leibniz-Zentrum für Agrarlandschaftsforschung (ZALF) e.V.* Müncheberg.

Siedentop, S. (2010). INNENENTWICKLUNG/AUSSENENTWICKLUNG". In D. Henckel, von Kuczkowski, K., Lau, P., Pahl-Weber, E., & Stellmacher, F. (Eds.), *Planen – Bauen – Umwelt* (pp. 235–240). Wiesbaden: VS Verlag für Sozialwissenschaften. https://doi.org/10.1007/978-3-531-92288-1_1.

Simons, H., & Weiden, L. (2016). Schwarmverhalten, Reurbanisierung und Suburbanisierung. *Informationen zur Raumentwicklung, 3,* 263–273.

Smit, J., Nasr, J., & Ratta, A. (2001). *Urban agriculture: Food, jobs and sustainable cities.* UNDP. New York: The Urban Agriculture Network, Inc.

Sondermann, M. (2014). Local cultures of urban gardening and planning in Germany. In *Riga event report: Urban allotment gardens in European cities – Future, challenges and lessons learned (COST Action TU1201).* Accessed on June 20, 2020. https://www.urbanallotments.eu/fileadmin/uag/media/D_Meetngs/Riga/Riga_Report_Final_NK.pdf

Specht, K., Reynolds, K., & Sanyé-Mengual, E. (2017). Community and social justice aspects of rooftop agriculture. In F. Orsini, M. Dubbeling, H. de Zeeuw, & G. Gianquinto (Eds.), *Rooftop urban agriculture* (pp. 277–290). Cham: Springer International Publishing. https://doi.org/10.1007/978-3-319-57720-3_17.

Specht, K., & Siebert, R. (2017). Städtische Landwirtschaft in, an und auf Gebäuden: Möglichkeiten für die Stadtentwicklung, Handlungsfelder und Akteure. In S. Kost & C. Kölking (Eds.), *Transitorische Stadtlandschaften: welche Landwirtschaft braucht die Stadt?* (pp. 95–114). Wiesbaden: Springer.

Stierand, P. (2012). Stadtentwicklung mit dem Gartenspaten. Umrisse einer Stadternährungsplanung. Dortmund. https://speiseraeume.de/downloads/SPR-Stadternaehrungsplanung-Stierand.pdf.

Thibert, J. (2012). Making local planning work for urban agriculture in the North American context. *Journal of Planning Education and Research, 32*(3), 349–357. https://doi.org/10.1177/0739456X11431692

UNDP (Ed.). (1996). *Urban Agriculture. Food, jobs and sustainable cities*. New York.

van der Jagt, A. P. N., Szaraz, L. R., Delshammar, T., Cvejić, R., Santos, A., Goodness, J., & Buijs, A. (2017). Cultivating nature-based solutions: The governance of communal urban gardens in the European Union. *Environmental Research, 159,* 264–275. https://doi.org/10.1016/j.envres.2017.08.013

van der Schans, J., Willem, W. L., Alfranca, Ó, Alves, E., Andersson, G., Branduini, P., et al. (2016). It is a business! Business models in urban agriculture. In F. Lohrberg, L. Lička, L. Scazzosi, & A. Timpe (Eds.), *Urban agriculture Europe* (pp. 82–91). Berlin: Jovis.

Vejre, H., Eiter, S., Hernández-Jiménez, V., Lohrberg, F., Loupa-Ramos, I., Recasens, X., et al. (2016). Can agriculture be urban? In F. Lohrberg, L. Lička, L. Scazzosi, & A. Timpe (Eds.), *Urban agriculture Europe* (pp. 18–21). Berlin: Jovis.

Viljoen, A., Bohn, K., & Howe, U. (Eds.). (2005). *Continuous productive urban landscapes: Designing urban agriculture for sustainable cities*. London: Elsiever Architectural Press.

Voicu, I., & Been, V. (2008). The effect of community gardens on neighboring property values. *Real Estate Economics, 36*(2), 241–283.

Zoll, F., Specht, K., Opitz, I., Siebert, R., Piorr, A., & Zasada, I. (2018). Individual choice or collective action? Exploring consumer motives for participating in alternative food networks. *International Journal of Consumer Studies*, 101–110. https://doi.org/10.1111/ijcs.12405.

Chapter 16
Integrating Ecosystem Services, Green Infrastructure and Nature-Based Solutions—New Perspectives in Sustainable Urban Land Management

Combining Knowledge About Urban Nature for Action

Dagmar Haase

Abstract Global urbanisation comprises both urban sprawl and increasing densification of existing cities. Along with the heat waves, floods and droughts associated with climate change, urbanisation challenges our cities, and thus the places where soon 60% of the world's population will live. In addition to human beings and their health, nature and biodiversity are under extreme pressure to function and to survive in these growing urban systems. More and more key biodiversity areas (KBAs) are becoming urbanised, and wetlands are being sealed. However, ecosystems are crucial for a healthy and safe life in cities. So how should we save urban nature as a habitat for humans, flora and fauna? This chapter presents three concepts that provide different perspectives for sustainable urban land management. They represent complementary paths to increased urban sustainability. Nonetheless, implementation is still a long way off, and moreover, unsolved issues still exist, such as the social inclusiveness of the three approaches.

Keywords Ecosystem services · Green infrastructure · Nature-based solutions · Complementary approaches for sustainable land use

16.1 Challenges in Urban Land Management: The Case of European Cities

Urbanisation and urban growth are two overarching phenomena in land use development affecting areas around the planet. Worldwide, more than 55% of the population lives and works in cities, and this trend does not seem to be subsiding (Haase et al.

D. Haase (✉)
Department of Geography, Humboldt Universität Zu Berlin, Unter den Linden 6, 10099 Berlin, Germany
e-mail: dagmar.haase@geo.hu-berlin.de; dagmar.haase@ufz.de

Department of Computational Landscape Ecology, Helmholtz Centre for Environmental Research—UFZ, Permoserstrasse 15, 04318 Leipzig, Germany

© The Author(s) 2021
T. Weith et al. (eds.), *Sustainable Land Management in a European Context*,
Human-Environment Interactions 8, https://doi.org/10.1007/978-3-030-50841-8_16

305

2018). Europe, a continent that became urbanised relatively early, is stagnating in population growth terms, but cities as such are becoming attractive places to move to (Scheuer et al. 2016; Wolff et al. 2018). Indeed, land take in and around cities is not only not subsiding in Europe—it is accelerating. In addition, when considering the per capita living space increase over the past few decades along with the average decrease in household sizes in Europe (Haase et al. 2013), land has become a scarce resource in cities. Recent construction activities are no longer exclusively concentrated on the urban periphery; on the contrary, densification of inner-city areas and infill development are high on the agenda (Wolff and Haase 2019).

Densification by infill development automatically leads to a decline and a partial complete disappearance of (spots of) nature in city centres (Haase et al. 2018), despite the fact that such areas are often high-value nature areas with a rich biodiversity, due to the wetland and riverine locations of many cities (Kühn et al. 2004). At the same time, we still find peri-urbanisation and land take outside the city cores on formerly arable ground, resulting in a decline in fertile land (Nilsson et al. 2014). Thus, the face of urban growth in European cities is multifaceted and does not include the considerable percentage of cities and towns in Europe that are shrinking (Wolff et al. 2018).

While growing and densifying, cities also face the direct consequences of ongoing climate change, such as long-lasting and early heat waves (as the summers of 2018 and 2019 recently demonstrated) including "tropical night" temperatures exceeding 20 °C. This is clearly a challenge for urban public health, in particular for an ageing urban population in a densely built area (Bosch and Sang 2017). At the same time, high daytime temperatures and continuous irradiance are a challenge for urban tree and shrub vegetation, which already suffers from the lack of rainfall. Therefore, heat has become one of the key challenges for entire urban systems in Europe, including the environment, public health and the economy, especially when considering cities that attract (mass) tourism (such as Vienna, Rome and Berlin).

In addition to heat, an increasing risk of flooding in lowland and coastal cities (Barcelona and Genoa after heavy rainfall, as well as Bosnia, Croatia or Germany after heavy rainfall and stationary depressions in the past decade) appears to be another key challenge for European cities (Scheuer et al. 2017). As cities increasingly accumulate economic value and, of course, human life, the frequency and degree of hazardous flood events need to be incorporated into a more sustainable and flood-proof urban land management (Krysanova et al. 2008). The case of drought and water shortages is similar, which have recently often alternated with floods: most European cities are not well prepared for longer-term water shortage and extreme irradiance.

However, cities in Europe are also places of great vestiges of nature (Haase and Gläser 2009). In addition to the above-mentioned wetlands and riverine land strips, cities harbour old forests, large parks, numerous gardens and green backyards. Recently, urban green ground infrastructure has been complemented by "vertical green" such as green rooftops and living walls (Pauleit et al. 2018). Moreover, we know about the positive effects of urban green and blue spaces in cities when dealing with high air temperatures and irradiance (Weber et al. 2014a). We know the positive effects of green space for public health and the prevention of heart or lung disease, as well as "lifestyle diseases", such as obesity and diabetes (very frequent in cities with a high poverty rate), and mental disease, such as depression or anxiety disorders (Gruebner and McCay 2019; Gruebner et al. 2017).

Thus, the major research question guiding this chapter of the book will be: How we can make use of urban nature and knowledge about nature to protect human life and, at the same time, protect nature from severe and hazardous conditions and events? Are there forms of urban land management that allow us to effectively and sensibly harness nature for human benefits, leading to more sustainable urban land use?

This chapter provides novel insights by discussing various concepts and the potential to integrate them into cities.

16.2 Three Concepts for One Goal

The next few pages will introduce three different approaches and concepts dealing with urban nature for sustainable cities:

- Urban ecosystem services (demand, flow, supply; Haase et al. 2014),
- Green infrastructure and green infrastructure types (Pauleit et al. 2018) and
- Nature-based solutions (Nesshoever et al. 2017).

All three concepts are interrelated and have a complementary character to a certain degree (Table 16.1 and Fig. 16.1). Urban ecosystem services (ES) focus on the processes and structure of urban nature and the beneficial effects of ecosystem process outcomes for people—in the case of this chapter, urban residents and urban society as a whole (Haase et al. 2014). Urban green infrastructure (UGI) can be understood as a strategic planning approach that takes these functional benefits of ES for "granted"; it thus aims to develop networks of green and blue spaces in urban areas, designed and managed to deliver a wide range of ES and other benefits at all spatial scales (Pauleit et al. 2018; EEA website). Finally, the concept of nature-based solutions (NBS) focuses on problems and challenges of an environmental or a social nature. NBS harnesses the ES functional approach and the design concept of green (blue) infrastructure to adapt both ES and UGI to the distinct and specific needs of cities. NBS, therefore, can be defined as living solutions that are inspired and supported by nature, which are cost-effective, whilst simultaneously providing environmental, social and economic benefits and helping to increase resilience and adaptation to climate change (Kabisch et al. 2017).

Table 16.1 Core properties of the three "green approaches" to sustainable urban land management (own conceptualisation and content compilation)

	Urban ecosystem services	Urban green infrastructure	Nature-based solutions for cities
Basic response or "working" units	Ecosystems (patterns and processes) and elements of them, such as soils, the water cycle and trees in an urban environment	Vegetation and vegetation types, their design and management in a city	Materials, structures and processes that function as, or like, ecosystems
How the approach works, or the idea behind it	Outcomes of ecosystem processes represent flows of material or energy that facilitate human life in cities, e.g. temperature cooling or water purification by soil sediment fixation	Elements of vegetation are planted and/or designed as well as maintained to make use of their ecosystem service flows for human well-being	Elements of nature are either used or constructed (mimicry) to produce ecosystem service flows to address issues related to climate change (solve the temperature problem) or facilitate human life in cities
Role of society	Beneficiaries of flows from ecosystem services at both individual and societal level; reduction of replacement costs	Users of the green infrastructure, whether as recreational users in parks or as urban gardeners (to provide two examples)	Active engagement in the (co-)development and (co-)design of nature (mimicry) and monitoring NBS success
State of implementation	Partly in implementation in cities; still criticism of the concept; ES indicators are in proper use in most urban planning departments across Europe	Widely implemented and refined in European cities; suffers from limited municipal budgets, but is also implemented through NGO and citizen-based activities and programmes	Novel approach, with most implementations in flood management and climate adaptation in bigger cities across Europe, less in food production or environmental education

16.3 Ecosystem Services, or the Benefits Nature Provides to Urban Populations

What are ecosystem services? Urban green and blue spaces deliver a number of ecosystem services (ES) that contribute to maintaining the physical and mental health of urban dwellers, improving their quality of life. Urban ecosystems in cities provide regulatory (air temperature and humidity regulation), cultural (recreation, tourism) and basic provisioning services (food, forage) to people (Haase et al. 2014; Fig. 16.2). Accordingly, healthy ecosystems deliver these services to a proper extent; degraded ones to a much lower extent, if at all (McPhearson et al. 2016).

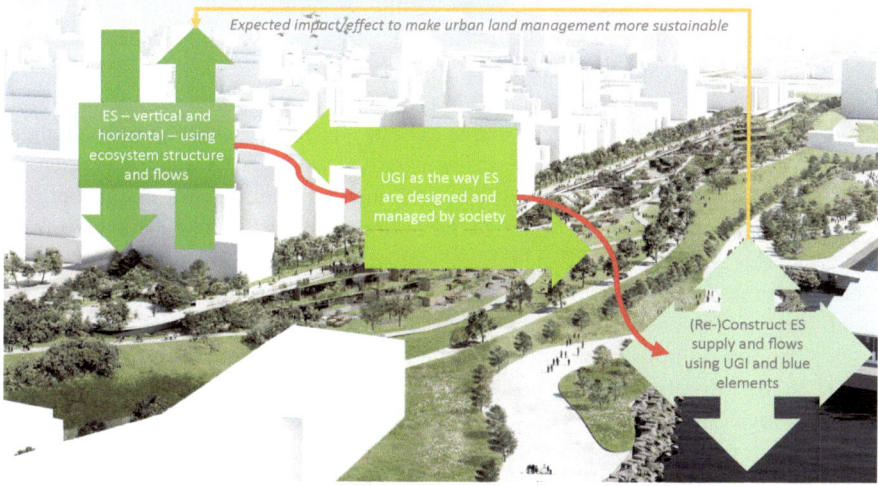

Fig. 16.1 Main links, partial overlaps and the differences between the three concepts (own sketch)

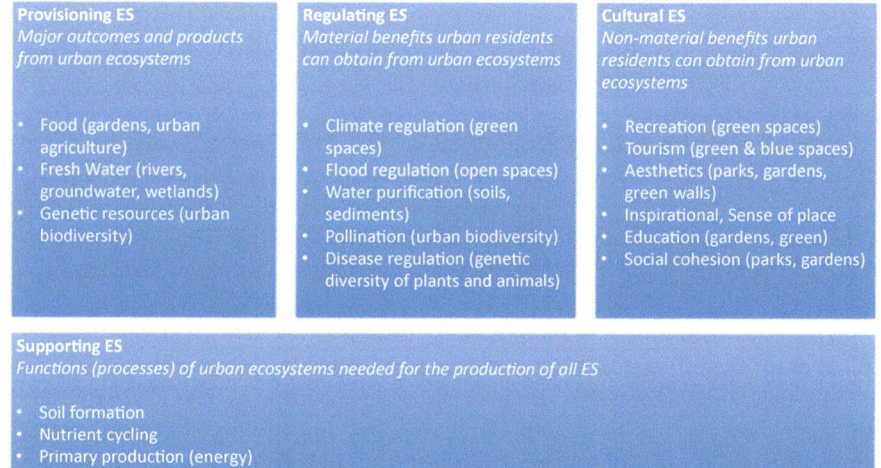

Fig. 16.2 Urban ES classifications with examples for typical urban infrastructures providing the respective services (own compilation)

The increasing frequency of heat waves have confronted Europe's urban residents with very high day and "tropical night" (>20 °C) temperatures (Weber et al. 2014a); moreover, they are exposed to particulate matter and traffic noise (Weber et al. 2014b). These environmental pressures can impair human health and result in higher illness, morbidity and mortality rates, as well as impaired mental health (Adli 2017). Europe's growing elderly population is particularly vulnerable to these problems (Gruebner et al. 2017). Illnesses caused by heat and pollution dramatically limit the quality of life in cities and incur major costs to urban society, especially for healthcare, as well as reducing labour capacity.

Regulatory ES provided by intact ecosystems definitely and effectively help to minimise these environmental pressures (TEEB Germany 2017): During spring and summer heat waves, such as Europe has experienced in 2018 and 2019, there is a significant increase in illness, morbidity and mortality rates (Gabriel and Endlicher 2011). For example, estimates of up to 5% of deaths in the city of Berlin are linked to heat (Gabriel and Endlicher 2011). Urban vegetation such as trees as well as various grasslands and meadows can significantly reduce peak summer temperatures (Weber et al. 2014a). Records show that a green space measuring 50 to 100 m wide is up to 3 °C cooler on hot, wind-still days than the surrounding developed area (Pauleit et al. 2018). Moreover, green spaces have a cooling impact on their direct urban surroundings (Andersson et al. 2019). In addition to heat relief, urban green spaces play a major role in air pollution control (Pauleit et al. 2018). Trees filter particulates by between 5 and 15%, depending on height, density and configuration (Weber et al. 2014a). In residential neighbourhoods, nature is especially beneficial to human health, as green spaces invite residents to spend time outdoors and to participate in active recreation such as sports, games, or even passive nature enjoyment and relaxation (Rall et al. 2017). A number of studies have provided very good and clear evidence that being outdoors supports reductions of aggression and anxiety, and, vice versa, raises concentration and performance levels across all age groups (Bosch and Sang 2017).

In terms of urban society and the social life in cities, which are also core concerns of urban land management, healthy ecosystems contribute to strengthening social cohesion by providing "aesthetic places for communication" (Kremer et al. 2016). When freely accessible, urban parks, gardens, rivers and lakes serve as refuges for urban residents to go to for multiple leisure and social activities with family and friends (Voigt et al. 2014). Allotments and community gardens facilitate encounters, joint activities and intercultural exchange (Pauleit et al. 2018). Growing local food in the city—be it in different types of gardens, on balconies or in abandoned cemeteries—increases urban self-sufficiency (Rodríguez-Rodríguez et al. 2015), and, at the same time, raises awareness about regional and healthy food (counteracting problems such as obesity among children and adults). Thus, recreational ecosystem services contribute to urban public health in multiple ways. However, these ES only arise if all groups of residents see these aforementioned green spaces as available, accessible, and attractive (Biernacka and Kronenberg 2018). With respect to the last of these, one key component of this attractiveness of green spaces is biodiversity— and is something that park users recognise (Fischer et al. 2018). This is a clear signal for more and better (more consistent) nature conservation in cities for ensuring the delivery of necessary ES.

Many of the aforementioned ES that nature delivers in cities are to a large extent neglected or simply ignored by urban planners and decision-makers dealing with land use and urban landscape/surface design (Kain et al. 2016; Kaczorowska et al. 2016; TEEB Germany 2017). Thus, the ES concept that is proposed here is a tool focusing

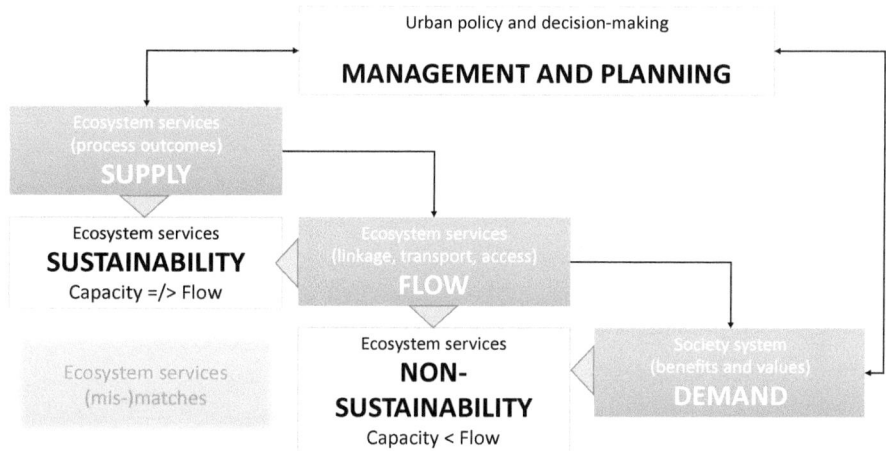

Fig. 16.3 Adapted cascade of urban ES supply, flow, and demand, and its incorporation into land management [building on earlier diagrams by Baró et al. (2017), Potschin and Haines-Young (2011), Villamagna et al. (2013) and Geijzendorffer et al. (2015)]

on the functional outcomes of nature's processes in urban areas; it can be used as both a planning and a monitoring tool in urban decision-making for fairer, more sustainable land use in our growing cities to balance density, social-environmental segregation and species loss (Fig. 16.3; McDonald et al. 2019).

16.4   Designing nature's Benefits into Green Urban Infrastructure in Cities

A second approach that appears promising for more sustainable urban land use through management and design is the urban green infrastructure (UGI) approach (Pauleit et al. 2018). The idea behind UGI is based on the principle that protecting and enhancing nature and natural processes are consciously integrated into urban spatial planning. UGI, in this sense, can be framed as a strategically planned network of (semi-)natural areas together with other natural features designed and managed to deliver a wide range of ES in the urban context (EEA 2019).

In contrast to common human-made, means-constructed, urban infrastructure approaches that often serve a single purpose, UGI's "living system" character entails multifunctionality; the elements or types of UGI can offer multiple benefits and flows of benefits—urban ecosystem services—provided that ecosystems are in a healthy state (Pauleit et al. 2018; Andersson et al. 2019): A single park supports not only climate change adaptation and mitigation, but also active and passive recreation, including educational benefits, and increases species biodiversity (Andersson et al. 2015; Rall et al. 2017). The multifunctional performance of such single infrastructure

Fig. 16.4 Types of UGI allocated by a multifunctional element of UGI—the urban tree. UGI can provide multiple benefits if it is healthy; if not, no flows of ES can be expected (tree by https://gun nisontree.com/tell-tree-dead-just-needs-water/)

units supports a more sustainable yet still resource-efficient urban land development process in European cities, where both space and resources are limited (Andersson et al. 2019).

UGI comprises a wide range of environmental features that operate at different scales—from the neighbourhood to the region—and in the best case these features form part of an interconnected ecological of new green infrastructure and other sustainability investments in cities have to accrue to positive outcomes for low-income and underprivileged residents as well, respecting their ideas and recreational needs equal to that of the wealthier part of urban society, which dominates discourse (Haase et al., 2017) (Fig. 16.4).

16.5   ES and UGI as Nature-Based Solutions to Urban Land Management Challenges?

A third approach has also started to emerge, making use of urban nature for more sustainable land management in cities and urban regions: nature-based solutions (NBS). According to the IUCN, NBS are defined as "actions to protect, sustainably manage, and restore natural or modified ecosystems, that address societal challenges effectively and adaptively, simultaneously providing human well-being and biodiversity benefits" (Cohen-Shacham et al. 2016). NBS are intended to support attaining society's development goals and safeguarding human well-being in ways that (a) reflect the cultural and societal values of a multi-origin urban society, and (b) enhance the resilience of urban ecosystems, and their capacity to provide the aforementioned ES (Kabisch et al. 2016a, b). NBS are designed nature—similar to UGI—that are implemented to address the urban challenges listed in the introduction of this chapter: food security, climate change, water shortage, human health, and disaster risk (Nesshoever et al. 2017).

NBS are based on both the ES and UGI concepts, but are novel in that they are conceptualised and implemented (Table 16.2): NBS always address a specific urban challenge, such as shown in Fig. 16.5, using the single planted tree as an example. NBS can be implemented as individual measures, or in an integrated manner combined with additional "grey" (i.e. technological, engineering or digital) solutions to urban challenges. Compared to city-wide ES flows and UGI networks,

Table 16.2 Classification of NBS in cities (modified from Cohen-Shacham et al. 2016)

Category of NBS Approaches	Examples from urban land management
Restoration NBS approaches	• Ecological restoration of wetlands, riparian forests and brownfields (including natural succession of grasslands) • Ecological engineering (co-creation of new parks at brownfield sites) • Forest landscape restoration (reforestation of former forest sites and afforestation of urban brownfields)
Adaptation NBS approaches	• Ecosystem-based adaptation (using functional adaptation and mutation properties of ecosystems, such as adapted species or populations) • Ecosystem-based mitigation • Climate adaptation ecosystem services (using the transpiration and evaporation functions of vegetation and soils) • Ecosystem-based disaster risk reduction (retention properties of open soil and natural wetlands)
Infrastructure NBS approaches	• Blue infrastructure (design of water-depending sites such as ponds or constructed wetlands) • Green infrastructure (design of parks, gardens, green roofs and walls)
Management NBS approaches	• Integrated coastal zone management (stormwater zones and coastal dune protection) • Integrated water resources management (constructed wetlands, bioswales, rain gardens at rooftop level, river revitalisation, floodplain de-sealing)
Conservation NBS approaches	• Locally based nature and biodiversity conservation approaches, including management of protected areas (urban national parks and biosphere reserves, nature playgrounds, beekeeping in cities, old tree maintenance)

Urban Green Infrastructure	Urban Ecosystem Services (supply)	Urban Ecosystem Services (flow)	Urban Ecosystem Services (demand)
	Evapotranspiration	*CC adaptation*	*Less air cond. Costs*
Tree	*Carbon sequestration*	*Air cooling (cool air)*	*CO_2 market: net win*
	Particle filtering	*Air quality (fresh air)*	*Less health costs*

Fig. 16.5 How an urban NBS works and how it can be related to the concepts of urban ES and UGI (own sketch)

NBS are often determined by site-specific natural and social-cultural contexts. NBS recognise and address existing trade-offs between the production of a few immediate health or economic benefits or risk reduction, and future (time-dependent) options for the production of the full range of ES flows and UGI network habitat and population-related effects, again as shown in Fig. 16.5, using the single planted tree as a multifunctional and long-living example (Nesshoever et al. 2017).

A recent review study reports, on the one hand, that, despite a lack of consensus about a single "final" definition of NBS, there is a shared understanding among European stakeholders that the NBS concept encompasses human and ecological benefits beyond the core objective of ecosystem conservation, restoration or enhancement. On the other hand, the study also reveals that resources are often limited in city municipalities, and each city has different needs. This makes it critical to prioritise the challenges NBS is to address during the urban land use planning process (Ershad Sarabi et al. 2019).

16.6 Conclusions for Sustainable Urban Land Management in the Future

The absolute strength of the three concepts and approaches introduced here lies in their combination and complementarity of functionality, design, management and straightforward implementation, as well as problem-based orientation to make urban land management more sustainable. Supply and demand as well as flows of nature are central in all three concepts. The complementary concepts link different disciplines and disciplinary strengths, bringing them all together towards a new approach in sustainable urban land management.

A clear weakness of all three approaches is that they neither include nor address one of the most crucial urban social and democracy-related questions of today: justice and fairness questions at the local—i.e. city—level are almost neglected. At the global level, telecouplings have not even been touched (Haase 2019), and thus urbanisation at the global level is difficult to tackle with any of the three concepts, although papers have already been published on global principles and upscaling from single cities and urban areas.

Acknowledgements Dagmar Haase's research was supported by Project ENABLE, funded via the 2015–2016 BiodivERsA COFUND call for research proposals. National funders of the project were the Swedish Research Council for Environment, Agricultural Sciences, and Spatial Planning, the Swedish Environmental Protection Agency, the German Aeronautics and Space Research Centre, the National Science Centre (Poland), the Research Council of Norway and the Spanish Ministry of Economy and Competitiveness.

References

Adli, M. (2017). *Stress and the City: Warum Städte uns krank machen.* Und warum sie trotzdem gut für uns sind. C: Bertelsmann Verlag.

Andersson, E., McPherson, T., Kremer, P., Frantzeskaki, N., Gomez-Baggethun, E., Haase, D., et al. (2015). Scale and context dependence of ecosystem service providing units. *Ecosystem Services, 12,* 157–164.

Andersson, E., Langemeyer, J., Borgström, S., McPhearson, T., Haase, D., Kronenberg, J., et al. (2019). Enabling green and blue infrastructure to improve contributions to human well-being and equity in urban systems. *BioScience*, in press. https://doi.org/10.1093/biosci/biz058.

Baró, F., Gómez-Baggethun, E., & Haase, D. (2017). Ecosystem service bundles from a supply-demand approach: Implications for landscape planning and management in an urban region. *Ecosystem Services, 24,* 147–159. https://doi.org/10.1016/j.ecoser.2017.02.021

Biernacka, M., & Kronenberg, J. (2018). Classification of institutional barriers affecting the availability, accessibility and attractiveness of urban green spaces. *Urban Forestry and Urban Greening, 36,* 22–33. https://doi.org/10.1016/j.ufug.2018.09.007

Bosch, M., & Sang, A. O. (2017). Urban natural environments as nature-based solutions for improved public health—A systematic review of reviews. *Environmental Research, 158,* 373–384.

Cohen-Shacham, E., Walters, G., Janzen, C., & Maginnis, S. (Eds.). (2016). *Nature-based solutions to address global societal challenges.* Gland, Switzerland: IUCN. https://doi.org/10.2305/IUCN.CH.2016.13.en.

EEA website. (2019). https://www.eea.europa.eu/themes/sustainability-transitions/urban-enviro nment/urban-green-infrastructure/what-is-green-infrastructure.

Ershad Sarabi, S., Han, Q., Romme, L. A. G., de Vries, B., & Wendling, L. (2019). Key enablers of and barriers to the uptake and implementation of nature-based solutions in urban settings: A review. *Resources, 8,* 121. https://doi.org/10.3390/resources8030121.

Fischer, L. K., Brinkmeyer, D., Honold, J., van der Jagt, A., Botzat, A., Lafortezza, R., et al. (2018). Recreational ecosystem services in European cities: Sociocultural and geographic context matters for park use. *Ecosystem Services.* https://doi.org/10.1016/j.ecoser.2018.01.015

Gabriel, K., & Endlicher, W. (2011). Urban and rural mortality rates during heat waves in Berlin and Brandenburg. *Germany. Environmental Pollution, 159*(8–9), 2044–2050. https://doi.org/10.1016/j.envpol.2011.01.016

Geijzendorffer, I. R., Martín-López, B., & Roche, P. K. (2015). Improving the identification of mismatches in ecosystem services assessments. *Ecological Indicators, 52,* 320–331.

Gruebner, O., & McCay, L. (2019). Urban design. In S. Galea, C. Ettman, & D. Vlahov, D. (Eds.), *Urban health*. Oxford University Academic Press, forthcoming.

Gruebner, O., Rapp, M. A., Adli, M., Kluge, U., Galea, S., & Heinz, A. (2017). Cities and mental health. *Deutsches Ärzteblatt International, 114,* 121–127. https://doi.org/10.3238/arztebl.2017.0121

Haase, D. (2019). Urban telecouplings. In C. Friis & J.Ø. Nielsen (Eds.), *Telecoupling*. Palgrave Studies in Natural Resource Management. https://doi.org/10.1007/978-3-030-11105-2_14.

Haase, D., & Gläser, J. (2009). Determinants of floodplain forest development illustrated by the example of the floodplain forest in the District of Leipzig. *Forest Ecology and Management, 258,* 887–894. https://doi.org/10.1016/j.foreco.2009.03.025

Haase, D., Kabisch, N., & Haase, A. (2013). Endless urban growth? On the mismatch of population, household and urban land area growth and its effects on the urban debate. *PLoS ONE, 8*(6), e66531. https://doi.org/10.1371/journal.pone.006653

Haase, D., Larondelle, N., Andersson, E., Artmann, M., Borgström, S., Breuste, J., et al. (2014). A quantitative review of urban ecosystem services assessment: Concepts, models and implementation. *Ambio, 43*(4), 413–433.

Haase, D., Kabisch, S., Haase, A., Larondelle, N., Schwarz, N., Wolff, M., et al. (2017). Greening cities—To be socially inclusive? About the paradox of society and ecology in cities. *Habitat International*. https://doi.org/10.1016/j.habitatint.2017.04.005.

Haase, D., Guneralp, B., Bai, X., Elmqvist, T., Dahiya, B., Fragkias, M., & Gurney, K. (2018). Different pathways of global urbanization. In T. Elmqvist, X. Bai, N. Frantzeskaki, C. Griffith, D. Maddox, T. McPhearson, et al. (Eds.), *The urban planet: Patterns and pathways to the cities we want*. Cambridge: Cambridge University Press.

Kabisch, N., Bonn, A., Korn, H., & Stadler, J. (Eds.). (2017). *Nature-based solutions to climate change in urban areas-linkages of science, society and policy*. Berlin: Springer.

Kabisch, N., Frantzeskaki, N., Pauleit, S., Naumann, S., Davis, M., Artmann, M., et al. (2016a). Nature-based solutions to climate change mitigation and adaptation in urban areas—perspectives on indicators, knowledge gaps, barriers and opportunities for action. *Ecology and Society, 21*(2), 39. https://doi.org/10.5751/ES-08373-210239.

Kabisch, N.; Strohbach, M.; Haase, D., & Kronenberg, J. (2016b). Green space availability in European cities. *Ecological Indicators, 70,* 586–596. https://doi.org/10.1016/j.ecolind.2016.02.029

Kaczorowska, A., Kain, J.-H., Kronenberg, J., & Haase, D. (2016). Ecosystem services in urban land use planning: Integration challenges in complex urban settings—Case of Stockholm. *Ecosystem Services, 22,* 204–212.

Kain, J. H., Larondelle, N., Haase, D., Rodríguez Rodríguez, D., & Kaczorowska, A. (2016). Land use scenarios exploring local consequences for supply of urban ecosystem services in Stockholm year 2050. *Ecological Indicators, 70,* 615–629. https://doi.org/10.1016/j.ecolind.2016.02.062

Kremer, P., Hamstead, Z., Haase, D., McPhearson, T., Frantzeskaki, N., Andersson, E., et al. (2016). Key insights for the future of urban ecosystem services research. *Ecology and Society, 21* (2), 29 [online]. https://www.ecologyandsociety.org/vol21/iss2/art29/.

Krysanova, V., Buiteveld, H., Haase, D., Hattermann, F. F., Van Niekerk, K.; Roest, K., et al. (2008). Practices and lessons learned in coping with climatic hazards at the river-basin scale: Floods and droughts. *Ecology and Society, 13*(2), 32. https://www.ecologyandsociety.org/vol13/iss2/art32/.

Kühn, I., Brandl, R., & Klotz, S. (2004). The flora of German cities is naturally species rich. *Evolutionary Ecology Research, 458*(6), 749–764.

McDonald, R, Mansur, A. V., Ascensão, F., Colbert, M., Crossman, K., Elmqvist, T., et al. (2019). The growing impacts of cities on biodiversity. Research gaps limit global decision-making. *Nature Sustainability*. https://doi.org/10.1038/s41893-019-0436-6.

McPhearson, T., Pickett, S., Grimm, N., Niemelä, J., Elmqvist, T., Weber, C., et al. (2016). Ecology for an urban planet: Advancing research and practice towards a science of cities. *BioScience.* https://doi.org/10.1093/biosci/biw002.

Nesshoever, C., Assmuth, T., Irvine, K. J., Rusch, G. M., Waylen, K. A., Delbaere, B., et al. (2017). The science, policy and practice of Nature-Based Solutions: An interdisciplinary perspective. *Science Total Environment, 579,* 1215–1227. https://doi.org/10.1016/j.scitotenv.2016.11.106

Nilsson, K., Nielsen, T. S., Aalbers, C., Bell, S., Boitier, B., Chery, J.-P., et al. (2014). Strategies for sustainable urban development and urban-rural linkages, research brief. *European Journal of Spatial Development.*

Pauleit, S., Olafsson, A. S., Rall, E., van der Jagt, A., Ambrose-Oji, B., Andersson, E., et al. (2018). Urban green infrastructure in Europe—Status quo, innovation and perspectives. *Urban Forestry and Urban Greening.* https://doi.org/10.1016/j.ufug.2018.10.006

Potschin, M. B., & Haines-Young, R. H. (2011). Ecosystem services: Exploring a geographical perspective. *Progress in Physical Geography, 35,* 575–594.

Rall, E., Bieling, C., Zytynska, S., & Haase, D. (2017). Exploring city-wide patterns of cultural ecosystem service perceptions and use. *Ecological Indicators, 77,* 80–95. https://doi.org/10.1016/j.ecolind.2017.02.001

Rodríguez-Rodríguez, D., Kain, J. H., Haase, D., Baró, F., Frantzeskaki, N., & Kaczorowska, A. (2015). Urban self-sufficiency through optimised ecosystem service demand. A utopian perspective from European cities. *Futures, 70,* 13–23. https://doi.org/10.1016/j.futures.2015.03.007

Scheuer, S., Haase, D., & Volk, M. (2016). On the nexus of the spatial dynamics of global urbanization and the age of the city. *PLoSONE, 11*(8). https://doi.org/10.1371/journal.pone.0160471.

Scheuer, S., Haase, D., & Volk, M. (2017). Fastest-growing urban areas as hotspots of change: 20th century climate trends and urbanization call for co-management of global change in cities. *PLoS ONE, 12*(12). https://doi.org/10.1371/journal.pone.0189451.

TEEB Germany. (2017). Ecosystem services in the city. Protecting health and enhancing quality of life. https://www.ufz.de/export/data/global/190507_TEEB_De_Broschuere_KF_Bericht3_Stadt_engl_web.pdf. Accessed on July 8, 2019.

Villamagna, A. M., Angermeier, P. L., & Bennett, E. M. (2013). Capacity, pressure, demand, and flow: A conceptual framework for analyzing ecosystem service provision and delivery. *Ecological Complexity, 15,* 114–121.

Voigt, A., Kabisch, N., Wurster, D., Haase, D., & Breuste, J. (2014). Structural diversity as a key factor for the provision of recreational services in urban parks—A new and straightforward method for assessment. *Ambio, 43*(4), 480–491.

Weber, N., Haase, D., & Franck, U. (2014a). Assessing traffic-induced noise and air pollution in urban structures using the concept of landscape metrics. *Landscape and Urban Planning, 125,* 105–116.

Weber, N., Haase, D., & Franck, U. (2014b). Zooming into the urban heat island: How do urban built and green structures influence earth surface temperatures in the city? *Science of the Total Environment, 496,* 289–298.

Wolff, M., & Haase, D. (2019). Mediating sustainability and liveability—Turning points of green space supply in European cities. *Frontiers in Environmental Science, Section Land Use Dynamics,* in press. https://doi.org/10.3389/fenvs.2019.00061.

Wolff, M., Haase, D., & Haase, A. (2018). Less dense or more compact? Discussing a density model of urban development for European urban areas. *PLOS ONE.* Published: February 28, 2018. https://doi.org/10.1371/journal.pone.0192326.

Chapter 17
Upcoming Challenges in Land Use Science—An International Perspective

Christine Fürst

Abstract This chapter provides an overview on relevant concepts, such as ecosystem services, sustainability, multifunctionality and social-ecological systems/frameworks applied in land use sciences. Current discussions, political debates and challenges in terms of methodological aspects, actor enrolment or project design are raised. Future research topics particularly related to the often non-coherent UN Sustainable Development Goals and their mutual trade-offs are raised and challenges in how to advance land use science are provided. An outlook is provided how co-development of knowledge and co-design of land use system research could be conceived in the future.

Keywords Land use science · Social-ecological systems · Co-development · Co-design · Integrated modelling and assessment

17.1 Introduction

Land use science or land system science turned out to be one of the key concepts in the recent years to integrate an inter- and transdisciplinary perspective in the sustainable management of our earth's natural resources and ecosystems (Müller and Monroe 2014; Rounsevell et al. 2012; Verburg et al. 2013, 2015). Disciplines that found entrance in this concept include geography, landscape ecology, environmental economics, behavioural sciences, social sciences, and biology, to name only few (Zscheischler and Rogga 2015). A key aspect of land use science consists in co-developing new and integrative methods in systemic modelling, multi-actor participation and in the appraisal of potential social-ecological impacts of changes in land use and land management (Fürst et al. 2017).

A number of *concepts* is closely related and meanwhile part of the methodological toolkit of land use science that should be shortly defined at the beginning. The concepts social-ecological systems and the major approaches for assessing the impact

---

C. Fürst (✉)
Institute for Geosciences and Geography, Dept. Sustainable Landscape Development, Martin Luther University Halle-Wittenberg, Von-Seckendorff-Platz 4, 06120 Halle (Saale), Germany
e-mail: christine.fuerst@geo.uni-halle.de

T. Weith et al. (eds.), *Sustainable Land Management in a European Context*,
Human-Environment Interactions 8, https://doi.org/10.1007/978-3-030-50841-8_17

of their changes, namely sustainability, ecosystem services, land use functions and multifunctionality have been selected since they are broadly referred and often used in parallel.

Social-ecological systems or system frameworks are one of the key approaches for enhancing the understanding of complex and multi-tiered human-environmental interactions at multiple scales and their outcomes (Ostrom 2007). The transition to a "framework" approach was suggested to form an umbrella for comparing in a meta-language different theories on the systemic interactions and cause-effect relationships in social-ecological models and thus contribute to highly generic systemic approaches (McGinnis and Ostrom 2014). Related to land use science, social-ecological systems or frameworks provide the theoretical background for identifying key system components and sub-systems and classify their relationships and interactions to come from a case-study and observation based understanding to generic system architectures that are a relevant basis for modelling land (use) systems (e.g. Tett et al. 2013).

For assessing the performance of land use systems, a number of assessment frameworks can be used (O'Farrell and Anderson 2010; Wu 2013). *Assessment approaches and frameworks* closely related to land use science are, among others, sustainability, ecosystem services (and synonymous terms), land use functions and multifunctionality. All of them are used in parallel, often with similar understanding but different relevance for land use sectors. By sustainability, we understand since the Brundtland report (1987) and the Rio Declaration in 1992 the "*development that meets the needs of the present without compromising the ability of future generations to meet their own needs*" (Brundtland report 1987) including the balancing of ecological, social and economic sustainability aspects. Anyhow, this understanding was developed more from a political and societal perspective (Lélé 1991) that missed in some aspects the relation to land use due to its high level of abstractness, even if it was broken down to sectors, using the Ministerial Conference for Pan-European Forestry (MCPFE) as an example (Mayer 2000). The novel concept of sustainability suggested by Von Carlowitz (1713) was much closer to land use since it was simply developed from a resource economic perspective to optimize over long time the harvesting of forest biomass to generate enough energy for ore smelting (Basiago 1995; Mebratu 1998; Wiersum 1995). There were manifold attempts to make the concept less abstract and break it down to indicators to support its implementation in practice (e.g.; forestry: Raison et al. 2001; agriculture: Harwood 1990; Zinck and Farshad 1995). A key criticism resulting from these attempts were the assessment efforts through too many, often redundant indicators and the high data demands (e.g. Ceron and Dubois 2003; Niemeijer and de Groot 2008; Hák et al. 2016).

The origin of the ecosystem services concept dates back to the 1970s, where Westman (1977) highlighted the social value of benefits provided by ecosystems to society (nature's services) as a basis for informed decisions. Subsequently, ecosystem services were mainstreamed in literature with a peak in the 1990s (e.g. Costanza et al. 1997; Daily 1997). Only little later, the Millennium Ecosystem Assessment (MEA 2005) became an important milestone in the conceptual development of ecosystem services and their relevance for policy consulting by synthesizing globally knowledge on the state of ecosystems. Since then, the number of publications addressing ecosystems and their services increased exponentially (Fisher et al. 2009; Gómez-Baggethun et al. 2010; Vihervaara et al. 2010). Ecosystem services found their entrance in political agenda setting through the UN science-policy process "Intergovernmental Panel for Biodiversity and Ecosystem Services (IBPBES), which was established in 2012 in Panama (Larigauderie and Mooney 2010; Larigauderie et al. 2012) and is currently (2019) ratified by 132 countries. Since the introduction of ecosystem services as an assessment framework e.g. in the context of the European Biodiversity Strategy (Maes et al. 2012a, b; Schägner et al. 2013), a multitude of definitions have been developed and discussed (Fisher et al. 2009). Definitions that reached highest interest, are, *"Conditions and Processes through which natural ecosystems, and the species that make them up, sustain and fulfill human life"* (Daily 1997); "Benefits human populations derive directly or indirectly from ecosystem functions" (Costanza et al. 1997); "Components of nature, directly enjoyed, consumed, or used to yield human well-being" (Boyd und Banzhaf 2007); "Aspects of ecosystems utilised actively or passively to produce human well-being" (Fisher et al. 2009); "Direct or indirect contributions of ecosystems to human well-being" (The Economics of Ecosystems and Biodiversity, see e.g. Ring et al. 2010). Still, the definition formulated in MEA (2005) is acknowledged as most important one, defining ecosystem services as *"The benefits people obtain from ecosystems"*. These include provisioning services such as food and water; regulating services such as regulation of floods, drought, land degradation, and disease; supporting services such as soil formation and nutrient cycling; and cultural services such as recreational, spiritual, religious and other nonmaterial benefits."

However, the division of ecosystem services as suggested by MEA (2005) is critically discussed and, for instance, The Common International Classification of Ecosystem services (CICES, Potschin and Haines-Young 2011; Haines-Young and Potschin 2012) suggest reducing the number of groups to regulating, provisioning and cultural services. They suggest assessing separately bio-geophysical structures, processes and ecosystem functions along a cascade that facilitates separating intrinsic values of nature from yielded services. Among the diverse suggestions how to best implement ecosystem services in support of policy consulting, the CICES cascade in combination with the DPSIR (Drivers-Pressures-State-Impact-Responses) framework (Müller and Burkhard 2012) and the IPBES framework (Díaz et al. 2015a, b) are promising and acknowledged suggestions. Both support strongly to separate the assessment of intrinsic values of nature and their benefits for human well-being, based on indicator sets, but also accounting for qualitative information. Anyhow, most recent terminological discussions on the use of *"Nature's contributions to people"* (Díaz et al. 2018) instead ecosystem services and resulting critical responses (Braat 2018; de Groot et al. 2018) indicate that the concept as such is not yet in its final stage to really support the assessment of outcomes of land use systems.

Consequently, there were attempts to make the concept suitable for a more integrative systemic perspective referring to landscapes as a holistic entity were multiple land uses are meeting (Termorshuizen and Opdam 2009; Termorshuizen et al. 2007) to address better human-nature interactions and landscape configuration as decisive

Table 17.1 Comparison of the concepts sustainability, ecosystem services, landscape services, land use functions and multifunctionality

Concept	Scale	Holistic?	Comprehensive?	Indicator availability?
Sustainability	Social-ecological systems	Yes	Partially	High
Ecosystem services	Ecosystems	Partially	Yes	Still in work
Landscape services	Landscapes	Partially	Partially	Not yet fully
Land use functions	Ecosystems/land use types	Reductionist	Yes	High
Multifunctionality	Landscapes (Ecosystems)	Partially	Yes	High

factors for services generation. Other attempts, such as the concept of land use functions (Pérez-Soba et al. 2008) suggest a more reductionist approach, focusing on those functions (or services) that are decisive in a regional context.

In contrast, multifunctionality was conceived as a concept to assess land use systems and landscapes in terms of their performance from an integrative transdisciplinary perspective (Antrop 2005; Fry 2001). Their relevance consists particularly in supporting a holistic (cultural) landscape (von Haaren 2002) and land use system planning (Selman 2009) and development including sectorial applications in agricultural (Renting et al. 2009) and forest management planning (Schmithüsen 2007) and in rural development (Knickel and Renting 2000). Table 17.1 summarizes the major aspects and differences in the applicability and usefulness of the assessment concepts.

17.2 Current International Debates and Political Discourses

Land use science is highly relevant to support consulting environmental and other policies, ensure their coherence and develop governance instruments at multiple scales for ensuring a sustainable development (Sterk et al. 2009; see ongoing discussions e.g. in: Weith et al. 2019). Most relevant international debates rank currently around the topics of Climate Change, Global Change and tele-coupling, the achievement of the UN Sustainable Development Goals and biodiversity losses considering their impacts on the resilience and vulnerability of our land use systems (Allen et al. 2016; Bahn et al. 2018; Olesen and Bindi 2002). Since Climate Change (CC) was brought into the policy and public perspective through the endorsement of the International Panel for Climate Change by the United Nations Environment Programme (UNEP) and the World Meteorological Organization (WMO) in 1988 (From Noordwijk 1990), it found most recently a culmination point in public perception through the grass root movement "Fridays for Future" (Wahlström et al. 2019). However, this public expression of needs to start quickly with instruments to realize the outcomes and of the Paris Agreement in 2015 (van den Bergh 2017) originates and is conducted mostly in the developed world and not in the Global South, where CC impacts are much more dramatic and relevant for peoples survival. Land use systems as an intrinsic factor in regulating the global climate through carbon sequestration are not really in the focus of the public perception and adapted behaviour (van de Ven et al. 2018) yet, even if certification systems that could further afforestation or halt losses of tropical forest areas are known and proven to be highly efficient (Brancalion et al. 2017; Kongsager et al. 2016). In contrast, policy changes, for instance in Brazil, foster the degradation and destruction of the Amazonian forest (Freitas et al. 2018), while in the boreal forests, large areas are burnt, for instance, in Russia in the recent years through missing monitoring and mismanagement, whose capacities to regulate the large subsurface Carbon resources in permafrost soils are now destroyed (Schaphoff et al. 2016; Shuman et al. 2017. The current debates on stopping CC are more focussing on replacing the critical use of fossil energy and related technologies by other critical technologies (e-Mobility) with increasingly disastrous social and ecological impacts on land use systems through the extraction of rare metals and particularly of lithium (Agusdinata et al. 2018; Lee and Wen 2017). In contrast, approved and traditionally developed strategies related to land use as a means to mitigate CC or to adapt to the not-anymore manageable impacts are lost in the societal and political discourse (Eguavoen et al. 2015; Knutti et al. 2016).

Global change and tele-coupling are not yet publicly perceived topics but anyhow found entrance in the political discourse on equitable polycentric international governance and the achievement of the UN Sustainable Development Goals (SDGs) (Bowen et al. 2017; Vasseur et al. 2017), but also in the discussion on global migration trends and how to co-manage them between Global North and Global South (Oberlack et al. 2018; Radel et al. 2019). Known interactions are the increasing consumption of meat in developed and transitioning economies, the globalization of production processes, land grabbing for resource extraction, the cultivation of renewable resources for energy production and the compensation of CO_2 emissions trough certificate trade (e.g. Fiske and Paladino 2016; Garrett and Rueda 2019; Harvey and Pilgrim, 2011; Hull and Liu 2018). These lead to highly negative impacts particularly on land use systems in the Global South and here particularly in weak economies in Africa or transitioning economies in Latin America, South-East and Central Asia (Liu et al. 2013; Gasparri et al. 2016). In result, highly valuable and native ecosystems are irreversibly degraded with huge consequences for the global loss of biodiversity (IPBES 2019) and its relevance for climate regulation, regulation of CC impacts and the intergenerational equity in the access to highly relevant natural resources (Cardinale et al. 2012; Redclift and Sage 1998; Tacconi and Bennett 1995). Since the role of all organisms threatened by the degradation of land use systems is not fully understood, negative consequences for the resilience not only of singular ecosystems, but the system earth and for the vulnerability of land use systems from local to global scale are not yet known (Bonan and Doney 2018).

The 17 UN Sustainable Development Goals (SDGs; UN SDGs 2015) address directly or indirectly land use systems, such as SDG 17 "*Life on Land*". Anyhow, the achievement of many of them provokes critical trade-offs for the sustainable development of land use systems, e.g. through soil sealing (SDG's 8 (Decent Work and Economic Growth), 9 (Industry, Innovation and Infrastructure), and 11 (Sustainable Cities and Communities)). This requires planning and policy instruments that reduce such trade-offs through spatial prioritization of areas foreseen to contribute to the SDGs and strategies to make them ecologically less problematic (e.g. through keeping green infrastructures, greening of facades and roofs, etc. (e.g. Ignatieva and Anrné 2013; Keesstra et al. 2016; Norton et al. 2015). An open question remains, how the land use systems in highly different political, cultural, social and economic surroundings can contribute not only locally but globally to the achievement of these goals and what kind of criteria need to be defined to correspond to the diverse contextual situations when judging how far adapted land use has contributed to or improved the achievement of the SDGs.

In this complex and interwoven areas of political and societal discourse, land use science has the potential to reveal dependencies of the different political areas and the interlinkages of the land use systems and subsystems from local to cross-continental scale referring to the theories of social-ecological systems and their tele-coupled connections (Friis et al. 2016; Liu et al. 2014; Schlueter et al. 2012). This theoretical frame can help to identify which intervention strategies, either through regulating or financial governance instruments or community-efforts can be most efficient in reducing CC, CC impacts and biodiversity losses (Carter et al. 2014; Pereira et al. 2012). Policy and planning recommendations need to stretch over all the above raised topics and reflect through an assessment of their impacts on the sustainable development by the use of approaches such as sustainability criteria or ecosystem services how relevant trade-offs, but also synergies are for local populations and in a global context (Costanza et al. 1991; Kumar et al. 2013).

17.3   Topics for Future Research Areas

Land use science and its methodological concepts can contribute largely to make societies aware of future challenges and risks in their development. Refocussing on social-ecological systems, for instance, health and medical treatments are at high risk to get lost including regulating and provisioning ES, taking waterborn diseases and not yet explored medical substances as examples (Alves and Rosa 2007; Romanelli et al. 2015). This poses also a problem for the long-term achievement of UN SDG 3 "*Health and Human Well-being*". Besides the public and political discourse on such highly relevant topics, there is a number of other research areas, where land use science is called to contribute from an inter- and transdisciplinary perspective.

One of them is urbanization; currently, roughly 55% of the world's population is living in a city or in an urban/metropolitan area. This is expected to be increased in the upcoming decades to an amount of 68% (UN World Populations Prospects 2019), which means that the major part of the global population has no direct access to natural resources and no direct relation to more natural land use systems such as agriculture or forestry. Concentration in urban areas might be sustainable from the perspective of health, education, access to clean water and energy and can boost urban and industrial development (SDGs 3 (Good Health and Well-being), 4 (Quality Education), 6 (Clean Water and Sanitation), 7 (Affordable and Clean Energy), 9 (Industry Innovation and Infrastructure) and 11 (Sustainable Cities and Communities); UN SDGs 2015) through more efficient bundling or resources and infrastructures. On the other hand, there might be huge trade-offs for poverty, hunger, gender equity, the relation between life and work, and sustainable consumption (SDGs 1 (No Poverty), 2 (Zero Hunger), 5 (Gender Equality), 8 (Decent Work and Economic Growth), 12 (Responsible Consumption and Production), 13 (Climate Action) and 15 (Life on Land); UN SDGs, 2015) through abandonment and lack of working power in rural areas as well as selective rural–urban migration of young people/young men, losses in cultural identity, and higher consumption needs in terms of changes in the diets (more meat) and energy consumption/mobility (Cutter 2017; Reckien et al. 2017; Springmann et al. 2018) (Fig. 17.1).

Related to this topic, rural–urban and cross-continental migration remains for the next decades one of the key research topics, where the question of how land use opportunities could counteract migration and how CC is driving migration are subject to recent research activities (Cattaneo et al. 2019; Pelling et al. 2018). "Attractive" land use opportunities require generally achieving a lower vulnerability and higher resilience towards climate and global change (Froese and Schilling 2019; Javadinejad et al. 2019). In many cases, cash-cropping instead of subsistence farming is considered to be an appropriate solution (e.g. Friend et al. 2019; Gentle et al. 2018). However, this provokes a higher dependency on markets considering the purchase of the seeds or seedlings, potentially higher efforts in fertilization and irrigation, and in selling the products (Amrouk et al. 2019; Robinson 2018). Globalized markets thus gain more and critical influence on highly vulnerable local systems and are prone to accelerate negative impacts on the ecological-economic resilience of the systems (Reyers and Selomane 2018; Rosa-Schleich et al. 2019). Failures in successfully cultivating cash crops due to either climate or market variabilities can destroy in very short time farming systems in the Global South and accelerate poverty-driven migration (e.g. McKeon 2018; Mustafa et al. 2019).

Fig. 17.1 Synergies and trade-offs of urbanization with the UN SDGs (2015)

A highly relevant integrative research area will thus be to focus less on agricultural adaptation strategies alone, but on the water-energy-food-biodiversity nexus (Fürst et al. 2017; Stoy et al. 2018; Venghaus and Hake 2018). The nexus approach as suggested, for instance, in the Future Earth initiative Food-Energy and Water (https://futureearth.org/networks/knowledge-action-networks/water-energy-food-nexus) delivers an integrative concept similar to the synergy-trade-off considerations in social-ecological systems, sustainability and ecosystem services assessments (e.g. Karabulut et al. 2018; Nhamo et al. 2019). Research in the sectors of water management, efficient energy production and accessibility, food systems and biodiversity is so far often highly segregated and solutions to overcome scarcity or losses in quality are missing to be coherent with each other sector (Fader et al. 2018) since causal interactions between these sectors are often not directly visible. The most recent calls for scoping as suggested by IPBES therefore enhance the system-overarching perspective and will trigger research that transitions from inter- and transdisciplinary approaches focussing on a specific topic or question to connecting similar disciplines and actor types across sectors regarding a multitude of questions. Part of these across-sector research demands will be the consideration of societal transformations concerning their capacities to contribute to a sustainable development and come up with related social and systemic innovations. Societal transformations are known to be key processes that impact land use systems in terms of their performance in providing a multitude of services and resources and equitable access to them (Ehrensperger et al. 2019; Long and Qu 2018) Expressions of social transformations in their land use context can range from changes in community living styles (e.g. from hunting to herding, Bergman et al. 2013), ecosystem management practices (Olsson et al. 2004), to changes from land use to land development rights (Zhu 2004). Land use planning as an integrative discipline needs to take into account such transformations concerning the prioritization of areas in regional spatial and urban planning for delivering the requested amounts of food, clean water, recreational areas and green infrastructure, areas for protecting settlements against CC driven extreme events and other demands. These can change dramatically along socio-cultural-economic transformations through rural–urban migration and informal settlements or changed diets and living styles (Bardsley and Hugo 2010; Lerner and Eakin 2011).

17.4   Challenges in Research Practice

An important challenge in land use science is an equal enrolment of social and natural sciences (Müller and Munroe 2014) and the development of an original set of inter- and transdisciplinary research methods (Rounsevell et al. 2012). Current research attempts are often either focussed on one of the disciplinary fields (natural/social sciences) and consider the other instead of coming to a fully integrative approach (Zscheischler and Rogga 2015). Reasons for this might consist in funding strategies that are in many countries worldwide still set up from a highly disciplinary point of view and train researchers that fail to understand the philosophies and theoretical-methodological backgrounds of other disciplines (Bromham et al. 2016). A huge hampering factors for land use sciences consists in the historically separately developed research in land use sectors, such as agriculture and forestry, that often have similar disciplinary approaches, but fail in the cross-sectoral integration and fail in cooperation from a systemic, landscape perspective that is essential for deriving suitable policy recommendations (e.g. Klein et al. 2005; Mickwitz et al. 2009). The lack in a systemic understanding is another challenge in further developing land use sciences. Many of the data acquisition, monitoring and modelling approaches are still oriented towards the micro-scale, miss spatial representativeness and thus fail to contribute to integrative assessments at superior scales and decision levels (e.g. Anderson 2018; De Palma et al. 2018). There is often no real valid relation between spot-oriented sampling or monitoring regarding up-scaling approaches to regional or global scales, but vice versa, also no real attempts to down-scale and validate outcomes from global assessments and modelling approaches with regard to their local reliability (e.g. Kolosz et al. 2018; Le Clec'h et al. 2018; Malenovský et al. 2019). Most of the modelling approaches in land use sciences are purely data driven and miss making use of such theoretical frameworks as provided by the social-ecological system concept (Colding and Barthel 2019). Sustainability and ecosystem services assessments focus often too narrowly on singular ecosystems or ecosystem types and thus do not contribute to holistic and integrative landscape-oriented planning and policy recommendations (von Haaren et al. 2019). Consequently, one of the most important challenges consists in an improved implementation of a systemic perspective and in the focussing of systemic architectures including all subsystems, subcomponents, their interrelations and the quality of these interrelations (e.g. Langhammer et al. 2019). Graph-node theory based approaches such as Bayesian Belief Networks (BBN) or Artificial Neural Networks (ANN) would provide adequate solutions that can either make use of local, indigenous and expert knowledge in drafting the system architecture (BBN), or make use of artificial intelligence algorithms to harvest data sets (ANN) (e.g. Marcot and Penman 2018; Schmidt et al. 2018). While ANN are reliant on the amount and quality of the available data, BBN and similar approaches hold the huge potential to serve also as a transdisciplinary method to approach the understanding of land use systems, integrate multiple knowledge types and data sets and combine qualitative and quantitative data sets. This is of even higher relevance, since challenges for land use science called by Future Earth are the co-design of research and subsequently the co-development of new knowledge (Liu et al. 2018). Using system architectures as a starting point in the discourse between science and practice reveals knowledge gaps and research needs in understanding specific land use systems and—in the sense of social-ecological frameworks—in generalizing their structure and functioning (Gu et al. 2018). The identified "nodes" (i.e. sub-systems) can be critically reflected considering the availability of data or methods to parameterize them. Finally, by the step-wise integration of knowledge to parameterize the subsystems and describe their interactions helps to co-develop knowledge on the system, but also knowledge on potential intervention scales or decision levels to accomplish sustainable development (see e.g. Kampelmann et al.

2018). A remaining challenge will however consist in the enrolment of actors from practice to bridge or potentially close the gap between land use science and land use practice (e.g. Partelow et al. 2019).

17.5   Outlook

Transdisciplinary methods alone will not solve the problem how to realize a permanent engagement of actors from practice. There is a need for a new understanding and conception in how land use systems research should be conceived for delivering relevant practical recommendations without losing its scientific character. If co-design of research and co-development of knowledge are taken serious, traditional "project-oriented" research might not be appropriate since the period between the start, where research questions and hypotheses are (co-) formulated and the end, where outputs are presented remains often non-transparent and inaccessible for actors from practice due to time, economic or methodological constraints (e.g. Hansson and Polk 2018). Also, their priorities and questions might undergo changes during the research project. An opportunity to overcome these discrepancies between science and practice would be to agree not on singular questions or projects, but on agenda-based approaches referring to the impact pathway strategy. This would include coming to a joint agreement on final impacts expected by policy, planning and practice (e.g. achievement of the UN SDGs) in a medium to long-term perspective and on outcomes that are perceived to be relevant to accomplish them (e.g. through societal transformations and the way how these can be initiated, supported and put into action). The agreed research agenda could then start with an initial set of outputs (i.e. singular projects) that are serving to succeed in the outcomes (e.g. scenario model-based recommendations that inform on optimal intervention scales and decision levels in land use systems regarding the UN SDGs). New relevant outputs could be added and existing ones should be subject to a critical review in reasonable time spans, so that a co-learning approach can be established. This could help to overcome the difficulties between day-to-day management in practice and the time lapse in conducting research. Certainly, the coordination of such an agenda-based approach will require a higher and very holistic coordination effort, which leaves enough decision and financial space for adjustment over the agenda process and that will require conducting some pre-assessments of financial ranges including uncertainties in which final budgets will range. On the other hand, it would be much more dynamic and adaptive, would offer much larger participation opportunities from practice, but also from scientific actors and would finally benefit both sides through more synergetic results instead of segregated results in a multitude of smaller (even joint research) projects. Such agenda-based processes could also be implemented on an international scale, calling for applications that deliver expected outputs from regions where these are perceived to be requested to fulfil the agenda, so that a synthesis across continents could be supported.

References

Agusdinata, D. B., Liu, W., Eakin, H., & Romero, H. (2018). Socio-environmental impacts of lithium mineral extraction: Towards a research agenda. *Environmental Research Letters, 13*(12), 123001.

Allen, C. R., Angeler, D. G., Cumming, G. S., Folke, C., Twidwell, D., & Uden, D. R. (2016). Quantifying spatial resilience. *Journal of Applied Ecology, 53*(3), 625–635.

Alves, R. R., & Rosa, I. M. (2007). Biodiversity, traditional medicine and public health: Where do they meet? *Journal of Ethnobiology and Ethnomedicine, 3*(1), 14.

Amrouk, E. M., Grosche, S. C., Heckelei, T. (2019). Interdependence between cash crop and staple food international prices across periods of varying financial market stress. Applied Economics, pp. 1–16.

Anderson, C. B. (2018). Biodiversity monitoring, earth observations and the ecology of scale. *Ecology letters, 21*(10), 1572–1585.

Antrop, M. (2005). From holistic landscape synthesis to transdisciplinary landscape management. In B. Tress, G. Tress, G. Fry, P. Opdam (Eds.), *From landscape research to landscape planning*.

Bahn, M., Erb, K., Harris, E., Hasibeder, R., Ingrisch, J., Mayr, S., Niedertschneidert, M., Oberhuber, W., Oberleitner, F., Tappeiner, U., Tasser, E., Viovy, N. (2018). ClimLUC–climate extremes and land-use change: Effects on ecosystem processes and services. Aspects of integration, education and application, pp. 27–50. https://epub.oeaw.ac.at/0xc1aa5576%200x003a37a5.pdf. Accessed online July, 22, 2019.

Bardsley, D. K., & Hugo, G. J. (2010). Migration and climate change: examining thresholds of change to guide effective adaptation decision-making. *Population and Environment, 32*(2–3), 238–262.

Basiago, A. D. (1995). Methods of defining 'sustainability.' *Sustainable development, 3*(3), 109–119.

Bergman, I., Zackrisson, O., & Liedgren, L. (2013). From hunting to herding: Land use, ecosystem processes, and social transformation among Sami AD 800–1500. *Arctic Anthropology, 50*(2), 25–39.

Bonan, G. B., Doney, S. C. (2018). Climate, ecosystems, and planetary futures: The challenge to predict life in Earth system models. *Science, 359*(6375), eaam8328.

Boyd, J., & Banzhaf, S. (2007). What are ecosystem services? The need for standardized environmental accounting units. *Ecological Economics, 63*(2–3), 616–626.

Bowen, K. J., Cradock-Henry, N. A., Koch, F., Patterson, J., Häyhä, T., Vogt, J., & Barbi, F. (2017). Implementing the "sustainable development goals": Towards addressing three key governance challenges—Collective action, trade-offs, and accountability. *Current Opinion in Environmental Sustainability, 26,* 90–96.

Braat, L. (2018). Five reasons why the science publication "assessing nature's contributions to people" (Díaz et al. 2018) would not have been accepted in Eco-system Services. *Ecosystem Services, 30,* A1–A2.

Brancalion, P. H., Lamb, D., Ceccon, E., Boucher, D., Herbohn, J., Strassburg, B., Edwards, D. P. (2017). Using markets to leverage investment in forest and landscape restoration in the tropics. *Forest Policy and Economics, 85,* 103–113.

Bromham, L., Dinnage, R., Hua, X. (2016). Interdisciplinary research has consistently lower funding success. *Nature, 534*(7609), 684.

Brundtland, G., Khalid, M., Agnelli, S., Al-Athel, S., Chidzero, B., Fadika, L., Hauff, V., Lang, I., Shijun, M., Morino de Botero, M., Singh, M., Oktito, S., (1987). Our Common Future (\'Brundtland report\').

Cardinale, B. J., Duffy, J. E., Gonzalez, A., Hooper, D. U., Perrings, C., Venail, P., et al. (2012). Biodiversity loss and its impact on humanity. *Nature, 486*(7401), 59.

Carter, N., Viña, A., Hull, V., McConnell, W., Axinn, W., Ghimire, D., Liu, J. (2014). Coupled human and natural systems approach to wildlife research and conservation. *Ecology and Society, 19*(3).

Cattaneo, C., Beine, M., Fröhlich, C. J., Kniveton, D., Martinez-Zarzoso, I., Mastrorillo, M., et al. (2019). Human migration in the era of climate change. *Review of Environmental Economics and Policy, 13*(2), 189–206.

Ceron, J. P., & Dubois, G. (2003). Tourism and sustainable development indicators: The gap between theoretical demands and practical achievements. *Current Issues in Tourism, 6*(1), 54–75.

Colding, J., Barthel, S. (2019). Exploring the social-ecological systems discourse 20 years later. *Ecology and Society, 24*(1).

Costanza, R., Daly, H. E., Bartholomew, J. A. (1991). Goals, agenda and policy recommendations for ecological economics. *Ecological Economics: The Science and Management of Sustainability,* (s 525).

Costanza, R., d'Arge, R., de Groot, R., Farber, S., Grasso, M., Hannon, B., Limburg, K, Naeem, S., O'Neill, RV., and l'aruelo, J. (1997). The value of the world's ecosystem services and natural capital. *Nature, 387*, 253–260.

Cutter, S. L. (2017). The forgotten casualties redux: Women, children, and disaster risk. *Global Environmental Change, 42*, 117–121.

Daily, G. C. (1997). *Nature's Services: Societal Dependence on Natural Ecosystems.* Washington DC: Island Press.

de Groot, R., Costanza, R., Braat, L., Brander, L., Burkhard, B., Carrascosa, J.L., Crossman, N., Egoh, B., Geneletti, D., Hansjuergens, B., Hein, L., Jacobs, S.J., Kubiszewski, I., Leimona, B., Li, B., Liu, J., Luque, S., Maes, J., Marais, C., Maynard, S., Montanarella, L., Moolenaar, S., Obst, C., Quintero, M., Saito, O., Santos-Martín, F., Sutton, P., van Beukering, P., van Weelden, M., Willemen, L. (2018). Ecosystem services are nature's contributions to people: Response to: Assessing nature's contributions to people. *Science Progress, 359* (6373).

De Palma, A., Sanchez-Ortiz, K., Martin, P. A., Chadwick, A., Gilbert, G., Bates, A. E., Börger, L., Contu, S., Hill, S.S.L., Purvis, A. (2018). Challenges with inferring how land-use affects terrestrial biodiversity: Study design, time, space and synthesis. In Advances in Ecological Research (Vol. 58, pp. 163–199). Academic Press.

Díaz, S., Demissew, S., Carabias, J., Joly, C., Lonsdale, M., Ash, N., Larigauderie, A., Adhikuri, J.R., Arico, S., Baldi, A., Bartuska, A., Baste, I.A., Bilgin, A., Brondizio, E., Chan, K.M.A., Figueroa, V.E., Duraiappah, A., Fischer, M., Hill, R., Koetz, T., Leadley, P., Lyver, P., Mace, G.M., Martin-Lopez, B., Okumura, M., Pacheco, D., Pascual, U., Perez, E.S. (2015a). The IPBES Conceptual Framework—connecting nature and people. *Current Opinion in Environmental Sustainability, 14*, 1–16.

Díaz, S., Demissew, S., Joly, C., Lonsdale, W. M., & Larigauderie, A. (2015b). A Rosetta stone for nature's benefits to people. *PLoS Biology, 13*(1), e1002040.

Díaz, S., Pascual, U., Stenseke, M., Martín-López, B., Watson, R. T., Molnár, Z., et al. (2018). Assessing nature's contributions to peopleAssessing nature's contributions to people. *Science, 359*, 270–272.

Eguavoen, I., Schulz, K., de Wit, S., Weisser, F., Müller-Mahn, D. (2015). Political dimensions of climate change adaptation: Conceptual reflections and African examples. In *Handbook of climate change adaptation*, pp. 1183–1199.

Ehrensperger, A., de Bremond, A., Providoli, I., & Messerli, P. (2019). Land system science and the 2030 agenda: exploring knowledge that supports sustainability transformation. *Current Opinion in Environmental Sustainability, 38*, 68–76.

Fader, M., Cranmer, C., Lawford, R., Engel-Cox, J. (2018). Toward an understanding of synergies and trade-offs between water, energy, and food SDG targets. *Frontiers in Environmental Science, 6*(NREL/JA-6A50-72168).

Fisher, B., Turner, R. K., & Morling, P. (2009). Defining and classifying ecosystem services for decision making. *Ecological Economics, 68*(3), 643–653.

Fiske, S. J., Paladino, S. (2016). Introduction: Carbon Offset Markets and Social Equity: Trading in Forests to Save the Planet. In *The Carbon Fix* (pp. 25–46). Routledge.

Freitas, F. L., Sparovek, G., Berndes, G., Persson, U. M., Englund, O., Barretto, A., & Mörtberg, U. (2018). Potential increase of legal deforestation in Brazilian Amazon after Forest Act revision. *Nature Sustainability, 1*(11), 665.

Friend, R. M., Thankappan, S., Doherty, B., Aung, N., Beringer, A. L., Kimseng, C., Cole, R., Immuong, Y., Mortensen, S., Nyunt, W.W., Paavola, J., Prombhakping, B., Salamanca, A., Soben, K., Win, S., Win, S., Yang, N. (2019). Agricultural and food systems in the Mekong region: Drivers of transformation and pathways of change. *Emerald Open Research, 1*(12), https://doi.org/10.12688/emeraldopenres.13104.1.

Friis, C., Nielsen, J. Ø, Otero, I., Haberl, H., Niewöhner, J., & Hostert, P. (2016). From teleconnection to telecoupling: Taking stock of an emerging framework in land system science. *Journal of Land Use Science, 11*(2), 131–153.

Froese, R., & Schilling, J. (2019). The Nexus of climate change, land use, and conflicts. *Current Climate Change Reports, 5*(1), 24–35.

From Noordwijk, R. (1990). United Nations Activities. In *Environmental Policy and Law*, (vol. 20, p. 71).

Fry, G. L. (2001). Multifunctional landscapes—Towards transdisciplinary research. *Landscape and Urban Planning, 57*(3), 159–168.

Fürst, C., Luque, S., & Geneletti, D. (2017). Nexus thinking–How ecosystem services can contribute to enhancing the cross-scale and cross-sectoral coherence between land use, spatial planning and policy-making. *International Journal of Biodiversity Science, Ecosystem Services and Management, 13*(1), 412–421.

Garrett, R., Rueda, X. (2019). Telecoupling and consumption in agri-food systems. In *Telecoupling* (pp. 115–137). Palgrave Macmillan, Cham.

Gasparri, N. I., Kuemmerle, T., Meyfroidt, P., Polain, Le., de Waroux, Y., & Kreft, H. (2016). The emerging soybean production frontier in Southern Africa: Conservation challenges and the role of south–south telecouplings. *Conservation Letters, 9*(1), 21–31.

Gentle, P., Thwaites, R., Race, D., Alexander, K., Maraseni, T. (2018). Household and community responses to impacts of climate change in the rural hills of Nepal. *Climatic Change, 147*(1–2), 267–282.

Gómez-Baggethun, E., De Groot, R., Lomas, P. L., & Montes, C. (2010). The history of ecosystem services in economic theory and practice: From early notions to markets and payment schemes. *Ecological Economics, 69*(6), 1209–1218.

Gu, Y., Deal, B., & Larsen, L. (2018). Geodesign processes and ecological systems thinking in a coupled human-environment context: an integrated framework for landscape architecture. *Sustainability, 10*(9), 3306.

Haines-Young, R., Potschin, M. (2012). *Common International Classification of Ecosystem Services (CICES, Version 4.1)*. European Environment Agency (vol. 33).

Hák, T., Janoušková, S., & Moldan, B. (2016). Sustainable development Goals: A need for relevant indicators. *Ecological Indicators, 60,* 565–573.

Hansson, S., & Polk, M. (2018). Assessing the impact of transdisciplinary research: The usefulness of relevance, credibility, and legitimacy for understanding the link between process and impact. *Research Evaluation, 27*(2), 132–144.

Harvey, M., & Pilgrim, S. (2011). The new competition for land: Food, energy, and climate change. *Food Policy, 36,* S40–S51.

Harwood, R. R. (1990). A history of sustainable agriculture. In: Edwards, C.I., Lal, R., Madden, P., Miller, R.H., House, G. (Eds.) *Sustainable Agricultural Systems*, pp. 3–19.

Hull, V., Liu, J. (2018). Telecoupling: A new frontier for global sustainability. *Ecology and Society, 23*(4).

Ignatieva, M., & Ahrné, K. (2013). Biodiverse green infrastructure for the 21st century: From "green desert" of lawns to biophilic cities. *Journal of Architecture and Urbanism, 37*(1), 1–9.

IPBES (Intergovernmental Panel for Biodiversity and Ecosystem services). (2019). *Global Assessment—Summary for Policy Makers*. https://www.ipbes.net/sites/default/files/downloads/spm_unedited_advance_for_posting_htn.pdf. Accessed online, July, 22, 2019.

Javadinejad, S., Eslamian, S., Ostad-Ali-Askari, K., Nekooei, M., Azam, N., Talebmorad, H., Hasantabar-Amiri, A., Mousavi, M. (2019). Relationship between climate change, natural disaster, and resilience in rural and urban societies. In *Handbook of Climate Change Resilience*, pp. 1–25.

Kampelmann, S., Kaethler, M., Hill, A. V. (2018). Curating complexity: An artful ap-proach for real-world system transitions. *Environmental innovation and societal transitions, 27*, 59–71.

Karabulut, A. A., Crenna, E., Sala, S., & Udias, A. (2018). A proposal for integration of the ecosystem-water-food-land-energy (EWFLE) nexus concept into life cycle assessment: A synthesis matrix system for food security. *Journal of cleaner production, 172*, 3874–3889.

Keesstra, S. D., Bouma, J., Wallinga, J., Tittonell, P., Smith, P., Cerdà, A., et al. (2016). The signifi-cance of soils and soil science towards realization of the United Nations Sustainable Development Goals. *Soil, 2,* 111–128.

Klein, R. J., Schipper, E. L. F., & Dessai, S. (2005). Integrating mitigation and adaptation into climate and development policy: Three research questions. *Environmental Science and Policy, 8*(6), 579–588.

Knickel, K., & Renting, H. (2000). Methodological and conceptual issues in the study of multifunctionality and rural development. *Sociologia Ruralis, 40*(4), 512–528.

Knutti, R., Rogelj, J., Sedláček, J., & Fischer, E. M. (2016). A scientific critique of the two-degree climate change target. *Nature Geoscience, 9*(1), 13.

Kolosz, B. W., Athanasiadis, I. N., Cadisch, G., Dawson, T. P., Giupponi, C., Honzák, M., et al. (2018). Conceptual advancement of socio-ecological modelling of ecosystem services for re-evaluating Brownfield land. *Ecosystem Services, 33*, 29–39.

Kongsager, R., Locatelli, B., & Chazarin, F. (2016). Addressing climate change mitigation and adaptation together: A global assessment of agriculture and forestry projects. *Environmental Management, 57*(2), 271–282.

Kumar, P., Brondizio, E., Gatzweiler, F., Gowdy, J., de Groot, D., Pascual, U., et al. (2013). The economics of ecosystem services: From local analysis to national policies. *Current Opinion in Environmental Sustainability, 5*(1), 78–86.

Langhammer, M., Thober, J., Lange, M., Frank, K., & Grimm, V. (2019). Agricultural land-scape generators for simulation models: A review of existing solutions and an outline of future directions. *Ecological Modelling, 393*, 135–151.

Larigauderie, A., & Mooney, H. A. (2010). The Intergovernmental science-policy Platform on Biodiversity and Ecosystem Services: Moving a step closer to an IPCC-like mechanism for biodiversity. *Current Opinion in Environmental Sustainability, 2*(1), 9–14.

Larigauderie, A., Prieur-Richard, A. H., Mace, G. M., Lonsdale, M., Mooney, H. A., Brus-saard, L., et al. (2012). Biodiversity and ecosystem services science for a sustainable planet: The DIVERSITAS vision for 2012–20. *Current Opinion in Environmental Sustainability, 4*(1), 101–105.

Le Clec'h, S., Sloan, S., Gond, V., Cornu, G., Decaens, T., Dufour, S., Grimaldi, M. Oszwald, J. (2018). Mapping ecosystem services at the regional scale: The validity of an upscaling approach. *International Journal of Geographical Information Science, 32*(8), 1593–1610.

Lee, J. C., & Wen, Z. (2017). Rare earths from mines to metals: Comparing environmental impacts from China's main production pathways. *Journal of Industrial Ecology, 21*(5), 1277–1290.

Lélé, S. M. (1991). Sustainable development: A critical review. *World Development, 19*(6), 607–621.

Lerner, A. M., & Eakin, H. (2011). An obsolete dichotomy? Rethinking the rural–urban interface in terms of food security and production in the global south. *The Geographical Journal, 177*(4), 311–320.

Liu, J. Q., Hull, V., Batistella, M., DeFries, R., Dietz, T., Fu, F., Hertel, T. W., Izaurralde, C. R., Lambin, E. F., Li, S., Martinelli, L. A., McConnell, W. C., Moran, E. F., Naylor, R., Ouyang, Z., Polenske, K. R., Reenberg, A., De Miranda Rocha, G., Simmons, C. S., Verburg, P. H., Vitousek, P. M., Zhang, F., Zhu, C. (2013). Framing sustainability in a telecoupled world.

Liu, J., Hull, V., Moran, E., Nagendra, H., Swaffield, S. R., Turner, B. (2014). Applications of the telecoupling framework to land-change science. In *Rethinking global land use in an urban era* (pp. 119–140). MIT Press.

Liu, J., Hull, V., Godfray, H. C. J., Tilman, D., Gleick, P., Hoff, H., et al. (2018). Nexus approaches to global sustainable development. *Nature Sustainability, 1*(9), 466.

Long, H., & Qu, Y. (2018). Land use transitions and land management: A mutual feedback perspective. *Land Use Policy, 74,* 111–120.

Maes, J., Paracchini, M. L., Zulian, G., Dunbar, M. B., & Alkemade, R. (2012a). Synergies and trade-offs between ecosystem service supply, biodiversity, and habitat conservation status in Europe. *Biological Conservation, 155,* 1–12.

Maes, J., Egoh, B., Willemen, L., Liquete, C., Vihervaara, P., Schägner, J. P., et al. (2012b). Mapping ecosystem services for policy support and decision making in the European Union. *Ecosystem Services, 1*(1), 31–39.

Malenovský, Z., Homolová, L., Lukeš, P., Buddenbaum, H., Verrelst, J., Alonso, L., Schaepman, M.E., Lauret, N., Gastellu-Etchegorry, J. P. (2019). Variability and uncertainty challenges in scaling imaging spectroscopy retrievals and validations from leaves up to vegetation canopies. *Surveys in Geophysics, 40*(3), 631–656.

Marcot, B. G., Penman, T. D. (2018). Advances in Bayesian network modelling: Integration of modelling technologies. *Environmental Modelling and Software.*

Mayer, P. (2000). Hot spot: Forest policy in Europe: achievements of the MCPFE and challenges ahead. *Forest Policy and Economics, 1*(2), 177–185.

McGinnis, M., Ostrom, E. (2014). Social-ecological system framework: Initial changes and continuing challenges. *Ecology and Society, 19*(2).

McKeon, N. (2018). 'Getting to the root causes of migration' in West Africa–whose history, framing and agency counts? *Globalizations, 15*(6), 870–885.

MEA. (2005). Millennium Ecosystem Assessment. https://www.millenniumassessment.org/. Accessed online July, 18, 2019.

Mebratu, D. (1998). Sustainability and sustainable development: historical and conceptual review. *Environmental Impact Assessment Review, 18*(6), 493–520.

Mickwitz, P., Aix, F., Beck, S., Carss, D., Ferrand, N., Görg, C., Jensen, A., Kivimaa, P., Kuhlicke, C., Kuindersma, W., Máñez, M., Melanen, M., Monni, S., Pedersen, A.B., Reinert, H., Van Bommel, S. (2009). Climate policy integration, coherence and governance (No. 2). Helsinki: PEER (PEER Report 2)—ISBN 9789521133794–92.

Müller, D., & Munroe, D. K. (2014). Current and future challenges in land-use science. *Journal of Land Use Science, 9*(2), 133–142.

Müller, F., & Burkhard, B. (2012). The indicator side of ecosystem services. *Ecosystem Services, 1*(1), 26–30.

Mustafa, M. A., Mayes, S., Massawe, F. (2019). Crop diversification through a wider use of under-utilised crops: A strategy to ensure food and nutrition security in the face of climate change. In *Sustainable Solutions for Food Security* (pp. 125–149). Springer, Cham.

Nhamo, L., Mabhaudhi, T., Mpandeli, S., Nhemachena, C., Senzanje, A., Naidoo, D., Liphadzi, S., Modi, A.T. (2019). Sustainability indicators and indices for the water-energy-food nexus for performance assessment: WEF nexus in practice–South Africa case study.

Niemeijer, D., & de Groot, R. S. (2008). A conceptual framework for selecting environmental indicator sets. *Ecological Indicators, 8*(1), 14–25.

Norton, B. A., Coutts, A. M., Livesley, S. J., Harris, R. J., Hunter, A. M., & Williams, N. S. (2015). Planning for cooler cities: A framework to prioritise green infrastructure to mitigate high temperatures in urban landscapes. *Landscape and Urban Planning, 134,* 127–138.

Oberlack, C., Boillat, S., Brönnimann, S., Gerber, J. D., Heinimann, A., Ifejika Speranza, C., Messerli, P., Rist, S., Wiesmann, U. M. (2018). Polycentric governance in telecoupled resource systems. *Ecology and Society, 23*(1).

O'Farrell, P. J., & Anderson, P. M. (2010). Sustainable multifunctional landscapes: a review to implementation. *Current Opinion in Environmental Sustainability, 2*(1), 59–65.

Olesen, J. E., & Bindi, M. (2002). Consequences of climate change for European agricultural productivity, land use and policy. *European Journal of Agronomy, 16*(4), 239–262.

Olsson, P., Folke, C., Hahn, T. (2004). Social-ecological transformation for ecosystem management: the development of adaptive co-management of a wetland landscape in southern Sweden. *Ecology and Society, 9*(4).

Ostrom, E. (2007). A diagnostic approach for going beyond panaceas. *Proceedings of the National Academy of Sciences, 104*(39), 15181–15187. https://doi.org/10.1073/pnas.0702288104.

Partelow, S., Fujitani, M., Soundararajan, V., & Schlüter, A. (2019). Transforming the social-ecological systems framework into a knowledge exchange and deliberation tool for comanagement. *Ecology and Society, 24*(1), 15.

Pelling, M., Leck, H., Pasquini, L., Ajibade, I., Osuteye, E., Parnell, S., et al. (2018). Africa's urban adaptation transition under a 1.5 climate. *Current Opinion in Environmental Sustainability, 31,* 10–15.

Pereira, H. M., Navarro, L. M., Martins, I. S. (2012). Global biodiversity change: The bad, the good, and the unknown. *Annual Review of Environment and Resources, 37.*

Pérez-Soba, M., Petit, S., Jones, L., Bertrand, N., Briquel, V., Omodei-Zorini, L., et al. (2008). Land use functions—a multifunctionality approach to assess the impact of land use changes on land use sustainability. In K. Helming, M. Pérez-Soba, & P. Tabbush (Eds.), *Sustainability impact assessment of land use changes* (pp. 375–404). Berlin Heidelberg: Springer.

Potschin, M. B., & Haines-Young, R. H. (2011). Ecosystem services Exploring a geographical perspective. *Progress in Physical Geography, 35*(5), 575–594.

Radel, C., Jokisch, B. D., Schmook, B., Carte, L., Aguilar-Støen, M., Hermans, K., et al. (2019). Migration as a feature of land system transitions. *Current Opinion in Environmental Sustainability, 38,* 103–110.

Raison, R. J., Brown, A. G., Flinn, D. W. (Eds.). (2001). Criteria and indicators for sustainable forest management IUFRO Research Series 7, p. 469.

Reckien, D., Creutzig, F., Fernandez, B., Lwasa, S., Tovar-Restrepo, M., McEvoy, D., & Satterthwaite, D. (2017). Climate change, equity and the Sustainable Development Goals: An urban perspective. *Environment and Urbanization, 29*(1), 159–182.

Redclift, M., & Sage, C. (1998). Global environmental change and global inequality: North/South perspectives. *International Sociology, 13*(4), 499–516.

Renting, H., Rossing, W. A. H., Groot, J. C. J., Van der Ploeg, J. D., Laurent, C., Perraud, D., et al. (2009). Exploring multifunctional agriculture. A review of conceptual approaches and prospects for an integrative transitional framework. *Journal of environmental management, 90,* S112–S123.

Reyers, B., Selomane, O. (2018). Social-ecological systems approaches: Revealing and navigating the complex trade-offs of sustainable development. In Ecosystem Services and Poverty Alleviation (Open Access) (pp. 39–54). Routledge.

Ring, I., Hansjürgens, B., Elmqvist, T., Wittmer, H., & Sukhdev, P. (2010). Challenges in framing the economics of ecosystems and biodiversity: The TEEB initiative. *Current Opinion in Environmental Sustainability, 2*(1–2), 15–26.

Rio Declaration. (1992). Rio declaration on environment and development. https://www.unesco.org/education/pdf/RIO_E.PDF. Accessed online July, 18, 2019.

Robinson, G. M. (2018). Globalization of agriculture. *Annual Review of Resource Economics, 10,* 133–160.

Romanelli, C., Cooper, D., Campbell-Lendrum, D., Maiero, M., Karesh, W.B., Hunter, D., Golden, C.D. (2015). Connecting global priorities: Biodiversity and human health: A state of knowledge review. WHO/CBD 344p. ISBN 978 92 4 150853 7. https://cgspace.cgiar.org/handle/10568/67397. Accessed online July, 22, 2019.

Rosa-Schleich, J., Loos, J., Mußhoff, O., & Tscharntke, T. (2019). Ecological-economic trade-offs of Diversified Farming Systems–A review. *Ecological Economics, 160,* 251–263.

Rounsevell, M. D., Pedroli, B., Erb, K. H., Gramberger, M., Busck, A. G., Haberl, H., et al. (2012). Challenges for land system science. *Land Use Policy, 29*(4), 899–910.

Schägner, J. P., Brander, L., Maes, J., & Hartje, V. (2013). Mapping ecosystem services' values: Current practice and future prospects. *Ecosystem Services, 4,* 33–46.

Schaphoff, S., Reyer, C. P., Schepaschenko, D., Gerten, D., & Shvidenko, A. (2016). Tamm Review: Observed and projected climate change impacts on Russia's forests and its carbon balance. *Forest Ecology and Management, 361,* 432–444.

Schlueter, M., McAllister, R. R. J., Arlinghaus, R., Bunnefeld, N., Eisenack, K., Hoelker, F., et al. (2012). New horizons for managing the environment: A review of coupled social-ecological systems modeling. *Natural Resource Modeling, 25*(1), 219–272.

Schmidt, A., Creason, W., & Law, B. E. (2018). Estimating regional effects of climate change and altered land use on biosphere carbon fluxes using distributed time delay neural networks with Bayesian regularized learning. *Neural Networks, 108,* 97–113.

Schmithüsen, F. (2007). Multifunctional forestry practices as a land use strategy to meet increasing private and public demands in modern societies. *Journal of Forest Science, 53*(6), 290–298.

Selman, P. (2009). Planning for landscape multifunctionality. Sustainability: Science, Practice, & Policy, 5(2).

Shuman, J. K., Foster, A. C., Shugart, H. H., Hoffman-Hall, A., Krylov, A., Loboda, T., et al. (2017). Fire disturbance and climate change: implications for Russian forests. *Environmental Research Letters, 12*(3), 035003.

Springmann, M., Clark, M., Mason-D'Croz, D., Wiebe, K., Bodirsky, B. L., Lassaletta, L., et al. (2018). Options for keeping the food system within environmental limits. *Nature, 562*(7728), 519.

Sterk, B., Carberry, P., Leeuwis, C., Van Ittersum, M. K., Howden, M., Meinke, H., et al. (2009). The interface between land use systems research and policy: Multiple arrangements and leverages. *Land Use Policy, 26*(2), 434–442.

Stoy, P. C., Ahmed, S., Jarchow, M., Rashford, B., Swanson, D., Albeke, S., et al. (2018). Opportunities and trade-offs among BECCS and the food, water, energy, biodiversity, and social systems nexus at regional scales. *BioScience, 68*(2), 100–111.

Tacconi, L., & Bennett, J. (1995). Economic implications of intergenerational equity for biodiversity conservation. *Ecological Economics, 12*(3), 209–223.

Termorshuizen, J. W., & Opdam, P. (2009). Landscape services as a bridge between landscape ecology and sustainable development. *Landscape Ecology, 24*(8), 1037–1052.

Termorshuizen, J. W., Opdam, P., & Van den Brink, A. (2007). Incorporating ecological sustainability into landscape planning. *Landscape and Urban Planning, 79*(3), 374–384.

Tett, P., Sandberg, A., Mette, A., Bailly, D., Estrada, M., Hopkins, T. S., Ribeira d'Alcala, M., McFadden, L. (2013). *Global Challenges in Integrated Coastal Zone Management.* In Mokness, E., Dahl, E. Støttrup, J. (eds.), Chichester: Wiley-Blackwell, pp. 229–243.

UN SDGs (Sustainable Development Goals). (2015). Sustainable Development Goals, Knowledge Platform. https://sustainabledevelopment.un.org/?menu=1300. Accessed online July, 22, 2019.

UN World Populations Prospects. (2019). UN DESA Population Division. https://population.un.org/wpp/. Accessed online, August, 25, 2019.

van de Ven, D. J., González-Eguino, M., & Arto, I. (2018). The potential of behavioural change for climate change mitigation: A case study for the European Union. *Mitigation and adaptation strategies for global change, 23*(6), 853–886.

van den Bergh, J. C. (2017). Rebound policy in the Paris Agreement: instrument comparison and climate-club revenue offsets. *Climate Policy, 17*(6), 801–813.

Vasseur, L., Horning, D., Thornbush, M., Cohen-Shacham, E., Andrade, A., Barrow, E., et al. (2017). Complex problems and unchallenged solutions: Bringing ecosystem governance to the forefront of the UN sustainable development goals. *Ambio, 46*(7), 731–742.

Venghaus, S., & Hake, J. F. (2018). Nexus thinking in current EU policies–The interdependencies among food, energy and water resources. *Environmental Science & Policy, 90,* 183–192.

Verburg, P. H., Erb, K. H., Mertz, O., Espindola, G. (2013). *Land system science: Between global challenges and local realities.*

Verburg, P. H., Crossman, N., Ellis, E. C., Heinimann, A., Hostert, P., Mertz, O., et al. (2015). Land system science and sustainable development of the earth system: A global land project perspective. *Anthropocene, 12,* 29–41.

Vihervaara, P., Kumpula, T., Tanskanen, A., Burkhard, B. (2010). Ecosystem services–A tool for sustainable management of human–environment systems. Case study Finnish Forest Lapland. *Ecological complexity, 7*(3), 410–420.

von Carlowitz, H.C. (1713). Sylvicultura oeconomica, oder haußwirthliche Nachricht und Natur-mäßige Anweisung zur wilden Baum-Zucht. Bibliothecia Regia Monacencis, Leipzig.

von Haaren, C. (2002). Landscape planning facing the challenge of the development of cultural landscapes. *Landscape and Urban Planning, 60*(2), 73–80.

von Haaren, C., Lovett, A. A., Albert, C. (2019). Landscape Planning and Ecosystem Services: The Sum is More than the Parts. In Landscape Planning with Ecosystem Services (pp. 3–9). Springer, Dordrecht.

Wahlström, M., Sommer, M., Kocyba, P., de Vydt, M., De Moor (2019, Eds.). Protest for a future: Composition, mobilization and motives of the participants in Fridays for future climate protests on March 15, 2019 in 13 European cities. https://eprints.keele.ac.uk/6536/1/Protest%20for%20a%20future_GCS%2015.03.19%20Descriptive%20Report-2.pdf. Accessed online August, 20, 2019.

Weith, T., Warner, B., & Susman, R. (2019). Implementation of international land use objectives-discussions in Germany. *Planning Practice and Research, 34*(4), 454–474.

Westman, W. (1977). How much are nature's services worth? *Science, 80*(197), 960–964.

Wiersum, K. F. (1995). 200 years of sustainability in forestry: Lessons from history. *Environmental management, 19*(3), 321–329.

Wu, J. (2013). Landscape sustainability science: ecosystem services and human well-being in changing landscapes. *Landscape Ecology, 28*(6), 999–1023.

Zhu, J. (2004). From land use right to land development right: institutional change in China's urban development. *Urban Studies, 41*(7), 1249–1267.

Zinck, J. A., & Farshad, A. (1995). Issues of sustainability and sustainable land management. *Canadian Journal of Soil Science, 75*(4), 407–412.

Zscheischler, J., & Rogga, S. (2015). Transdisciplinarity in land use science–a review of concepts, empirical findings and current practices. *Futures, 65,* 28–44.

Part IV
Outlook

Chapter 18
Conclusions and Research Perspectives

Thomas Weith, Tim Barkmann, Nadin Gaasch, Sebastian Rogga, Christian Strauß, and Jana Zscheischler

18.1 Problems and Challenges

The previous chapters and articles in this book present not only a wide variety of challenges with regard to the sustainable use of land, but also a vast array of possible solutions. Without claiming to give an exhaustive overview, these possibilities are pooled and further developed by the editors in this concluding chapter. On the whole, emphasis is given to the aspiration described in the introduction for further developing current forms of land management as governance of land in the European context,

T. Weith (✉)
Institute of Environmental Science and Geography, University of Potsdam, Campus-Golm, Karl-Liebknecht-Str. 24-25, 14476 Potsdam, Germany
e-mail: Thomas.Weith@zalf.de

T. Barkmann · C. Strauß
Research Area 'Land Use and Governance', Leibniz Centre for Agricultural Landscape Research, Eberswalder Str. 84, 15374 Müncheberg, Germany
e-mail: timbarkmann@gmx.de

C. Strauß
e-mail: Christian_Strauss@gmx.de

N. Gaasch
Science Management and Transfer, Potsdam Institute for Climate Impact Research, Telegraphenberg A31, 14473 Potsdam, Germany
e-mail: nadin.gaasch@pik-potsdam.de

S. Rogga
Leibniz Centre for Agricultural Landscape Research, Eberswalder Str. 84, 15374 Müncheberg, Germany
e-mail: Sebastian.Rogga@zalf.de

T. Weith · J. Zscheischler
Working Group "Co-Design of Change and Innovation", Leibniz Centre for Agricultural Landscape Research, Eberswalder Str. 84, 15374 Müncheberg, Germany
e-mail: Jana.Zscheischler@zalf.de

© The Author(s) 2021   339
T. Weith et al. (eds.), *Sustainable Land Management in a European Context*,
Human-Environment Interactions 8, https://doi.org/10.1007/978-3-030-50841-8_18

simultaneously taking up and implementing the conceptual approach of co-design formulated at the start of the book in a variety of forms.

All of the authors who contributed to this volume stress that the current use of land, and combined functions and ecosystem services, cannot be described as sustainable. The limited space and the services provided by land for society are being handled in a way that will make it difficult for future generations in particular to use them sustainably in a better way (García-Martín et al.). As urbanisation processes progress, more and more land is used for settlement and infrastructure, leading to an irreversible loss of soils (Haber). In addition, agricultural and forestry land is also being damaged by erosion and leaching (Haber; Nuissl & Siedentop). At the same time, there is a large-scale loss of biodiversity, and landscapes lose their character as multifunctional cultural landscapes, due in part to fragmentation and to focusing on a one-sided use of land (García-Martín et al.).

Yet changes and burdens do not occur consistently across the whole area. Rather, they are characterised by a patchwork of the considerable intensification of use in the context of agricultural intensification, urbanisation and a strong demand for land in specific places, with a simultaneous abandonment of land and extensification elsewhere (García-Martín et al.; Doernberg & Weith.) It is not only this concurrence of opposing developments that makes it so difficult to develop suitable responses. There are also diverse functional and spatial linkages (see also "telecoupling") that have an impact beyond the regional, national and European context, up to the international scale. These interrelations show connections to global developments such as economic globalisation processes and digitalisation. Added to this are often unforeseeable impacts of urbanisation, as well as shrinking processes, which cannot be represented by means of simple cause-and-effect relationships (Nuissl & Siedentop). The previously and currently emerging competition for land use and land use conflicts are therefore often very difficult to analyse fully due to their complexity, and it is difficult to draw distinctions between them, particularly with regard to spatial designation (e.g. urban, rural) (Kanning et al.).

While single conflicts such as between settlement development and protection of open space appear to be obvious, it is more difficult to identify interconnections in the case of agricultural biodiversity conservation (agriculture versus nature conservation?) or the promotion of sustainable energy use (Kirschke et al., Nuissl & Siedentop). In heterogeneous fields of impact such as demographic change, in fact, they are sometimes even impossible to determine (Hoffmann).

This is also the case because various contemporary governance activities that influence land use are responsible for creating land use conflicts to a large extent in the first place. For example, besides mitigating climate change, current energy transition policies also cause land use conflicts to a considerable extent, whether between different forms of use (wind energy) or within a particular use such as agriculture (food versus fuel) (García-Martín et al.). Often, these conflicts are only recognised early on in part due to the limited knowledge in connection with a lack of impact assessment (processes relating to land use conflicts). This is compounded by the fact that integrative perspectives reflecting the complexity of the matter are rarely implemented due to persistent interest-led perspectives focusing on specific sectors (agricultural production, housing, optimising transport connections). For instance, there is a lack of adequate governance approaches for land that take into account the complexity of the challenges as given specific expression in Sustainable Development Goals such as Life on Land (15), and that integrate the diversity of the ecosystem services to be observed. For agricultural practices, for example, this would mean no longer prioritising particular economic products (commodity outputs), but simultaneously taking adequate account of the demand for additional ecosystem services (non-commodity outputs), taking full consideration of the diversity of the demand for land.

18.2 Sustainable Land Management—New Approaches

According to the authors who contributed to this book, previous paths have so far been unable to provide an adequate solution to the multitude of challenges and problems occurring in land use. For this reason, they propose new approaches that can be translated into a concept of sustainable land management that specifically integrates science and practice.

A central starting point in this regard is the early and comprehensive recognition of land use competition and conflicts, enabling synergetic solutions (in particular) to be found. The approaches developed in the process represent modules of sustainable land management as a conceptual idea that should relate to each other and be further elaborated. In spatial terms, both developments in urban and rural areas need to be considered (Kanning et al.). Such consideration often deeply contests how actors have managed land in the past.

In concrete terms, six linked approaches can be detected

(1) Refining the focuses of analysis
(2) Enhancing process and knowledge orientation
(3) Redesigning processes of implementation
(4) Naming and implementing concrete objectives
(5) Defining institutional frameworks and, in particular, governance approaches
(6) Further developing research activities in the interdisciplinary and transdisciplinary context

Refining the focuses of analysis
According to the contributors to this book, the first step to sustainable land management requires problem analysis that does sufficient justice to the complex challenges. Such analysis must particularly capture the wide variety of drivers behind land use demands and land use conflicts (García-Martín et al.). Besides general and supraregional factors such as the general demand for settlement areas, regional contexts must also be captured adequately. Place-based analysis (García-Martín et al.) is specifically requested. Such analysis should also include the views of local stakeholders

from the outset (Kanning et al., Zscheischler & Rogga), enabling the co-production of knowledge. Moreover, greater attention should be paid to indirect effects (Kirschke et al., Nuissl & Siedentop). For instance, the effects of urbanisation processes go far beyond urban areas, due to the induced traffic, the consumption of resources and the demographic/spatial pull effects, and can often only be recognised and detected after a considerable time lag (Hoffmann). For this reason, analyses should, if possible, be undertaken also empirically for prolonged periods of time, comparing regions (Hoffmann). In some cases, researchers are totally uncertain as to the key factors driving these developments, owing to the complexity of the cause-and-effect relationships. Long-term analyses would then at least enable the uncovering of blind spots, e.g. concerning the consequences of the impact of demographic change on land use that often cannot yet be assessed (Hoffmann).

In combination with follow-up activities undertaken in science and practice, this would also lead to an overall improvement of knowledge in the medium term. In general, however, there are often complaints about a lack of basic statistical data, as well as a lack of data on current land use and land use change (Hoffmann). Although geodata inventories and infrastructures have improved considerably in recent years, recording and monitoring options should be further improved in this area (Nuissl). At the same time, there is still, on the one hand, a lack of key basic data such as comprehensible surveys on ownership of land and property (Kirschke et al.), while, on the other, there is also a greater need for the stronger aggregation of empirical knowledge, culminating in enhanced impact models and concepts, also in combination with socio-economic analyses (e.g. lifestyle analysis: Hoffmann, Doernberg & Weith).

Enhancing process and knowledge orientation
The authors note that the development and implementation of sustainable land management should consider the specific design of change processes, paying particular attention to the generation and provision of different knowledge bases (co-production of knowledge).

Transdisciplinary processes in particular have a high potential for dealing with complex land use issues, where normative discourses, conflicting interests, sectoral as well as disciplinary viewpoints and different knowledge types are to be increasingly integrated in the search for co-designed sustainable solutions (Zscheischler; Zscheischler & Rogga). Such transdisciplinary processes can be supported by various conceptual and methodological approaches. The real world labs (as a research concept) and serious games (as a method) addressed in this book are just two possible approaches for generating better informed conclusions (Maaß, Kanning et al.). It is particularly important to enable practitioners to gain access to knowledge sharing processes and to view matters from the viewpoint of other stakeholders (Kanning et al.).

Since the demand for knowledge related to land use diverges, the forms of knowledge provision and sharing should also be diverse, but also manageable for the stakeholders involved (Pütz & Brassel). In particular, knowledge sharing in a common problem area is often described as a great benefit in research and development processes. In this case, digitalisation may create possibilities for generating and sharing knowledge, and promote multiperspectivity (Schulz et al.). The development of digital infrastructures and learning formats enables this to take place anywhere (Schulz et al.).

Substantial bottlenecks are currently emerging in this connection, which must be overcome specifically in science and practice. In the world of science, in particular, there are few opportunities for the development of supra-disciplinary transformative research, due to the forces of inertia in discipline-based systems of reputation and communication (Zscheischler; Rogga). By contrast, practice often lacks the possibilities to test experimentally different options for action (protected experimental space). Another important aspect in this connection is external stimulating support, not only financial and organisational, but also with regard to competencies in process management.

Redesigning processes of implementation
The early involvement of different stakeholder groups also plays a vital role in the implementation of approaches towards sustainable land management with a new direction in content. In light of this, implementation and transfer should not only be "co-planned and thought out" from the very outset (Rogga), but additional resources should also be earmarked for this purpose. This calls for a changed understanding of co-dissemination that goes beyond the conventional paradigm of the loading dock (Rogga). That is to say, in line with a transdisciplinary basic approach, the different stakeholders and their views are included in the problem-solving process from the very beginning. In this connection, digitalisation processes may be understood as the creation of enabling spaces (Schulz et al.) that not only facilitate local and regional knowledge generation, but also support—globally and flexibly—the dissemination and adoption of knowledge. Several authors believe that this will result in a greater likelihood of the use and impact of new land management approaches, without being able to guarantee this, however (Rogga, Schulz et al.).

Naming and implementing concrete objectives
As is the case with any political process, the implementation of sustainable land management requires clear objectives. One particularly important aspect in this regard is linking efficiency and sufficiency goals that additionally include regionalised sustainable development goals. In the area of settlement development, for example, it is not only important to promote more inner urban development to optimise existing settlements, but also to stop, or at least reduce, the development of further land on the outskirts of the city (Nuissl & Siedentop). Failed incentive structures must also be addressed in this connection (Nuissl & Siedentop). Another important aspect is the systematic search for synergies of supposedly different land use demands in order to achieve multifunctional land use. Rewetted peatlands, for instance, need not automatically exclude economic function. Paludiculture is a way of merging the aspects of protection and utilisation. Intensified efforts should be made to achieve such combined solutions. They are easier to achieve in co-designed processes

in which supposedly opposing stakeholder groups assume joint responsibility for the intended integrated use at an early stage.

More importantly, land use conflicts will likely be managed much more efficiently on the basis of co-designed use and development processes than conventional decision processes in which conflicts are detected at a later stage, or specifically addressed at a later stage for the public.

Defining institutional frameworks and, in particular, governance approaches

In the context of European institutional frameworks, the outlined handling of different demands to land and of land use conflicts simultaneously suggests a specific understanding of governance. Building on regional political agenda setting (Nuissl & Siedentop; Fürst), the interaction of different stakeholder groups (collaboration) plays an important role. In this connection, the early coordination of activities as co-design, the involvement of civil society stakeholders (Nuissl & Siedentop) and the nomination of persons with regional responsibility (minders) are of particular importance.

System solutions at different scales should be sought (Heck). Where possible, the approaches take up existing production and governance systems, and take into account their inherent logics and structures. Different regulatory approaches (state-driven hierarchical, market-based, cooperative) interact in the process. It is also important that the specific governance approaches are suitable for establishing and developing decentralised network systems (Heck). In the process, local stakeholders must also be able to possess or acquire the relevant skills required for decision processes.

The conceptual approaches named include nature-based solutions, focusing on green infrastructures, greater consideration of ecosystem services, and a combination of these approaches (Haase). In light of this, consideration must be taken of developments in all communities of the city, on the urban fringe, and in rural regions, including also interlinkages on the regional scale (Doernberg & Weith) as well as the global scale (telecoupling) (Haase, Schulz et al.). This then also facilitates de facto sharing and even fair burden-sharing (Specht et al.). In consequence, this results in much more resilient solutions (Heck). Care should be taken that the different regulatory approaches do not hinder or obstruct each other (Kirschke et al.). Each approach should be assessed for effectiveness, particularly with regard to implementation, not least because of the complexity of the challenges and the spatial differentiation (Heck).

Knowledge management processes play a significant role in this respect. In the context of co-design processes, it is not only the involvement of regional know-how that plays a major role, but also regional values and the advancement of existing practices by initiating and continuing social learning processes (Nuissl & Siedentop) that address practical recommendations (Fürst). At the same time, it is evident that solutions based purely on the development and use of information and communication technologies, while often helpful and expandable, fall short of what is required, and only take effect in the process-related interactive interplay with key players (Fürst).

In addition, it is highly important to broach the subject of power structures when seeking to achieve sustainable land management. In land management processes, it is essential to illustrate which stakeholders currently have the opportunities to significantly manage land uses at present. It is also equally important to realise which changes may be necessary to make adaptations in keeping with sustainable land use in a bid to strike a better balance between different land use interests.

Current normative values are the core foundation of this. Overarching norms and values, such as justice, are the core foundation of target setting, assessment and development in land management. They need to be disclosed and presented for debate (Doernberg & Weith; Haase). The same should also apply to the question of ownership structures. However, this issue has only been raised rarely in the past, not least because of poor data availability. Yet it is considered an increasingly important component for sustainable land management in the current literature (above all Davy 2012).

Further developing research activities in the interdisciplinary and transdisciplinary context

In addition to the aforementioned approaches for implementing and advancing sustainable land management, the authors, with research in mind, make additional recommendations for action that go beyond the approaches addressed in the individual chapters.

First and foremost, they address the generation of additional knowledge bases with change processes in mind. The development of what is referred to as transformation knowledge (Zscheischler & Rogga) should accompany the further development of models and concepts for social innovations and transition processes. In the process, the gap between theoretical requirements and practical implementation, described repeatedly in the literature (e.g. Pohl and Hirsch-Hadorn 2008), should also be addressed at an early stage. This can be achieved by linking various approaches, where possible, to concrete real-world use cases, particularly in transdisciplinary settings (Zscheischler).

To ensure the design of such an approach is feasible in practice, scientific skills should focus more strongly on this aspect, whilst also proposing potential practical applications in collaboration with the various stakeholder groups (roadshows, Heck) or actively initiating them (*Dorfkern* companies, Heck).

18.3 Final Conclusion

The approaches to sustainable land management compiled in this book provide key building blocks for a new form of land governance that

- considers the use of land resources in an integrative rather than a sectoral manner, i.e. that is derived from concrete problems with reference to land, which society recognises as a problem, formulating a need for action accordingly,

- features strong process orientation, by way of co-design, since a state of sustainability will never finally be achieved, and the frameworks and normative objectives constantly change,
- is simultaneously oriented to options for change and innovation,
- strongly requires the use and further development of interdisciplinary and transdisciplinary approaches, due to the diversity of stakeholder perspectives,
- is designed in an evidence-based manner, due to its knowledge orientation, or in an evidence-informed manner with stakeholders in society in mind.

The aspects compiled here also reflect some of the research and development activities promoted in the Global Land Programme Science Plan (GLP 2015), not least because of the activities undertaken by the authors involved, and now provide their own contribution to the debate. Rather than constituting the one and only way forward in the European context, they provide building blocks for a continuously adaptive system of responses that can—and must—be adapted, depending on the problem, the location, and the type of land use concerned. They constitute a kind of mindset that requires the capacity for individual and collective reflection and learning beyond methodological and conceptual forms of access. Only then can multiple perspectives and approaches to problem solutions be further developed into new thought styles that enable negotiation processes to be undertaken on equal terms and socially robust solutions to be supported, gradually breaking old path dependencies.

Clearly, there cannot be a single stakeholder who is solely responsible for the implementation of such an approach. Even though the need for intermediaries, change agents and integration experts remains undisputed, especially with regard to professional process design, all those who have a direct or indirect influence on land use shall take greater responsibility for the development and implementation of sustainable land management within the meaning of co-management. The editors simultaneously view this as a form of practical implementation of the imperative of responsibility, as called for, philosophically speaking, by Hans Jonas more than three decades ago, and which is ultimately reflected today in the principle of intergenerational justice.

This book discusses issues and responses in a decidedly European context. From the editors' point of view, the analyses, concepts and methods presented can—and should—be increasingly examined with regard to their relevance and implementation capacity in a non-European context in the future. The non-sustainable use of land as practiced today is by no means an exclusively European problem. A number of responses have explicitly been developed within the "Sustainable Land Management" research programme, funded by the German Federal Ministry of Education and Research (BMBF), referred to several times in this book. A promising option for the future would be to gradually develop these responses into an integrated framework for assessing sustainable land management for policies, plans and programmes.

References

Davy, B. (2012). *Land Policy*. New York: Routledge.

Global Land Programme (GLP). (2015). Science plan and implementation strategy 2016–2021. Online: https://glp.earth/sites/default/files/uploads/glpscienceplan_25_10_16.pdf. Last accessed on January 30, 2020

Pohl, C., & Hirsch Hadorn, G. (2008). Methodological challenges of transdisciplinary research. *Natures Sciences Societies, 16,* 111–121.